Electrical Science for Technicians

An indispensable resource for electrical technicians and trainees, *Electrical Science for Technicians* walks readers through the subject in a logical order, providing a historical overview alongside modern electrical theory and practice. You will be guided through the subject in a topic by topic manner with each section building upon the one that came before it. By adding context to the principles of electrical science they become easier to both understand and remember, providing a background in the subject that will remain with you for life.

▶ Fully aligned to the 17th edition of the wiring regulations
▶ Topic-based approach ensures suitability for both technicians and students
▶ Clear objectives outlined at the start and revisited at the end of each chapter as a checklist allow readers to check their learning before moving on.

Adrian Waygood currently works as a freelance lecturer and instructional designer, but in the past has worked for organisations as diverse as the South Wales Electricity Board, the Royal Navy, Royal Navy of Oman, British Aerospace, the Northern Alberta Institute of Technology, the British Columbia Institute of Technology, the UAE's Higher Colleges of Technology, and the Apprenticeship Branch of British Columbia's Ministry of Labour.

Electrical Science for Technicians

Adrian Waygood

Routledge
Taylor & Francis Group

LONDON AND NEW YORK

First published 2015
by Routledge
2 Park Square, Milton Park, Abingdon, Oxon OX14 4RN

and by Routledge
711 Third Avenue, New York, NY 10017, USA

First issued in hardback 2017

Routledge is an imprint of the Taylor & Francis Group, an informa business

British Library Cataloguing-in-Publication Data
A catalogue record for this book is available from the British Library

Library of Congress Cataloging in Publication Data
Waygood, Adrian.
Electrical science for technicians / Adrian Waygood.
pages cm
ISBN 978-1-138-84926-6 (pbk. : alk. paper) -- ISBN 978-1-315-72574-1 (ebook) 1. Electrical engineering--Textbooks. I. Title.
TK146.W374 2015
621.3--dc23
2014042155

ISBN 13: 978-1-138-42208-7 (hbk)
ISBN 13: 978-1-138-84926-6 (pbk)

Typeset in Univers LT Std by
Servis Filmsetting Ltd, Stockport, Cheshire

Dedication

To my wife, Anita, for putting up with those long periods of silence, while our garden suffered and a long list of unfinished jobs accumulated, as I sat in front of my computer preparing the manuscript and panicking to meet the publisher's deadline!

To my friend and former colleague, Anne Wafaa, who kindly edited much of the original material on which this book is based.

To my old chum, Doug Craig, for kindly reviewing sample chapters.

Contents

Foreword

It is with a great deal of pleasure that I write this foreword to **Electrical Science for Technicians** by Adrian Waygood, whom I have known for over thirty years and have the greatest respect for.

Adrian and I worked together for many years in the Department of Electrical Engineering Technology, at the Northern Alberta Institute of Technology (NAIT), in Edmonton, Canada, where we were assigned to both the Electrical Apprenticeship and Electrical Technology programs. I knew him then as an outstanding instructor, and he created a vast amount of electrical learning materials that I, together with many of our colleagues, used on a regular basis. Some elements of this book are based on these learning materials.

I also very much admired his devotion to his students who, in turn, treated him with a great deal of well-earned respect.

Unusually, in addition to his experience as an electrical instructor, Adrian also has a great deal of international experience as an instructional designer, and this is evident from the way he has constructed the content and sequence of each of his two books.

I like to believe that I was instrumental in encouraging him to write his earlier book, **An Introduction to Electrical Science**, which was published in 2012, intended for apprentices and tradespersons in the electrotechnology industry.

I also believe that this new book achieves the same very high standard of his earlier work. This book continues on from where his previous book left off and, although intended for technology students, it may also prove a useful introduction to the subject for those entering a degree program in electrical engineering.

I thoroughly recommend **Electrical Science for Technicians** to anyone who wants an interesting and easily assimilated introduction to what is often perceived to be a difficult subject.

George Brain, P.Eng.
Texada Island, British Columbia, Canada

The author

Adrian Waygood is a retired engineer, lecturer and instructional designer, and holds a Higher National Certificate in Electrical Engineering together with a Master's Degree in Education Technology.

He has held commissions in both the Royal Navy and the Royal Navy of Oman, specialising in training engineering artificer apprentices in the RN, and head of curriculum design in the RNO.

In Canada, he taught electrical engineering technology at the Northern Alberta Institute of Technology and worked as an instructional design consultant at the British Columbia Institute of Technology. While at NAIT and BCIT he consulted on curriculum-design projects at technical teachers' training colleges in Indonesia and India.

He has also worked as a manager with the Government of British Columbia apprenticeship programme, Senior Training Adviser to the Royal Saudi Air Force, and Head of Instructional Media Production for the Higher Colleges of Technology in the United Arab Emirates.

Introduction

Welcome to **Electrical Science for Technicians**. This book continues from where its companion book, **An Introduction to Electrical Science**, left off.

Although this book is principally aimed at students undertaking technology-level courses, some of its coverage of electrical machines may prove useful for electrical apprentices. I hope, too, that it may be of use as an introductory text for students embarking on engineering degrees, and to hobbyists who are interested in developing a greater understanding of the science behind their particular hobby.

As with its companion book, I have tried to make the topics as interesting as possible by relating, where appropriate, their historical background. I have also tried to develop topics in a conversational, easy-to-understand, logical sequence – for example, in the chapters on alternating current, I show how phasor diagrams are constructed, step-by-step, rather than by simply providing the reader with a completed phasor diagram.

I have also attempted to *explain* the science behind each topic rather than simply treating it as an exercise in mathematics. My approach has been, '*This is how it works . . .*', rather than '*From the following equation, you will see . . .*'!

Each chapter in this book starts with a list of '**objectives**'. You can think of objectives as being the required 'learning outcomes' for a particular chapter. *Avoid reading them at your peril*!

It's well worth your while familiarising yourselves with these objectives, as they list the *essential* things you will need to learn as you read through the chapter. They allow you to identify material that is 'essential' from material that is 'nice to know' or material that has been included to try and make the chapter more interesting.

For example, as already explained, to make the topics more interesting, most chapters include some historical background to the subject matter, but you are *not* expected to learn this information, so you will *not* find it listed in the chapter's objectives.

Another useful feature of the list of objectives is that, if you were to add a *question mark* at the end of each, then they become **self-test items**!

For example, an objective may read, '*explain what is meant by the term, "luminous efficacy"*'. By adding a question mark, you can you then ask yourself, '***Can I*** *. . . explain what is meant by the term, "luminous efficacy"?*' If you *can*, then you have achieved that particular objective; if you *cannot*, then you should re-read the part of the chapter that explains what 'luminous efficacy' means!

Understanding purpose of **objectives**, then, *is the key to efficiently learning the essential material within any particular chapter.*

Thank you for purchasing this book. I sincerely hope you will derive as much pleasure from reading it as I had in writing it.

Adrian Waygood, MEd (Tech)

CHAPTER 1

Solving complex circuits: network theorems

Objectives

On completion of this chapter you should be able to:

1. recognise a complex circuit.

2. apply 'sense arrows' to complex circuits.

3. identify junctions ('nodes') and loops ('meshes') in complex circuits.

4. recognise complex circuits to which the following 'network analysis' techniques may be applied:

   a. Kirchhoff's Loop Equation method.

   b. Superposition Theorem method.

   c. Delta-Star/Star-Delta Transformation method.

   d. Thévenin's Theorem method.

5. define **Kirchhoff's 'Current Law'** and **'Voltage Law'**.

6. solve complex circuits using Kirchhoff's 'Loop Analysis' method.

7. solve complex circuits, using the 'Superposition Theorem'.

8. solve complex circuits, using the 'Delta-Star/Star-Delta Transformation' method.

9. solve complex circuits, using 'Thévenin's Theorem'.

Introduction

In the companion book, *__An Introduction to Electrical Science__*, we learnt that electrical **circuits** are categorized according to the way in which their various components are connected relative to each other.

And we learnt that there are *four* such categories as shown in Figure 1.1. We also learnt that the term, '**complex circuit**', is nothing more than a *category* into which we can place any circuit that isn't connected in 'series', 'parallel', or 'series-parallel'. Used in this context, then, the term 'complex' *doesn't* mean 'complicated', although many complex circuits are indeed complicated! 'Complex circuits' are often referred to as '**networks**', so the analysis of complex circuits is often referred to as '**network analysis**'.

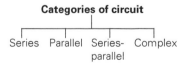

Categories of circuit

Series Parallel Series-parallel Complex

Figure 1.1

A basic example of a circuit within each of these categories is shown in Figures 1.2a–d. We need to bear in mind, however, that there are countless variations of the examples of *series-parallel* and *complex* connections having various degrees of difficulty.

We also learnt how to solve **series**, **parallel**, and **series-parallel** circuits. By 'solve', we mean determining the *equivalent resistance* of the circuit, as well as the *currents through*, and the *voltage drops across*, each of their components.

Electrical Science for Technicians. 978-1-138-84926-6 © Adrian Waygood.
Published by Taylor & Francis. All rights reserved.

Unfortunately, the techniques we use to solve series, parallel, and series-parallel circuits *cannot* be used to solve complex circuits. Well . . . this is only *partly* true, because we *may* need to use those techniques to simplify *parts* of a complex circuit, but they certainly cannot be used alone to solve the circuit as a whole.

To solve complex circuits, we need access to a wider range of techniques.

In this chapter, then, we are going to learn various new techniques for solving complex circuits or networks.

But, before we do so, let's briefly remind ourselves of how we determine the 'equivalent' (or 'total') resistance of a series circuit (Figure 1.3) and of a parallel (Figure 1.4) circuit:

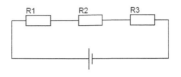

Series circuit

Figure 1.2a

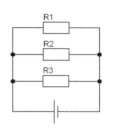

Parallel circuit

Figure 1.2b

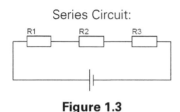

Series Circuit:

Figure 1.3

$$R = R_1 + R_2 + R_3 + etc.$$

Parallel Circuit:

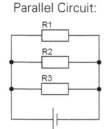

Figure 1.4

$$\frac{1}{R} = \frac{1}{R_1} + \frac{1}{R_2} + \frac{1}{R_3} + etc.$$

For a circuit with just *two* resistors, R_1 and R_2, in parallel, we can use this useful **'product-over-sum'** shortcut:

$$R = \frac{R_1 R_2}{R_1 + R_2} \left(= \frac{\text{product}}{\text{sum}} \right)$$

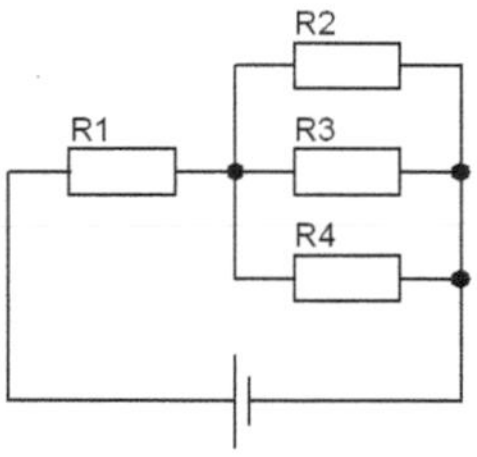

The *'product-over-sum'* shortcut method of determining the resistance of two resistances connected in parallel is particularly useful to know, as we shall be using it elsewhere in this chapter.

Returning, now, to **complex circuits**: it's obviously important to be able to *recognise* one when we see one.

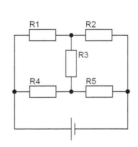

Series-parallel circuit

Figure 1.2c

For example, if we examine the circuit shown in Figure 1.2d (a common type of complex circuit that we call a 'bridge' circuit), we should quickly realise that, because of the way that resistor R_3 is connected, *none* of the resistors is either in series or in parallel with another: thus confirming that it *cannot* be a series-parallel circuit and, so, must be an example of a complex circuit.

Many complex circuits *include more than one source of e.m.f.*, such as the example shown in Figure 1.5:

Complex circuit

Figure 1.2d

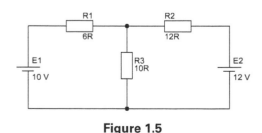

Figure 1.5

So complex circuits or 'networks' simply *cannot* be solved using only those techniques we've learnt to solve series, parallel, or series-parallel circuits.

Instead, we must use a process called '**network analysis**', which uses one or other of various techniques that are collectively known as '**network theorems**'.

There are *numerous* network theorems; *far* too many to be adequately dealt with in this chapter. So, we'll restrict ourselves to a selection of some of the more commonly used theorems. These are:

▶ **Loop Analysis** ('**Kirchhoff's Laws**' method)
▶ **Superposition Theorem**
▶ **Star/Delta Transformation**
▶ **Thévenin's Theorem.**

Before we start, however, we need to remind ourselves of the significance of **sense arrows**, **nodes**, and **closed loops**.

Understanding sense arrows

The term, '**sense arrows**', describes the use of arrows in circuit diagrams to represent the 'sense' or 'direction' in which potential differences act and currents flow. Sense arrows represent, if you like, a 'snapshot' of what is actually going on, or what we *assume* is going on, in a circuit *at any given instant*.

Even though we now know that, in metal conductors at least, electric current is a movement of free electrons from a negative potential to a positive potential, it has been a long traditional to use '**conventional flow**' instead.

With 'conventional flow', it's traditional to consider a *positive* potential as having a 'higher' potential than a *negative* potential, and that current therefore flows *from* this 'higher potential' *to* a 'lower potential' – in other words, 'conventional flow' takes place *through the load from the positive to the negative terminal.*

'**Conventional flow**' was established long before there was any understanding of the atom and its structure. At that time physicists, such as Benjamin Franklin, believed that current was a flow of some mysterious 'fluid' that moved through a wire from a higher (positive) pressure to a lower (negative) pressure. Despite our present knowledge of electron flow, it has been a well-established tradition to use conventional flow to indicate current direction circuit diagrams.

Sense arrows apply to *both* direct current (d.c.) and alternating current (a.c.) but, in this book, we will only be considering d.c. circuits.

Sense arrows for sources of electromotive force *(E)*

For *sources* of electromotive force *(E)*, such as batteries or d.c. generators, a single-headed sense arrow is used. The arrow head itself *always* represents the higher (**positive**) potential of that source – as illustrated in Figure 1.6:

Sense arrows for current

For **current**, an arrow superimposed over a conductor in a circuit diagram *always* points *in the direction of the assumed conventional current flow: i.e. from a higher (positive) potential to lower (negative) potential.*

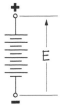

Figure 1.6

Sense arrows for voltage drops

For current to flow through an individual circuit component, such as resistors for example, we have learnt that there must be a difference in potential across each of those individual components. We call this difference in potential, a '**voltage drop**', which we label: U_1, U_2, U_3, etc. Again, the arrow-head for a sense arrow that represents a voltage drop *always* points towards the higher *(more positive)* potential – that is, in the *opposite direction* to the current sense arrow. This is shown in Figure 1.7.

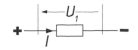

Figure 1.7

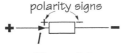

Figure 1.8

Whenever we have a particularly complicated circuit diagram, to avoid cluttering that diagram with lots of voltage sense arrows, it's often clearer, instead, to simply label voltage sources and voltage drops with positive (+) and negative (–) polarity signs, *instead* of sense arrows, as shown in Figure 1.8.

Examples of sense arrows in a series circuit

Figure 1.9 shows a **voltage-source** *(E)*, and the corresponding **current** *(I)* and **voltage drop** *(U₁, U₂,* and *U₃)* sense arrows (together with the polarity signs, mentioned above) in a straightforward series resistive circuit:

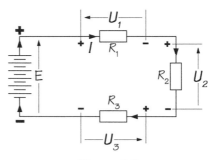

Figure 1.9

You will notice that the direction of each of the voltage drop sense arrows, U_1, U_2, and U_3, acts in the *opposite* sense to that of the electromotive force, E, provided by the battery. This agrees with **Kirchhoff's Voltage Law**, which states that *'the algebraic sum of the potential differences (voltages) around any closed loop equals zero'*. We can express this, as follows:

$$E - U_1 - U_2 - U_3 = 0$$

Sometimes, it's not clear in *which* direction a particular potential difference or current is acting within some parts of a complex circuit. In this case, we have to make an 'educated guess' and insert sense arrows according to that educated guess.

For example, in the circuit shown in Figure 1.10, the directions of the currents through resistors R_1, R_2, R_3, and R_4 are quite obvious, but we don't really have enough information to judge whether the direction of the current through resistor R_5 is upward or downward.

So, we must simply *guess* a direction, and insert a sense arrow in that particular direction. Then, if the *calculated* value of that current turns out to be a *positive* value, then our guess was correct, and it must be *acting in the same direction as our*

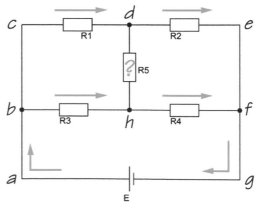

Figure 1.10

assumed sense arrow; on the other hand, should the calculated current turn out to be a *negative* value, then our guess was wrong and, instead, it must be *acting in the opposite direction to the assumed sense arrow.*

So even our 'educated guess' is *wrong*, the resulting polarity of our answer will always correct us!

'Nodes' and 'meshes'

It's important to understand what we mean by a '**node**' and by a '**mesh**', because many textbooks use these terms and, as we shall see, they play very important rôles in our calculations – particularly when we solve complex circuits using the '**Loop Equations**' method.

▶ A '**node**' is simply another term for a '**junction**' – that is, those points at which conductors are connected together. In Figure 1.11, for example, there are *four* 'nodes', or junctions, located at points **b**, **d**, **f**, and **h**.

▶ A '**mesh**' is also known as a '**closed loop**' or, simply, a '**loop**', and describes *any* complete route or pathway around an electric circuit, which has the same start and end point. A mesh *may*, or *may not*, include a voltage source such as a battery.

In the case of a series circuit, there is obviously just *one* mesh or closed loop. But for parallel, series-parallel, and complex circuits there can be *numerous* meshes. The exact number of such loops, of course, will depend entirely upon the configuration of any particular circuit.

For example, Figure 1.11 shows a series of schematic diagrams that illustrate the various meshes or loops (shown in bold) that exist in a typical complex circuit. For the complex circuit shown in **Figure 1.11a**, we can identify *seven* meshes:

▶ (**Figure 1.11b**):   mesh *a-c-e-g-a*
▶ (**Figure 1.11c**):   mesh *a-b-f-g-a*
▶ (**Figure 1.11d**):   mesh *a-c-d-h-f-g-a*
▶ (**Figure 1.11e**):   mesh *a-b-h-d-e-g-a*
▶ (**Figure 1.11f**):   mesh *b-c-e-f-b*
▶ (**Figure 1.11g**):   mesh *b-c-d-h-b*
▶ (**Figure 1.11h**):   mesh *h-d-e-f-h.*

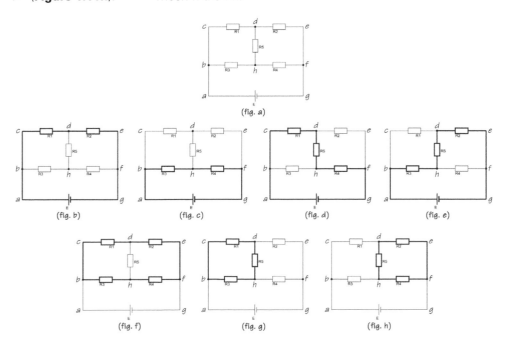

Figure 1.11

Meshes or loops don't necessarily include a source of potential difference. For example, in Figure 1.11, you will notice that, in *some* cases, the meshes *include* the voltage source (Figures *b, c, d,* and *e*) while, in *other* cases, they *don't*.

Kirchhoff's Laws

Kirchhoff's Laws are named in honour of the Prussian physicist, Gustav Kirchhoff (1824–1887), who formulated these laws while he was still a student.

His two laws apply to *all* categories of circuit, and it is very important to understand these laws if we are to understand the behaviour of voltages and currents in *any* circuit, whether simple or complicated.

There are *two* laws called, respectively, '**Kirchhoff's Current Law**' and '**Kirchhoff's Voltage Law**', as outlined below.

Kirchhoff's Current Law

Kirchhoff's Current Law may be expressed as follows:

> **Kirchhoff's Current Law** states that *'the algebraic sum of the currents at any junction is zero'.*

Let's apply this law to the circuit shown in Figure 1.12:

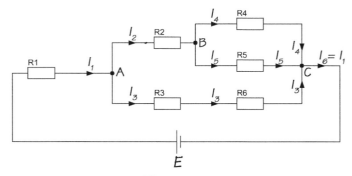

Figure 1.12

In this particular circuit, there are three 'junctions' or 'nodes', labelled **A**, **B**, and **C**.

▶ At junction **A**, the current (I_1) *approaching* that junction must be equal to the sum of the two currents, I_2 and I_3, *leaving* that junction. i.e.:

$$(I_1 = I_2 + I_3) \quad \text{or} \quad (I_1 - I_2 - I_3 = 0)$$

▶ At junction **B**, the current (I_2) *approaching* that junction must be equal to the sum of the two currents, I_4 and I_5, *leaving* that junction. i.e.:

$$(I_2 = I_4 + I_5) \quad \text{or} \quad (I_2 - I_4 - I_5 = 0)$$

▶ Finally, at junction **C**, the sum of the three currents ($I_3 + I_4 + I_5$) *approaching* that junction must be equal to the current, I_6, *leaving* that junction. i.e.:

$$(I_3 + I_4 + I_5 = I_6) \quad \text{or} \quad (I_3 + I_4 + I_5 - I_6 = 0)$$

And, of course, current I_6 is exactly the same current as I_1. So, in the above expression, if we wished, we could simply replace current label I_6 with I_1.

It is a misconception to think that the current *approaching* a junction 'splits up' at that junction. It's far more accurate to say that the current 'approaching' a junction is a combination of the two currents 'leaving' that junction. In other words, the magnitude of the current approaching a junction is determined by the sum of the currents leaving that junction – not the other way around!

Kirchhoff's Voltage Law

Kirchhoff's Voltage Law may be expressed as follows:

> **Kirchhoff's Voltage Law** states that *'for any closed loop, the algebraic sum of the voltages is zero'.*

If we examine the circuit, shown in Figure 1.12, again, we should be able to identify *six* closed loops or meshes. As examples, let's examine just *three* of these to learn how Kirchhoff's Voltage Law is applied.

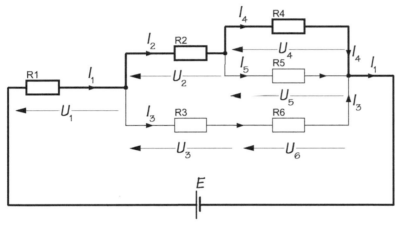

Figure 1.13

Let's start by applying Kirchhoff's Voltage Law to the loop shown in bold, in Figure 1.13, above.

Working from the battery's positive terminal, around the circuit, to its negative terminal (i.e. in a clockwise direction), we can write the following expression:

$$E - U_1 - U_2 - U_4 = 0$$

These voltage drops can also be expressed in terms of their relevant currents and resistances, as follows:

$$E - (I_1 R_1) - (I_2 R_2) - (I_4 R_4) = 0$$

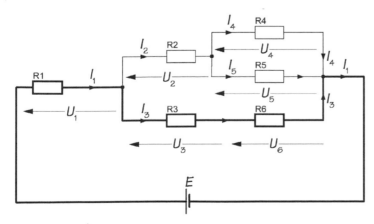

Figure 1.14

Let's repeat this for the loop shown in bold, in Figure 1.14.

$$E - U_1 - U_3 - U_6 = 0$$

Which, again, can also be written in terms of the currents and resistances, as follows:

$$E - (I_1 R_1) - (I_3 R_3) - (I_3 R_6) = 0$$

. . . which, if we wish, we could simplify to:

$$E - (I_1 R_1) - I_3 (R_3 + R_6) = 0$$

Finally, let's look at the loop shown in bold in Figure 1.15. This particular loop *doesn't contain any voltage sources*, so it's a little more complicated.

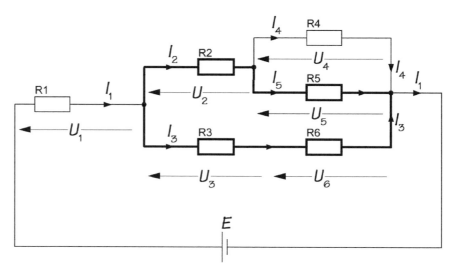

Figure 1.15

There's no voltage source in this particular loop, so we can work either *clockwise* or *counterclockwise*, around the loop in order to write a Kirchhoff's Voltage Law equation. As we worked *clockwise* in the other two examples, for consistency (after all, there's no point in making life more difficult than it already is!) we'll do so again: in which case we see that the voltage drops U_2 and U_5 will be working in *the opposite direction* to the voltage drops U_3 and U_6:

$$-U_2 - U_5 + U_6 + U_3 = 0$$

. . . which we can also write as:

$$-(I_2 R_2) - (I_5 R_5) + (I_3 R_6) + (I_3 R_3) = 0$$

(Had we chosen to work *counterclockwise*, instead of clockwise, the signs in the above equation would change but, as pointed out earlier, our error of judgment will come to light as we work through the problem.)

Having cleared away all the preparatory work, let's now move on to examine the network theorems themselves . . .

'Loop analysis' method for solving complex circuits

The '**Loop Analysis**' method of solving complex circuits is also widely known as the '**Kirchhoff's Laws Method**'. But by whichever name we choose to call it, it's exactly the same technique. So, perhaps, it would be a good idea to call it the '**Kirchhoff's Loop Analysis Method**'.

Strictly speaking, this method of solving complex circuits *isn't* really a 'network theorem' (all network theorems involve *simplifying* complex circuits by redrawing them as 'equivalent circuits') but, rather, is simply an extension of what we have already learnt about Kirchhoff's Current and Voltage Laws. Nevertheless, this technique is widely used to solve complex circuits, which is why we've bundled it under the heading 'network theorems'.

In this section, then, we will learn how this method makes use of both **Kirchhoff's Current** and **Voltage Laws** to solve complex circuits, by generating **simultaneous equations** in order to determine the various unknown quantities within that circuit.

Simultaneous equations are used to represent conditions in which two or more unknown variables are related to each other through an equal number of equations.

For relatively straightforward complex circuits with, say, just *two* or *three* unknown currents, Kirchhoff's Loop Analysis method is reasonably straightforward. But for more difficult complex circuits, with a correspondingly larger number of unknowns, the resulting number of simultaneous equations becomes unwieldy, and the necessary 'number crunching' makes them both difficult and time consuming to solve longhand.

A free MS Windows software download that may be used for solving simultaneous equations with up to six unknowns, is **Microsoft *Mathematics***, which is available from the Microsoft website (see the **Appendix** to this chapter for more information).

So, for more complicated complex circuits with, say, four or more unknowns, it is usually far easier to use one or other of the alternative network theorems *instead* of Kirchhoff's Loop Analysis method described in this section.

For this reason, therefore, we will restrict our explanation of the Loop Analysis method to relatively straightforward complex circuits with no more than three unknowns.

There are *two* different techniques for applying Kirchhoff's Loop Analysis and, for this reason, we shall solve the same worked example, twice – using each of the two approaches.

Worked example 1

Calculate (a) the currents through each of the resistors in the circuit shown in Figure 1.16, together with (b) the potential difference appearing across resistor R_3:

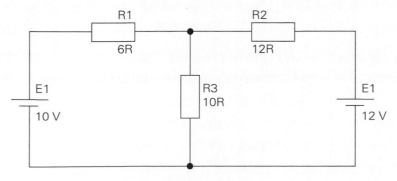

Figure 1.16

Solution (using technique 1)

The first step is to *label* the diagram, so that the junctions ('nodes') and loops ('meshes') can be clearly identified – this is shown in Figure 1.17, using the letters *a–f*. After that, we can insert *polarity signs* (rather than sense arrows, to minimise clutter) to represent the circuit's e.m.f.s, voltage drops, and sense arrows for the currents. In this example, the directions of the (conventional) current sense arrows appear to be obvious but, remember, even if we should get it wrong, the *calculated* currents would bring this error to light by being a negative quantity.

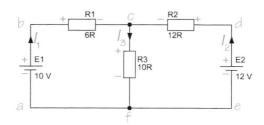

Figure 1.17

Applying **Kirchhoff's Current Law** to node (junction) *c*, we can write:

sum of currents *leaving* junction c = sum of currents *approaching* junction c

$$I_3 = I_1 + I_2 \quad \text{or} \quad I_3 - I_1 - I_2 = 0$$

In this example, although there are *three* different currents, there are actually only *two* unknowns (I_1 and I_2), because current I_3 is simply the sum of I_1 and I_2. *This technique should be applied to all complex circuits, to minimum the number of unknowns.*

So, in this example, we need to generate just *two* simultaneous equations in order to determine the values of all three currents.

Next, for each of the batteries, E_1 and E_2, we apply **Kirchhoff's Voltage Law** around the closed loops that include the resistor, R_3. So, let's start by considering loop **b-c-f-a-b**, shown in Figure 1.18:

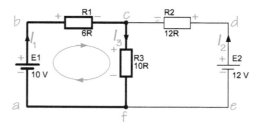

Figure 1.18

Taking the voltage polarity marks (+ and −) into account:

Algebraic-sum of potential differences around any closed loop is zero:

$$E_1 - I_1 R_1 - I_3 R_3 = 0$$
$$E_1 = I_1 R_1 + I_3 R_3$$
$$\text{(since } I_3 = I_1 + I_2)$$
$$E_1 = I_1 R_1 + (I_1 + I_2) R_3$$
$$10 = 6I_1 + 10(I_1 + I_2)$$
$$10 = 6I_1 + 10I_1 + 10I_2$$
$$10 = 16I_1 + 10I_2 \quad ---\text{(equation 1)}$$

Next, let's consider loop **d-c-f-e**, shown in Figure 1.19:

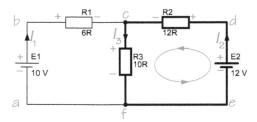

Figure 1.19

Again, taking the voltage polarity marks (+ and −) into account:

Algebraic-sum of potential differences around any closed loop is zero:

$$E_2 - I_2R_2 - I_3R_3 = 0$$
$$E_2 = I_2R_2 + I_3R_3$$
$$\text{(since } I_3 = I_1 + I_2)$$
$$E_2 = I_2R_2 + (I_1 + I_2)R_3$$
$$12 = 12I_2 + 10(I_1 + I_2)$$
$$12 = 12I_2 + 10I_1 + 10I_2$$
$$12 = 10I_1 + 22I_2 \quad \text{---(equation 2)}$$

We now have *two* simultaneous equations with *two* unknowns (I_1 and I_2). To eliminate I_2, we can start by multiplying equation (2) by 1.6:

$$1.6 \times (12 = 10I_1 + 22I_2)$$
$$19.2 = 16I_1 + 35.2I_2 \quad \text{---(equation 3)}$$

Now, we can *subtract* equation (1) from equation (3) in order to eliminate current I_1:

$$
\begin{array}{rcl}
19.2 = & 16I_1 & +35.2I_2 \text{ (equation 3)} \\
10 \;\; = & 16I_1 & +10\;\;I_2 \text{ (equation 1)}- \\
\hline
9.2 = & & +25.2I_2 \\
\hline
\end{array}
$$

$$I_2 = \frac{9.2}{25.2} = \textbf{0.365 A (Answer)}$$

So, now, we can insert the value of I_2 into either equation (1) or equation (2), to determine the value of I_1:

$$10 = 16I_1 + 10I_2 \quad \text{---(equation 1)}$$
$$10 = 16I_1 + (10 \times 0.365)$$
$$10 = 16I_1 + 3.65$$
$$16I_1 = 10 - 3.65$$
$$16I_1 = 6.35$$
$$I_1 = \frac{6.35}{16} = \textbf{0.397 A (Answer)}$$

Now we can find the value of I_3, by simply *adding* the values of I_1 and I_2:

$$I_3 = (I_1 + I_2) = (0.365 + 0.397) = \textbf{0.762 A (Answer)}$$

Finally, we can find out the potential difference across resistor R_3, as follows:

$$U_3 = I_3R_3 = 0.762 \times 10 = \textbf{7.62 V (Answer)}$$

Solution (using technique 2)

An *alternative* technique for solving the circuit in worked example 1 involves us not worrying about labelling the individual currents through the various components or making any assumptions at all about their directions. *Instead*, we simply consider two loop currents, *both* acting in an arbitrary direction: in this example, *clockwise* (we could have chosen a counterclockwise direction had we wished, as it doesn't really matter!), and labelled I_1 and I_2 respectively.

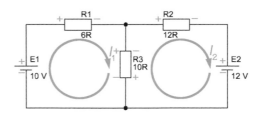

Figure 1.20

Before proceeding any further, we should add polarity signs for the supply voltages as well as for the voltage drops appearing across the resistors, *based on the loop current directions*.

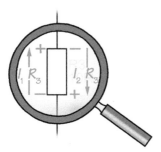

Figure 1.21

> We need to take *extra care* when we consider resistor R_3, because there are actually *two* voltage drops to worry about: one (I_1R_3) due to loop current I_1 and the other (I_2R_3) due to loop current I_2, each acting in *opposite directions!*
>
> We *have* to take *both* these voltage drops into account, *whether we are considering the loops taken by current I_1 or current I_2* (Figure 1.21).

Notice that loop current I_1 passes *downward* through resistor R_3, while loop current I_2 passes *upward* through that *same* resistor. So, we must take into account *two* voltage drops appearing across R_3: one (acting *upwards*) caused by loop current I_1, *together with* another (acting *downwards*) caused by loop current I_2. To emphasise these two voltage drops, in the following equations we'll highlight one of them in blue.

So, bearing this in mind, and applying Kirchhoff's Voltage Law to loop current I_1,

$$E_1 = I_1R_1 + I_1R_3 - (I_2R_3)$$
$$10 = 6I_1 + 10I_1 - (10I_2)$$
$$10 = 16I_1 - 10I_2 \; - - - \text{(equation 1)}$$

Applying Kirchhoff's Voltage Law to loop current I_2,

$$-E_2 = I_2R_3 - (I_1R_3) + I_2R_2$$
$$-12 = 10I_2 - (10I_1) + 12I_2$$
$$-12 = -10I_1 + 22I_2 \; - - - \text{(equation 2)}$$

We now have *two* simultaneous equations with *two* unknowns (I_1 and I_2). To eliminate I_2, we can start by multiplying equation (2) by 1.6:

$$1.6 \times (-12 = -10I_1 + 22I_2)$$
$$-19.2 = -1.6I_1 + 35.2I_2 \; - - - \text{(equation 3)}$$

Now, we can *add* equations (1) and (3) in order to eliminate current I_1:

$$
\begin{array}{rcl}
10 & = & 16I_1 - 10 \;\; I_2 \;\text{(equation 1)}\\
-19.2 & = & -16I_1 + 35.2I_2 \;\text{(equation 3)} +\\
\hline
-9.2 & = & 25.2I_2
\end{array}
$$

$$I_2 = \frac{-9.2}{25.2} = \underline{\mathbf{-0.365\ A\ (Answer)}}$$

The *negative* sign, of course, means that the current is actually flowing in the *opposite* direction to the direction that we assumed in the diagram.

We can now insert this value for I_2 into equation (1) to determine a value for current I_1:

$$10 = 16I_1 - 10I_2$$
$$10 = 16I_1 - [10 \times (-0.365)]$$
$$10 = 16I_1 + 3.65$$
$$16I_1 = 10 - 3.65 = 6.35$$
$$I_1 = \frac{6.35}{16}$$
$$= \underline{\mathbf{0.397A\ (Answer)}}$$

Either of these two techniques can be used to solve complex circuits using **Kirchoff's Loop Analysis** method. Both are equally valid, but some students find one technique easier to use than the other.

Worked example 2

In this next example, we'll look at a somewhat more difficult complex circuit: that shown in Figure 1.22. Once again, we want to determine the values of the currents through, and the voltage drops, across each of the five resistors. For this worked example, we'll use the first technique we applied to worked example 1.

With just a single source of e.m.f., this circuit might *appear* to be less complicated than the circuit in **worked example 1** but, in fact, it *isn't* because, as we shall see, this circuit will have *three* unknowns and, so, it will require us to generate and solve *three* simultaneous equations!

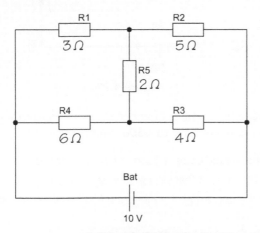

Figure 1.22

Solution

Once again, the first step is to label the diagram, so that the junctions and loops can be clearly identified. After that, we can insert sense arrows to represent the circuit's e.m.f.s, voltage drops and currents. In this example, the directions of the (conventional) currents are, for the most part, obvious – *but we'll have to make a guess regarding the current direction through R_5* (we'll *assume* that it acts downwards). If we should get it wrong, it doesn't really matter, because the calculated current would indicate our error by being a negative quantity.

The key to solving *any* circuit using the **Loop Analysis** method is to do our best to *minimise the number of unknown currents*.

> Just how to minimise the number of unknown currents is not always obvious, and to do this requires a lot of practice and, sometimes, some 'lateral thinking', so don't be put off if you experience difficulty at first – that's perfectly normal and you will *not* be alone in finding it difficult! But, as they say, 'practice makes perfect'!

For example, in the above circuit, we *could* assign a different label to the various currents flowing through each of the resistors (e.g. I_1, I_2, I_3, I_4, and I_5), as well as to the supply current *(I)* – but that gives us *six* different currents to solve!

It will make the problem *very* much easier if we can express some of these currents in terms of some of the others. In fact, with this particular circuit, we can reduce the number of unknown currents (six) to just *three* – as we have done in Figure 1.23.

Carefully examine the current labelled, below, to understand how we managed to achieve this. *Being able to do this is an important skill, and it only comes with practice.*

Based on these assumed current directions, we can also apply potential difference sense arrows or, to reduce any clutter, apply positive and negative signs to the resistors.

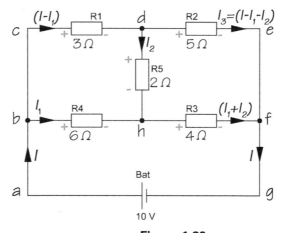

Figure 1.23

This is how we reduced the number of unknown currents from six to just *three:*
Applying Kirchhoff's Current Law to node (junction) *b:*

> let the supply current *approaching* junction ***b*** = I
> let the current *leaving* junction ***b***, through $R_4 = I_1$
> and the current *leaving* junction ***b***, through $R_1 = I_1 + (I - I_1)$

Applying Kirchhoff's Current Law to node (junction) *d:*

> current *approaching* junction ***d***, through $R_1 = (I - I_1)$
> current *leaving* junction ***d***, through through $R_5 = I_2$
> current *leaving* junction ***d***, current through $R_2 = (I - I_1 - I_2)$

Applying Kirchhoff's Current Law to node (junction) *h:*

> current *approaching* junction ***h***, through $R_4 = I_1$
> current *approaching* junction ***h***, through through $R_5 = I_2$
> current *leaving* junction ***h***, current through $R_3 = (I_1 + I_2)$

Applying Kirchhoff's Current Law to node (junction) f:

current *approaching* junction **f**, through $R_3 = (I_1 + I_2)$

current *approaching* junction **f**, through $R_5 = (I - I_1 - I_2)$

supply current *leaving* junction **f** $= (I - I_1 - I_2) + (I_1 + I_2) = I$

So, as you can see, we have managed to define all *six* currents in terms of just *three* – that is: **I**, **I₁**, and **I₂**. This means we will only have to solve *three*, not six, simultaneous equations!

Next, we apply **Kirchhoff's Voltage Law** around the closed loops until we achieve three equations with three unknowns. So, let's start by considering loop **a-b-f-g-a** – as highlighted in blue in Figure 1.24.

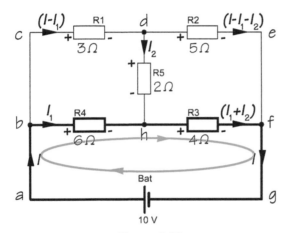

Figure 1.24

For loop **a-b-f-g-a**:

Algebraic-sum of potential differences around any closed loop is zero:

$$E - I_1 R_4 - (I_1 + I_2)R_3 = 0$$
$$E = I_1 R_4 + (I_1 + I_2)R_3$$
$$10 = 6I_1 + 4(I_1 + I_2)$$
$$10 = 6I_1 + 4I_1 + 4I_2$$
$$10 = 10I_1 + 4I_2 \ {-}{-}{-} \text{(equation 1)}$$

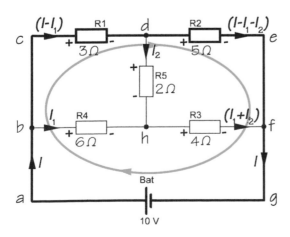

Figure 1.25

For loop **a-c-e-g-a**, in Figure 1.25:

Algebraic-sum of potential differences around any closed loop is zero:

$$E - (I - I_1)R_1 - (I - I_1 - I_2)R_2 = 0$$
$$E = (I - I_1)R_1 + (I - I_1 - I_2)R_2$$
$$10 = 3(I - I_1) + 5(I - I_1 - I_2)$$
$$10 = (3I - 3I_1) + (5I - 5I_1 - 5I_2)$$
$$10 = 8I - 8I_1 - 5I_2 \ \text{---(equation 2)}$$

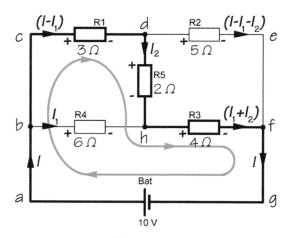

Figure 1.26

For loop **a-c-d-h-f-g-a**, in Figure 1.26:

Algebraic-sum of potential differences around any closed loop is zero:

$$E - (I - I_1)R_1 - I_2R_5 - (I_1 + I_2)R_3 = 0$$
$$E = (I - I_1)R_1 + I_2R_5 + (I_1 + I_2)R_3$$
$$10 = 3(I - I_1) + 2I_2 + 4(I_1 + I_2)$$
$$10 = (3I - 3I_1) + 2I_2 + (4I_1 + 4I_2)$$
$$10 = 3I + I_1 + 6I_2 \ \text{---(equation 3)}$$

Let's list these three loop equations, together:

$$10 = 10I_1 + 4I_2 \ \text{---(equation 1)}$$
$$10 = 8I - 8I_1 - 5I_2 \ \text{---(equation 2)}$$
$$10 = 3I + I_1 + 6I_2 \ \text{---(equation 3)}$$

If we multiply equation (2) by 3 and multiply equation (3) by 8, and subtract the two equations, we can eliminate current *I:*

$$(10 = 8I - 8I_1 - 5I_2) \times 3: \quad 30 = 24I - 24I_1 - 15I_2 \ \text{---(equation 4)}$$
$$(10 = 3I + I_1 + 6I_2) \times 8: \quad 80 = 24I + 8I_1 + 48I_2 \ \text{---(equation 5)}$$

Subtracting equation (5) from equation (4):

$$-50 = -32I_1 - 63I_2 \ \text{---(equation 6)}$$

Now, if we multiply equation (1) by 3.2, and add it to equation (6), we can eliminate current **I_1**:

$$(10 = 10I_1 + 4I_2) \times 3.2: \quad 32 = 32I_1 + 12.8I_2 \ \text{---(equation 1)}$$
$$-50 = -32I_1 - 63I_2 \ \text{---(equation 6)}$$

Adding equations (1) and (6):

$$-18 = -50.2I_2$$

$$I_2 = \frac{18}{50.2} = 0.36 \text{ A}$$

We can now insert this value for I_2 into equation (1) in order to determine the value of current I_1:

$$10 = 10I_1 + 4I_2 ---\text{(equation 1)}$$
$$10 = 10I_1 + 4(0.36)$$
$$10 = 10I_1 + 1.44$$
$$10I_1 = 10 - 1.44$$
$$I_1 = \frac{8.56}{10} = 0.86 \text{ A}$$

Finally, we can insert the values of currents I_1 and I_2 into equation (2), in order to determine the value of the supply current, I:

$$10 = 8I - 8I_1 - 5I_2 ---\text{(equation 2)}$$
$$10 = 8I - 8(0.86) - 5(0.36)$$
$$10 = 8I - 6.88 - 1.80$$
$$10 = 8I - 8.643$$
$$8I = 10 + 8.88$$
$$I = \frac{18.88}{8} = 2.36 \text{ A}$$

So, to summarise:

current through resistor $R_1 = (I - I_1) = 2.36 - 0.86 =$ **1.50 A (Answer)**

current through resistor $R_2 = (I - I_1 - I_2) = 2.36 - 0.86 - 0.36 =$ **1.14 A (Answer)**

current through resistor $R_3 = (I_1 + I_2) = 0.86 + 0.36 =$ **1.22 A (Answer)**

current through resistor $R_4 = I_1 =$ **0.86 A (Answer)**

current through resistor $R_5 = I_2 =$ **0.36 A (Answer)**

Now, we can determine the voltage drop across each of the resistors:

voltage drop across $R_1 = 1.50 \times 3 =$ **4.50 V (Answer)**

voltage drop across $R_2 = 1.14 \times 5 =$ **5.70 V (Answer)**

voltage drop across $R_3 = 1.22 \times 4 =$ **4.88 V (Answer)**

voltage drop across $R_4 = 0.86 \times 6 =$ **5.16 V (Answer)**

voltage drop across $R_5 = 0.36 \times 2 =$ **0.72 V (Answer)**

Summary

By now, we will have realised how time consuming (and how open to errors) **Kirchhoff's Loop Analysis** method of solving complex circuits can be, *even for just two or three simultaneous equations!*

Even though the complex circuits in **worked examples 1** and **2** were *relatively* straightforward, the amount of number crunching involved to solve them was extensive, time consuming, and open to all sorts of inadvertent mistakes. So, *surely* there *must* be simpler ways of solving complex circuits than by using the Loop Analysis method?

Fortunately, there *are* simpler methods of solving complex circuits, as we shall now learn, as we move on to examine various network theorems!

Although these methods are, perhaps, *more intuitive*, or *easier to understand*, or *mathematically simpler*, they are still **time consuming** – *so it would be a mistake to think that these methods will necessarily save us any time!* **Network theorems** differ from the Loop Analysis method because they *all* involve *redrawing a complex circuit* in order to create a simpler, '**equivalent circuit**' or, in some cases, two or more equivalent circuits that are easier to solve than the *actual* circuit itself.

Superposition Theorem

The **Superposition Theorem** offers us an instinctively logical method for solving the type of complex circuit shown in **worked example 1**. That is, *for complex circuits with more than one voltage source.*

The **Superposition Theorem** states that, '*in a network of fixed resistances containing more than one source of e.m.f., the resultant current in any branch is the algebraic sum of the currents that would be produced by each e.m.f., acting alone, with all other sources of e.m.f. being replaced by their respective internal resistances'.*

Figures 1.27a–c summarise how we apply the **Superposition Theorem** to a complex circuit having two source e.m.f.s:

In Figure 1.27b, we have replaced battery E_2 with a short-circuit (or, if given, its internal resistance). And, in Figure 1.27c, we have replaced battery E_1 with a short-circuit (or, if given, its internal resistance).

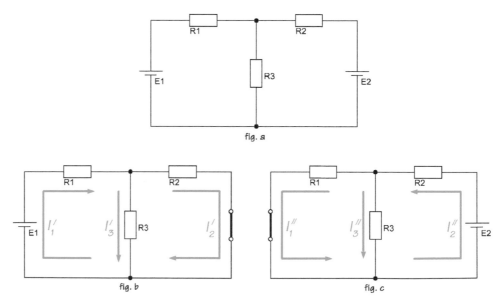

fig. a

fig. b   fig. c

Figure 1.27

These change the configuration of the circuit from being a *complex* circuit into a pair of simple *series-parallel* circuits, and . . .

...the current flowing through resistor R_1 = algebraic-sum of I_1' and I_1''.

...the current flowing through resistor R_2 = algebraic-sum of I_2' and I_2''.

...the current flowing through resistor R_3 = algebraic-sum of I_3' and I_3''.

So now let's apply this theorem to **worked example 1**, which we tackled earlier using the Kirchhoff's Loop Analysis method.

Worked example 3

Using the **Superposition Theorem**, calculate (a) the currents through each of the resistors shown in the circuit shown in Figure 1.28a, together with (b) the potential difference appearing across resistor R_3:

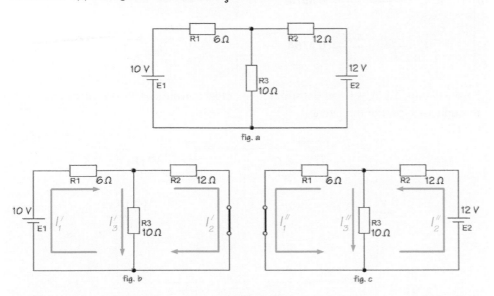

Figure 1.28

In Figure 1.28b, we've removed the battery E_2, and replaced it with a short-circuit, and the resulting currents, I_1', I_2', and I_3' are shown. In Figure 1.28c, we've removed the battery E_1, and replaced it with a short-circuit, and the resulting currents, I_1'', I_2'', and I_3'' are shown.

According to the **Superposition Theorem**, the current through resistance, R_1, will be the *difference* between currents I_1' and I_1'', the current through resistance R_2 will be the *difference* between currents I_2' and I_2'', while the current through R_3 will be the *sum* of currents I_3' and I_3''.

Referring to Figure 1.28b	**Referring to Figure 1.28c**

Referring to Figure 1.28b

Let the equivalent resistance of R_2 and R_3 be R_A:

$$R_A = \frac{R_2 R_3}{R_2 + R_3} = \frac{12 \times 10}{12 + 10}$$

$$= \frac{120}{22} = 5.45\ \Omega$$

$$R_{total} = R_1 + R_A = 6 + 5.45 = 11.45\ \Omega$$

$$I_1' = \frac{E_1}{R_{total\,b}} = \frac{10}{11.45} = \textbf{0.873 A}$$

Voltage drop across R_3 and R_2:

$$U_{R3/2} = E_1 - \left(I_1' R_1\right) = 10 - \left(0.873 \times 6\right)$$

$$= 10 - 5.24 = 4.76\ \text{V}$$

Referring to Figure 1.28c

Let the equivalent resistance of R_1 and R_3 be R_B:

$$R_B = \frac{R_1 R_3}{R_1 + R_3} = \frac{6 \times 10}{6 + 10}$$

$$= \frac{60}{16} = 3.75\ \Omega$$

$$R_{total\,c} = R_2 + R_B = 12 + 3.75 = 15.75\ \Omega$$

$$I_2'' = \frac{E_2}{R_{total\,b}} = \frac{12}{15.75} = \textbf{0.762 A}$$

Voltage drop across R_3 and R_1:

$$U_{R3/1} = E_2 - \left(I_2'' R_2\right) = 12 - \left(0.762 \times 12\right)$$

$$= 12 - 9.14 = 2.86\ \text{V}$$

Now, we can determine the currents through R_2 and R_3:

$$I'_2 = \frac{U_{R3/2}}{R_2} = \frac{4.76}{12} = \textbf{0.397A}$$

$$I'_3 = \frac{U_{R3/2}}{R_3} = \frac{4.76}{10} = \textbf{0.476A}$$

Now, we can determine the currents through R_1 and R_3:

$$I''_2 = \frac{U_{R3/1}}{R_1} = \frac{2.86}{6} = \textbf{0.477A}$$

$$I''_3 = \frac{U_{R3/1}}{R_3} = \frac{2.86}{10} = \textbf{0.286A}$$

Finally (Figure 1.29), we can determine the actual currents and their directions through each part of the circuit:

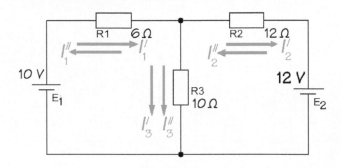

Figure 1.29

current through $R_1 = I'_1 - I''_1 = 0.873 - 0.477 = \textbf{0.396 A}$ (left to right) **Answer**

current through $R_2 = I'_2 - I''_2 = 0.397 - 0.762 = \textbf{0.365 A}$ (right to left) **Answer**

current through $R_3 = I'_3 + I''_3 = 0.476 - 0.286 = \textbf{0.762 A}$ (right to left) **Answer**

These answers correspond to **worked example 1**, *in which the Loop Equation method was used to solve the circuit.*

Conclusion

Although the amount of 'number crunching' used in this example of using the **Superposition Theorem** might not be a great deal less than we used in example 1, the technique is certainly far more intuitive and simpler than the Loop Analysis method.

Delta-Star/Star-Delta Transformation Theorems

As the name suggests, the '**Delta-Star/Star-Delta Transformation Theorems**' offer us a way of converting a delta connection to an equivalent star connection, and *vice versa*. And, as we shall see, this gives us a somewhat easier way of solving problems of the type we saw in worked example 2.

We first met the **delta** and **star** (wye) connections in the companion book, *An Introduction to Electrical Science*, where we learnt that they were two standard connections used in three-phase a.c. systems. However, the connections themselves are by no means unique to three-phase a.c. systems, and *any* three resistors can be connected in delta or star configuration, as shown in Figure 1.30.

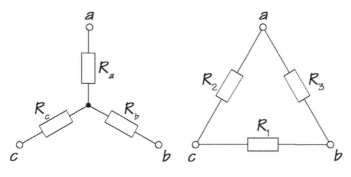

Figure 1.30

Of course, these connections do *not* have to be arranged to form the above *shapes*. For example, the schematics shown in Figure 1.31 represent resistors connected in *exactly* the same configurations.

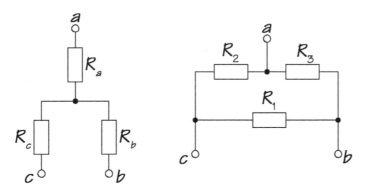

Figure 1.31

In electronics engineering, these configurations are widely used, but are more commonly known as a '**Tee Network**' (equivalent to 'star' configuration) and a '**Pi (π) network**' (equivalent to 'delta' configuration), and are generally illustrated as shown in Figure 1.32.

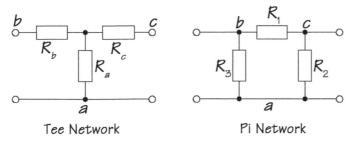

Tee Network   Pi Network

Figure 1.32

So *where* are you likely to see these sort of connections, and *why* should it be useful to be able to convert from one to the other?

Well, consider the example in Figure 1.33. The left-hand illustration shows a complex circuit and even a brief examination will reveal that three of its resistors (R_1, R_2, and R_3) form a *delta* connection (centre). If we then apply delta-star transformation to this, we end up with an equivalent circuit (right) that is an easily solved *series-parallel* circuit *instead* of the original *complex* circuit.

(We could, of course, apply *exactly* the same transformation, instead, to the right-hand side of this particular circuit.)

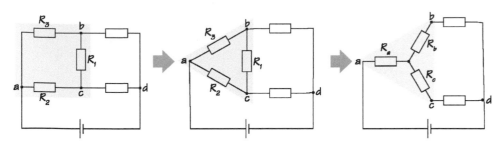

Figure 1.33

Returning, now, to the delta and star connections, to prevent any potential confusion, it's very important to apply consistent labelling to both the delta and star connections. In this case, as illustrated in the following figures.

▶ we start by assigning the letters *a-b-c*, in a *clockwise* direction, to the terminals and
▶ for the **star connection**, we then label the resistors to match the terminals to which they are connected, i.e. R_a, R_b, and R_c
▶ for the delta connection, we use numerals to label the resistors, i.e. R_1, R_2, and R_3, where resistor R_1 is directly *opposite* terminal *a*, resistor R_2 is directly *opposite* terminal *b*, and resistor R_3 is directly *opposite* terminal *c*.

Now let's move on to the two transformations. We're not going to derive the following equations; in fact, as we shall see, it's not even necessary to learn the equations themselves as it's much easier to identify the *patterns* these equations form. In other words, we can treat the equations in much the same way that some people memorise their bank PIN numbers by the *pattern* formed when they key in those numbers into a keyboard at an ATM machine.

'Delta to star' transformation

This describes how, given a *delta* connection (the greyed-out resistors, in Figure 1.34), we can produce an equivalent *star* (or *'wye'*) connection.

*The equivalent star resistance connected to any given terminal is equal to the **product** of the two delta resistances connected to the same terminal, divided by the **sum** of all three delta resistances.*

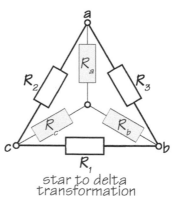

delta to star transformation

Figure 1.34

$$R_a = \frac{R_2 R_3}{R_1 + R_2 + R_3} \qquad R_b = \frac{R_3 R_1}{R_1 + R_2 + R_3} \qquad R_c = \frac{R_1 R_2}{R_1 + R_2 + R_3}$$

'Star to delta' transformation

This describes how, given a *star* (or *'wye'*) connection (the greyed-out resistors, in Figure 1.35), we can produce an equivalent *delta* connection.

*The equivalent delta resistance between any two terminals is the **sum** of the two star resistances connected to those terminals, plus the **product** of the same two resistances **divided** by the third star resistance.*

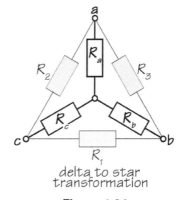

star to delta transformation

Figure 1.35

$$R_1 = R_b + R_c + \frac{(R_b R_c)}{R_a} \qquad R_2 = R_c + R_a + \frac{(R_c R_a)}{R_b} \qquad R_3 = R_a + R_b + \frac{(R_a R_b)}{R_c}$$

Once again, it's worth emphasising that the trick for remembering these transformations is *not* to attempt to memorise the equations themselves but, instead, to recognise the 'pattern' of the relationship between the resistances at any one corner of the triangle, in much the same way as we might remember our bank account's PIN number by the *pattern* formed when we punch-in the numbers at an ATM machine, rather than by actually memorizing the PIN number itself. Once we have learnt that pattern for these transformations for one corner, we can apply the same pattern to the other two corners of the triangle.

Now, let's apply this technique to **worked example 2**. We want to determine the current through each resistor, and the voltage drop across them. So let's redraw the circuit, with all the given values shown:

Worked example 4

Let's repeat worked example 2 but, this time, using the delta-star transformation method – as illustrated in Figure 1.36.

As it stands, it's a *complex* circuit, as there is no resistor in series or in parallel with another. But, by applying the delta-star transformation, we can create an equivalent series-parallel circuit that then becomes relatively easy to solve using the techniques that we learnt for series-parallel circuits.

So, if we consider the 3-Ω, 2-Ω and 6-Ω resistors, we see that they form a *delta* configuration.

From this, we can determine the equivalent resistances for a star configuration:

$$R_a = \frac{R_2 R_3}{R_1 + R_2 + R_3} = \frac{6 \times 3}{2+6+3} = \frac{18}{11} = 1.64\,\Omega$$

$$R_b = \frac{R_3 R_1}{R_1 + R_2 + R_3} = \frac{3 \times 2}{2+6+3} = \frac{6}{11} = 0.55\,\Omega$$

$$R_c = \frac{R_1 R_2}{R_1 + R_2 + R_3} = \frac{2 \times 6}{2+6+3} = \frac{12}{11} = 1.09\,\Omega$$

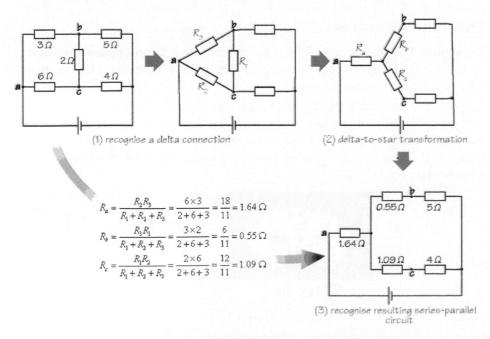

(1) recognise a delta connection

(2) delta-to-star transformation

$$R_a = \frac{R_2 R_3}{R_1 + R_2 + R_3} = \frac{6 \times 3}{2+6+3} = \frac{18}{11} = 1.64\,\Omega$$

$$R_b = \frac{R_3 R_1}{R_1 + R_2 + R_3} = \frac{3 \times 2}{2+6+3} = \frac{6}{11} = 0.55\,\Omega$$

$$R_c = \frac{R_1 R_2}{R_1 + R_2 + R_3} = \frac{2 \times 6}{2+6+3} = \frac{12}{11} = 1.09\,\Omega$$

(3) recognise resulting series-parallel circuit

Figure 1.36

This circuit has now been changed into a straightforward series-parallel circuit, enabling us to easily determine its total resistance:

$$R_{total} = Ra + \frac{(0.55+5)(1.09+4)}{(0.55+5)+(1.09+4)} = 1.64 + \frac{5.55 \times 5.09}{5.55+5.09} = 1.64 + \frac{28.25}{10.64} = 4.30\,\Omega$$

So the resulting supply current must be:

$$I = \frac{E}{R_{total}} = \frac{10}{4.30} = 2.33\text{ A}$$

This allows us to determine the voltage drop across **R_a**:

$$U_a = IR_a = 2.33 \times 1.64 = 3.82\text{ V}$$

So, the voltage drop across the rest of the circuit is

$$U = E - U_a = 10 - 3.82 = 6.18\text{ V}$$

. . . enabling us to determine the current through the 5Ω resistor:

$$I = \frac{U}{R_b+5} = \frac{6.18}{0.55+5} = \frac{6.18}{5.55} = 1.11\text{ A}$$

. . . and to determine the current through the 4-Ω resistor:

$$I = \frac{U}{R_c+4} = \frac{6.18}{1.09+4} = \frac{6.18}{5.09} = 1.21\text{ A}$$

So the voltage drop across the 5 Ω resistor must be:

$$= 1.11 \times 5 = 5.55\text{ V}$$

. . . and the voltage drop across the 4 Ω resistor must be:

$$= 1.21 \times 4 = 4.84\text{ V}$$

So the potential difference between points **b** and **c**, must be:

$$= 5.55 - 4.84 = \mathbf{0.71\text{ V (Answer)}}$$

Any minor differences between these answers and those obtained in worked example 2 are due to rounding-up or rounding-down.

Thévenin's Theorem

Thévenin's Theorem is named in honour of Frenchman **Léon Charles Thévenin** (1857–1926), a telegraph engineer with the *Postes et Télégraphes*, then the French national communications company.

Suppose that we have a '**black box**' (Figure 1.37) containing voltage source together with other electrical components, but the *only* access to whatever is *inside* that black box is by means of a pair of external terminals, **x**–**y**. How can we determine the essential characteristics of whatever circuitry is within the box without opening it up and taking a peek?

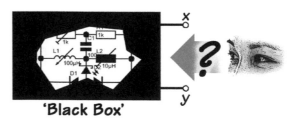

'Black Box'

Figure 1.37

Figure 1.38

Well, first, we could find the **open-circuit voltage** (its electromotive force) appearing across the terminals, by using a high-impedance voltmeter (Figure 1.38). A high-impedance voltmeter draws so little current, that any internal voltage drop within the voltage source or other circuit components can be ignored and, so, we really are measuring the open-circuit voltage.

Next, we could short-circuit the terminals, using a low-impedance ammeter (Figure 1.39) to measure the resulting **short-circuit current**. As this is only a theoretical exercise, there is no fear of harming whatever circuit components are actually inside the black box!

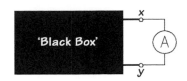

Figure 1.39

Let's suppose the value of the open-circuit voltage was, say, 100 V, and the value of the short-circuit current was, say, 5 mA.

Even though we have absolutely no knowledge whatsoever of the *actual* circuit within the black box (and regardless of how complicated that circuit actually is), we could conclude that it *behaves* as though it contained a 100 V source in series with a resistance of:

$$R = \frac{100}{5 \times 10^{-3}} = 20 \times 10^3 = 20 \text{ k}\Omega$$

So, whatever is inside the 'black box' can always be reduced to a simple **equivalent circuit** comprising, in this case, *a **single-source e.m.f.** of 100 V*, in **series** with a **single resistance** of 20 kΩ.

Thévenin's equivalent circuit is illustrated in Figure 1.40. We've labelled the single-source e.m.f. as E_{Thevenin}, and the series resistance as R_{Thevenin}. We call these, '**Thévenin's Equivalent Voltage**' and '**Thévenin's Equivalent Resistance**', respectively.

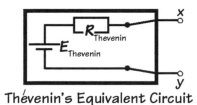

Thévenin's Equivalent Circuit

Figure 1.40

Now, suppose we have a resistive circuit as shown in Figure 1.41, and we wish to determine the current through, say, resistor R_x:

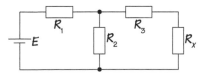

Figure 1.41

We could now consider resistor R_x as being an *external load*, connected to a 'black box' *containing the rest of the circuit*, as shown in Figure 1.42:

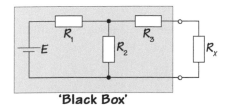

'Black Box'

Figure 1.42

As explained previously, the circuit elements contained within the black box, regardless of how complicated they *really* are, can be reduced to an *equivalent circuit*, comprising *a single source e.m.f.* in series with *a single internal resistance* – as shown in Figure 1.43.

We call this '**Thévenin's Equivalent Circuit**':

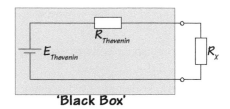

Figure 1.43

Now, it is a relatively simple matter to determine the current flowing through resistor, R_x – *providing* we can first of all determine values for $E_{Thevenin}$ and $R_{Thevenin}$!

This, then, is the principle behind **Thévenin's Theorem**, which is defined as follows:

> ***Thévenin's Theorem*** *states that, 'Any two-terminal network of fixed resistances and voltage sources may be replaced by a single voltage source having an equivalent voltage equal to the open-circuit voltage across the terminals of the original network, and having an internal resistance equal to the resistance looking back into the network from the two terminals, with all the voltages sources replaced by their internal resistances.'*

Thévenin's Theorem, itself, is *not* complicated. But what *can* be complicated is rearranging the circuit within our imaginary 'black box' into a more understandable or recognisable form.

So, to apply Thévenin's Theorem, we simply follow these steps:

step 1: Remove the resistor under consideration, and calculate the value of the voltage ($E_{Thevenin}$) appearing across the 'gap' left between the terminals.

step 2: Replace all existing voltage sources with their internal resistances (unless otherwise stated, we can assume that these are zero ohms). Then determined the equivalent resistance ($R_{Thevenin}$), as 'seen' from the open terminals.

step 3: Draw the Thévenin equivalent circuit, with the resistor under consideration placed in series with a constant-voltage generator (***$E_{Thevenin}$***) determined in step 1 and the internal resistance (***$R_{Thevenin}$***) determined in step 2.

So let's demonstrate the use of Thévenin's Theorem, using a simple series-parallel circuit before moving on and applying it to a complex circuit. By applying Thévenin's Theorem to this simple problem, we are using a 'sledgehammer to crack a nut'; we're *only* doing it in order to demonstrate the steps required to apply the theorem.

Worked example 5

Use Thévenin's Theorem to determine the value of current passing through resistor R_4 in the circuit illustrated in Figure 1.44.

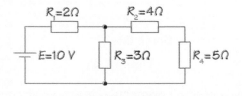

Figure 1.44

Solution

So Thévenin's Theorem assumes that resistor R_4 is the *load*, and the rest of the circuit is *enclosed inside a 'black box'* – as shown in Figure 1.45:

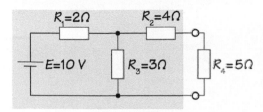

Figure 1.45

Step 1

Remove the load (resistor R_4), and 'look back' into the black box to determine the voltage that will appear across the terminals:

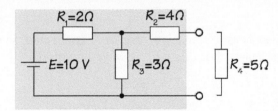

Figure 1.46

With the 'load' disconnected, no current is being drawn from the black box, so there can be no voltage drop across resistor R_3 and, so, the voltage appearing across the terminals ($E_{Thevenin}$) must be *exactly* the same as the voltage drop (U_{R3}) appearing across R_3.

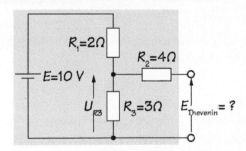

Figure 1.47

Note how, in Figure 1.47, we have redrawn the circuit that's inside the 'black box', to make it clearer what is going on.

So the current *(I)* produced by the battery will only flow around the loop formed by resistors R_1 and R_3:

$$I = \frac{E}{\left(R_1 + R_3\right)} = \frac{10}{2+3} = 2 \text{ A}$$

From which we can determine $\boldsymbol{E}_{\textbf{Thevenin}}$:

$$\boldsymbol{E}_{\textbf{Thevenin}} = U_{R3} = IR_3 = 2 \times 3 = \textbf{6 V}$$

Step 2

Now, we remove the battery, and replace it with a short-circuit, and determine the resistance ($R_{Thevenin}$) of the components inside the 'black box':

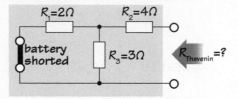

Figure 1.48

It should be clear that, from Figure 1.48, with the battery replaced by a short-circuit, resistors R_1 and R_3 are in parallel with each other, and this combination is in series with resistor R_2, so we can find $R_{Thevenin}$ as follows:

$$R_{Thevenin} = \left(\frac{R_1 R_3}{R_1 + R_3}\right) + R_4 = \left(\frac{2 \times 3}{2+3}\right) + 4 = \mathbf{5.2\ \Omega}$$

Step 3

Draw Thévenin's equivalent circuit with resistor R_4 connected, as shown in Figure 1.49, and calculate the current (I_{R4}) flowing through R_4.

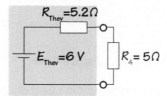

Figure 1.49

$$I_{R4} = \frac{E_{Thevenin}}{\left(R_{Thevenin} + R_4\right)} = \frac{6}{(5.2+5)} \approx \mathbf{0.59\ A\ (Answer)}$$

Now that we've solved a simple circuit, let's apply Thévenin's Theorem to worked example 3 that we completed earlier.

Worked example 6

Repeat worked example 2 but, this time, using Thévenin's Theorem to calculate (a) the current through, and (b) the voltage drop across resistor, $\mathbf{R_5}$, in the complex circuit shown in Figure 1.50:

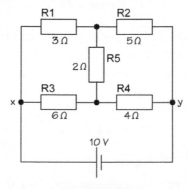

Figure 1.50

Solution

Step 1

Remove resistor, R_5, (the 'load'), and 'look back' into the circuit from terminals a–b, to determine the voltage (Thévenin's equivalent voltage, $E_{Thevenin}$) appearing across those terminals.

As you can see from Figure 1.51, this is no longer a 'complex circuit' because, with resistor, R_5, removed, *it has become a simple series-parallel circuit:*

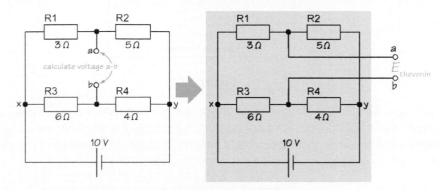

Figure 1.51

To find the potential difference between terminals a and b, we start by finding the potential differences or voltage drops across resistors R_1 and R_2. Since junction x is common to *both* resistors, all we then need to do is to subtract these two voltage drops to find the *potentials* at terminals a and b:

So, first, we need to find the currents flowing through both the upper and lower branches:

$$I_{upper} = \frac{E}{R_1 + R_2} = \frac{10}{3+5} = \frac{10}{8} = 1.25\,A$$

$$I_{lower} = \frac{E}{R_3 + R_4} = \frac{10}{6+4} = \frac{10}{10} = 1.0\,A$$

This enables us to calculate the voltage drops across resistors R_1 and R_3:

$$U_1 = I_{upper}R_1 = 1.25 \times 3 = 3.75\,V$$

$$U_3 = I_{lower}R_3 = 1.0 \times 6 = 6.00\,V$$

So the potential at point a with respect to junction x must be **–3.75 V**, and the potential at point b with respect to junction x must be **–6.00 V** and, therefore, the *potential difference* ($E_{thevenin}$) between terminals a–b will be the *difference* between these two potentials:

$$E_{thevenin} = -3.75 - (-6) = \textbf{2.25 V}$$

Step 2

Now, we short-circuit the battery (or replace it with its internal resistance, if given), and 'look back' into the circuit from terminals a–b, to determine the resistance (Thévenin's equivalent resistance, $R_{thevenin}$).

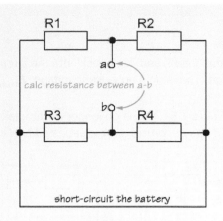

Figure 1.52

Although the resulting circuit shown in Figure 1.52 is actually a simple series-parallel circuit, at first glance, this isn't necessarily obvious. So it's often necessary to *redraw the circuit* so that it becomes a little *more* obvious, as shown in Figure 1.53 (Figure 1.55 explains *how* we did this, step-by-step):

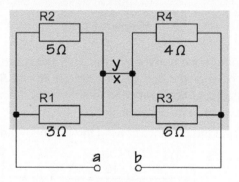

Figure 1.53

When redrawn, it becomes obvious that the circuit is a series-parallel circuit and, so, finding its resistance is straightforward:

Let R_a be the equivalent resistance of R_2 and R_1 in parallel, and R_b be the equivalent resistance of R_4 and R_3 in parallel:

$$R_a = \frac{R_2 R_1}{R_1 + R_2} = \frac{5 \times 3}{3+5} = \frac{15}{8} = 1.88\ \Omega \quad \text{and} \quad R_b = \frac{R_4 R_3}{R_3 + R_4} = \frac{4 \times 6}{6+4} = \frac{24}{10} = 2.4\ \Omega$$

So, the total resistance ($R_{thevenin}$) is the sum of R_a and R_b:

$$R_{thevenin} = R_a + R_b = 1.88 + 2.4 = \mathbf{4.28\ \Omega}$$

Step 3

Now, we can draw the **Thévenin's equivalent circuit**, as shown in Figure 1.54:

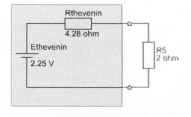

Figure 1.54

. . . from which, we can now calculate the current through R_5, as follows:

$$I_5 = \frac{E_{thevenin}}{(R_{thevenin} + R_5)} = \frac{2.25}{4.28 + 2} = \frac{2.25}{6.28} = 0.359 \text{ A (Answer a.)}$$

The voltage drop across resistor, R_5, will, of course, be:

$$U_5 = I_5 \times R_5 = 0.359 \times 2 = 0.718 \text{ V (Answer b.)}$$

If we compare this with the answers in worked example 2, we will see that the answers are identical.

We can use exactly the same technique to determine the current through, and the voltage drop across, *any* of the resistors in the complex circuit shown in this example. You might like to try this, and compare your answers with those provided in worked example 1.

How we simplified the circuit in worked example 6

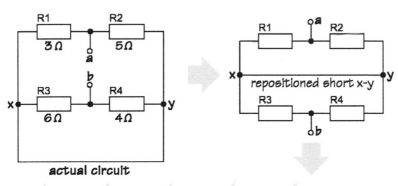

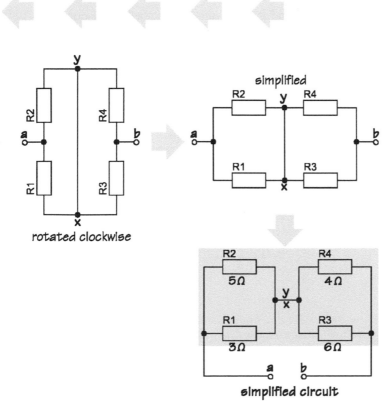

Figure 1.55

Self-test problems

1   Calculate the value of current passing through resistor, R_5, in the circuit shown in Figure 1.56, using (a) the *Kirchhoff's Loop Equation* method, and (b) *Thévenin's Theorem*. (**Answer**: 0.32 A)

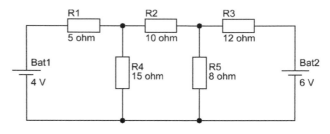

Figure 1.56

2   Calculate the voltage across resistor R_2 in Figure 1.57 using (a) *Kirchhoff's Loop Equation* method, and (b) *Thévenin's Theorem*. (**Answer**: 2 V)

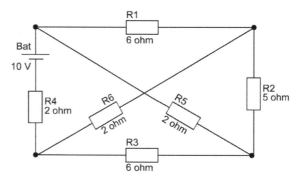

Figure 1.57

3   Two batteries, *A* and *B*, are connected as shown in Figure 1.58. Battery *A* has an e.m.f. of 100 V with an internal resistance of 0.5 Ω, and battery *B* has an e.m.f. of 80 V with an internal resistance of 0.4 Ω. Calculate the value of the currents passing through each battery and through resistor R_2, using (a) the *Kirchhoff's Loop Equation* method, and (b) the *Superposition Theorem*. (**Answer**: battery *A*: 9.03 A discharge; battery *B*: 6.45 A discharge; resistor R_2: 15.49 A)

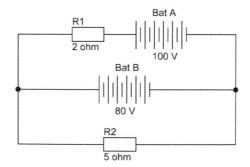

Figure 1.58

4   Use (a) the *Kirchhoff's Loop Equation* method, (b) *Thévenin's Theorem*, and (c) the *Delta-Star Transformation Theorem*, to calculate the currents passing through each resistor. (**Answer**: resistor R_1: 0.372 A; R_2: 0.251 A; R_3: 0.896 A; R_4: 0.776 A; R_5: 0.120 A)

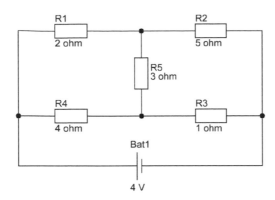

Figure 1.59

Conclusion

In this chapter, we have learnt that the techniques used to solve *series*, *parallel*, and *series-parallel* circuits *cannot be applied to complex circuits or 'networks'*.

Instead, we have to use what are generally called '**network theorems**'.

There are *numerous* network theorems, but we have limited our study of them to just four: '**Kirchhoff's Loop Equations**' (not strictly a 'theorem'). the '**Superposition Theorem**', '**Delta-Star/Star-Delta Transformation Theorem**', and '**Thévenin's Theorem**'.

With the exception of the **Kirchhoff's Loop Equation** technique, all the others require us to *simplify the original circuit diagram in such a way that we can reduce it to an 'equivalent' series-parallel circuit*, thus making it *relatively* easy to solve. We should emphasise the word, 'relatively' because there is *no 'truly simple' way of solving complex circuits!*

It would be foolish to suggest that *any* of these techniques are 'easy' but, unfortunately, that is the nature of complex circuits! Each technique requires practice, which is why the self-test problems have been provided.

Now that we've completed this chapter, we need to examine its **objectives** listed at its start. Placing a question mark at the end of each objective turns that objective into a **test item**. If we can answer those test items, then we've met the objectives of this chapter.

Appendix

Using Microsoft 'Mathematics' to solve simultaneous equations

Microsoft Mathematics is a free software download for Windows' PCs (unfortunately, it's currently not available for the Macintosh OSX operating system), which provides a set of very useful mathematical tools that, amongst other things, will allow you to solve simultaneous equations with up to *six* unknowns. It's useful, therefore, when solving complex circuits using the **Loop Equations** method.

Microsoft Mathematics is available from the **Microsoft** website. You can download the software file, and install it on your Windows PC, following the on-screen instructions.

Figure 1.60 shows the Microsoft Mathematics' opening screen:

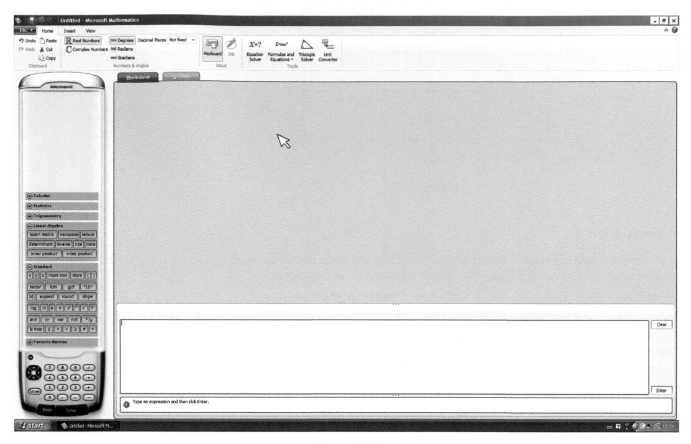

Figure 1.60

Click on **Formulas and Equations**, in the top menu bar, to open the **Equation Solver** (Figure 1.61):

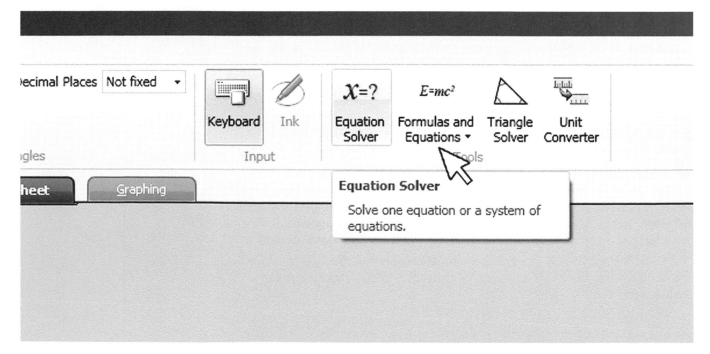

Figure 1.61

For, say, *two* simultaneous equations, in the **Equation Solver** dialogue box, select **Solve a System of 2 Equations** from the drop-down list (Figure 1.62):

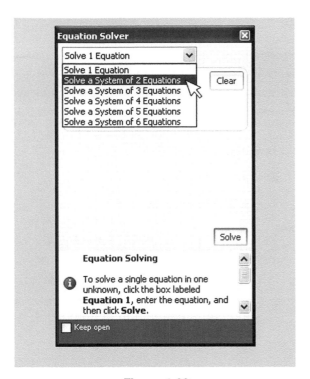

Figure 1.62

Enter each equation into each of the two, **Equation 1** and **Equation 2** boxes (Figure 1.63). You can use **x** to represent current I_1 and **y** to represent current I_2:

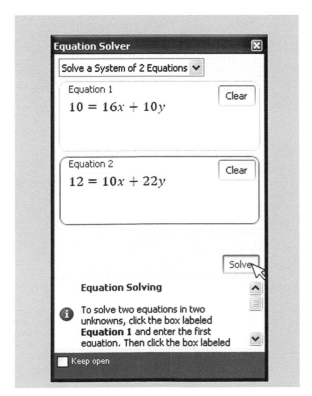

Figure 1.63

Finally, click on the **Solve** button and the software will do all the work necessary to present the answers (Figure 1.64):

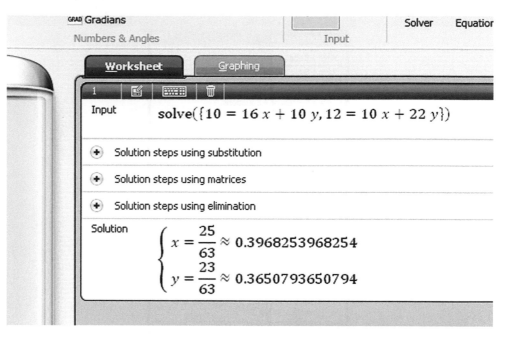

Figure 1.64

CHAPTER 2

Review of series and parallel a.c. circuits

Objectives

On completion of this chapter you should be able to:

1. sketch a current phasor diagram, representing, a
   a. series *R-L* circuit.
   b. series *R-C* circuit.
   c. series *R-L-C* circuit.
2. convert a series circuit phasor diagram into an impedance diagram.
3. convert a series circuit phasor diagram into a power diagram.
4. apply basic geometry and/or trigonometry, to derive equations for
   a. impedance,
   b. resistance,
   c. reactance,
   d. apparent power,
   e. true power,
   f. reactive power,
   g. phase angle, and
   h. power factor.

5. sketch a voltage phasor diagram, representing, a
   a. parallel *R-L* circuit.
   b. parallel *R-C* circuit.
   c. parallel *R-L-C* circuit.
6. convert a parallel circuit phasor diagram into an admittance diagram.
7. convert a parallel circuit phasor diagram into a power diagram.
8. apply basic geometry and/or trigonometry, to derive equations for
   a. admittance,
   b. conductance,
   c. susceptance,
   d. impedance,
   e. resistance,
   f. reactance,
   g. apparent power,
   h. true power,
   j. reactive power,
   k. phase angle, and
   l. power factor.

Electrical Science for Technicians. 978-1-138-84926-6 © Adrian Waygood.
Published by Taylor & Francis. All rights reserved.

Introduction

In this chapter, we are going to review the behaviour of **series** and **parallel a.c. circuits**, which were covered in detail in the companion to this book, *An Introduction to Electrical Science*.

In particular, we are going to remind ourselves how *easy* it is to generate *any* equation required for solving series or parallel a.c. circuit problems, *through the application of basic geometry and/or trigonometry to right-angled triangles*.

And, *most importantly*, we are going to remind ourselves that **none of the equations derived in this chapter need be committed to memory!**

Quite literally, the *only* two equations that we need to remember throughout this chapter are those for inductive reactance and capacitive reactance.

To understand the behaviour of series and parallel a.c. circuits, we must be able to

▶ represent the voltages and currents in a circuit, by constructing an appropriate phasor diagram
▶ memorise and apply the following equations for inductive and capacitive reactance:

   ▷ inductive reactance: $\qquad X_L = 2\pi f L$

   ▷ capacitive reactance: $\qquad X_c = \dfrac{1}{2\pi f C}$

▶ memorise and apply the mnemonic, '**CIVIL**' (or '**ELI** the **ICE** man') to inductive and capacitive circuits
▶ memorise and apply Pythagoras's Theorem to right-angled triangles
▶ memorise and apply the following ratios to right-angled triangles:

   ▷ $\sin\phi = \dfrac{\text{opposite}}{\text{hypotenuse}}; \quad \cos\phi = \dfrac{\text{adjacent}}{\text{hypotenuse}}; \quad \tan\phi = \dfrac{\text{opposite}}{\text{adjacent}}.$

Armed with the above knowledge, we will be able to solve *any* series or parallel a.c. circuit . . . *without memorising any of the equations that follow!*

Before we proceed, though, a word on the alternative mnemonics: '**CIVIL**' and '**ELI the ICEman**'. Each of these is intended to remind us of the phase relationships between currents and voltages in inductive and capacitive circuits:

▶ **CIVIL** reminds us that, in a capacitive *(C)* circuit, current *leads* voltage (**CIV**IL) whereas, in an inductive *(L)*, current *lags* voltage (C**IVIL**).
▶ **ELI the ICEman** reminds us that, in an inductive *(L)*, current *lags* voltage (**ELI**) whereas, in a capacitive *(C)* circuit, current *leads* voltage (**ICE**).

The mnemonic, '**ELI the ICEman**' is particularly popular with Canadian students because, no doubt, they can relate 'him' to their country's severe winters!

Purely resistive, inductive, and capacitive circuits

Let's start, then, by realising that **purely resistive**, **purely inductive**, and **purely capacitive** circuits are all 'ideal' circuits. By 'ideal', we mean they are circuits that exist theoretically but *not in real life!* However, by understanding how these 'ideal' circuits behave theoretically, then we can extend that understanding to 'real' circuits: i.e. **resistive-inductive (R-L), resistive-capacitive (R-C),** and **resistive-inductive-capacitive (R-L-C)** circuits.

We must remember that the circuit symbols we use are representing the *quantities*, resistance, inductance, and capacitors; they do *not* represent resistors, inductors, or capacitors!

Purely resistive circuit

If we apply a sinusoidal voltage to a **purely resistive circuit** (Figure 2.1), the resulting current will be **in phase** with the applied voltage, as illustrated in Figure 2.2:

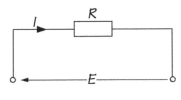

Figure 2.1

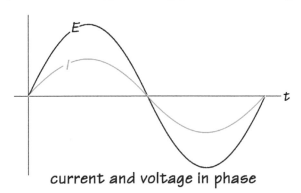

current and voltage in phase

Figure 2.2

We can represent this condition using a **phasor diagram**, in which the voltage and current phasors both 'point' in the same direction, normally drawn along the real (positive) horizontal axis, as illustrated in Figure 2.3.

Figure 2.3

The *lengths* of the voltage and current phasors are unrelated to each other, as (a) they represent two different quantities and (b) are drawn to completely different scales. What is important is the *angle between the two phasors*.

The opposition to the movement of current in a purely resistive circuit is, of course, **resistance (R),** which is the ratio of voltage to current:

$$R = \frac{\bar{E}}{I} \text{ , expressed in ohms.}$$

Purely inductive circuit

If we apply a sinusoidal voltage to a purely inductive circuit (Figure 2.4), the resulting current would **lag** the applied voltage by 90 electrical degrees, as illustrated in Figure 2.5. We can remember this relationship from the mnemonic, '**CIVIL**' (or, if you prefer, '**ELI** the ICE man'):

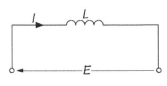

Figure 2.4

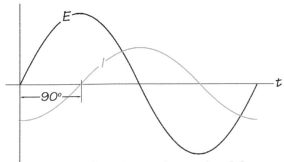

current lagging voltage by 90°

Figure 2.5

Figure 2.6

We can represent this condition using a **phasor diagram**, in which the voltage and current phasors are displaced from each other by 90°. As phasors are considered to 'rotate' in a *counterclockwise* direction (and positive angles are measured, counterclockwise, from the horizontal positive axis), the voltage phasor is drawn 90° counterclockwise ('leading') relative to the current phasor as shown in Figure 2.6.

(It would be equally correct to have drawn the voltage phasor horizontally, and the current phasor 90° clockwise.)

We call the opposition to current in a purely inductive circuit '**inductive reactance**' (X_L), which is directly proportional to the inductance (L) of the circuit and the frequency (f) of the supply, i.e.:

$$X_L = 2\pi fL, \text{ expressed in ohms.}$$

The inductive reactance of any purely inductive circuit can also be determined from the ratio of voltage to current, i.e.:

$$X_L = \frac{\overline{E}}{\overline{I}}, \text{ expressed in ohms.}$$

Purely capacitive circuit

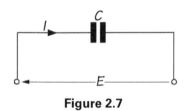

Figure 2.7

If we apply a sinusoidal voltage to a purely capacitive circuit (Figure 2.7), the resulting current would **lead** the applied voltage by 90 electrical degrees, as illustrated in Figure 2.8. Again, we can remember this relationship from the mnemonic, '**CIV**IL' (or, if you prefer, 'ELI the **ICE** man'):

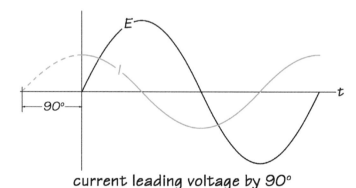

current leading voltage by 90°

Figure 2.8

Figure 2.9

We can represent this condition using a **phasor diagram**, in which the voltage and current phasors are displaced from each other by 90°. As phasors are considered to 'rotate' in a counterclockwise direction, the voltage phasor is drawn 90° clockwise ('lagging') relative to the current phasor as shown in Figure 2.9.

(It would be equally correct to have drawn the voltage phasor horizontally, and the current phasor 90° counterclockwise.)

We call the opposition to current in a purely capacitive circuit, '**capacitive reactance**' (X_C), which is inversely proportional to the capacitance (C) of the circuit and the frequency (f) of the supply, i.e.:

$$X_C = \frac{1}{2\pi fC}, \text{ expressed in ohms.}$$

The capacitive reactance of a purely capacitive circuit can also be determined from the ratio of voltage to current, i.e.:

$$X_C = \frac{\overline{E}}{\overline{I}}, \text{ expressed in ohms.}$$

R-L, R-C, and R-L-C series circuits

'Real', a.c. circuits comprise *combinations* of resistance, inductance, and capacitance. Whenever we draw schematic diagram for a series **R-L**, **R-C**, or **R-L-C** circuit, we must once again remind ourselves that the circuit symbols we use are representing the *quantities*, resistance, inductance, and capacitors; they do *not* represent resistors, inductors, or capacitors!

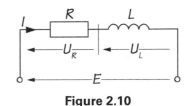

Figure 2.10

> **VITAL !** The key to understanding and solving *any* a.c. circuit, series or parallel, is **the ability to draw its phasor diagram**, *and* **to apply basic geometry and trigonometry to it**.
>
> Remember, a phasor diagram is little more than a **right-angled triangle** for which we need to be able to find the lengths of its three sides and the angles between them.
>
> If you learn nothing else from this chapter, learn this!

Figure 2.11a

R-L series circuit

To draw the phasor diagram for an **R-L** series circuit, we *always* start by drawing the reference phasor along the positive, horizontal, axis (Figure 2.10).

Figure 2.11b

Because **current** is common to each component of a series circuit, the current is *always* used as the **reference phasor** (Figure 2.11a).

Next, we draw the phasor representing the voltage drop across the resistive component $(\bar{U}_R)$ in phase with the reference phasor (Figure 2.11b). Then, we draw the phasor representing the voltage drop across the inductive component $(\bar{U}_L)$, leading the reference current by 90° (Figure 2.11c).

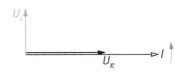

Figure 2.11c

Finally, we vectorially add $\bar{U}_R$ and $\bar{U}_L$ to show the supply voltage, $\bar{E}$ (Figure 2.11d).

The mnemonic, **CIVIL**, will always help you remember whether a voltage drop leads or lags the current.

So, for a series **R-L circuit**, the supply voltage is shown *leading* the current by some angle, ϕ, which we call the 'phase angle'.

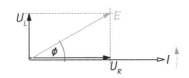

Figure 2.11d

R-C series circuit

The phasor diagram for an **R-C** series circuit (Figure 2.12) is drawn in just the same way as it was for an *R-L* series circuit, *except* that the voltage drop across the capacitive component $(\bar{U}_C)$ is drawn *lagging* the reference phasor, as shown in Figure 2.13.

So, for a series **R-C circuit**, the supply voltage is shown *lagging* the current by some angle, ϕ, which we call the 'phase angle'.

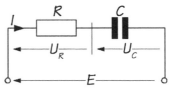

Figure 2.12

R-L-C series circuit

To draw the phasor diagram for an **R-L-C** series circuit (Figure 2.14), we *always* start by drawing the reference phasor along the positive, horizontal, axis (Figure 2.15a). Because current is common to each element of a series circuit, the reference phasor always represents the current (Figure 2.15a). Next we draw the phasor representing the voltage drop across the resistive component $(\bar{U}_R)$ *in phase* with the reference phasor (Figure 2.15b). Then we draw the phasor representing the voltage drop across the inductive component $(\bar{U}_L)$, *leading* the reference current by 90° (Figure 2.15c). Next, we drawn the phasor representing the voltage drop across the capacitive component $(\bar{U}_C)$, *lagging* the reference current by 90° (Figure 2.15d).

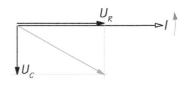

Figure 2.13

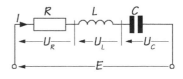

Figure 2.14

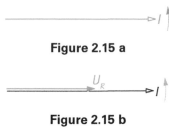

Figure 2.15 a

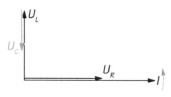

Figure 2.15 b

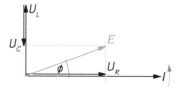

Figure 2.15 c

Figure 2.15 d

Figure 2.15 e

> **Important**! We *always* make phasor $\bar{U}_L$ *bigger* than phasor $\bar{U}_C$, or *vice versa* – it really doesn't matter *which* is longer, provided one is longer than the other (the purpose of the phasor diagram is to *generate equations*, rather than to accurately depict the circuit)! If we should make them the *same* length, then they will represent the circuit at **resonance**, which is a unique condition that *we want to avoid* when drawing a general phasor diagram!

Finally, we vectorially add $\bar{U}_R$, $\bar{U}_L$, and $\bar{U}_C$ to show the supply voltage, $\bar{E}$ (Figure 2.15e). So, for a series **R-L-C circuit**, the supply voltage can *either* lag or lead the current. If voltage drop, $\bar{U}_L$, is *larger* than voltage drop, $\bar{U}_C$ (as shown), then the supply voltage will be *leading* the current; however, if voltage drop, $\bar{U}_C$, is *larger* than voltage drop, $\bar{U}_L$, then the supply voltage will be *lagging* the current.

Solving series a.c. circuits by generating equations

Now that we know how to draw a circuit's phasor diagram, solving an 'electrical' problem becomes a simple matter of applying either *basic geometry* (Pythagoras's Theorem) or *basic trigonometry* (sine, cosine, or tangent ratios) involving simple right-angled triangles.

For example, by applying **Pythagoras's Theorem** to each of the following phasor diagrams, which are really just simple **right-angled triangles**, the following relationships are revealed:

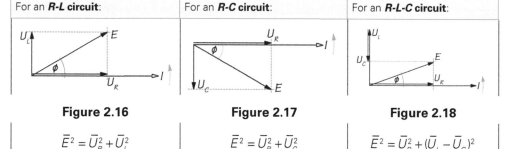

For an **R-L circuit**:	For an **R-C circuit**:	For an **R-L-C circuit**:
Figure 2.16	**Figure 2.17**	**Figure 2.18**
$\bar{E}^2 = \bar{U}_R^2 + \bar{U}_L^2$	$\bar{E}^2 = \bar{U}_R^2 + \bar{U}_C^2$	$\bar{E}^2 = \bar{U}_R^2 + (\bar{U}_L - \bar{U}_C)^2$

From which we can derive the following equations:

$\bar{E} = \sqrt{\bar{U}_R^2 + \bar{U}_L^2}$	$\bar{E} = \sqrt{\bar{U}_R^2 + \bar{U}_C^2}$	$\bar{E} = \sqrt{\bar{U}_R^2 + (\bar{U}_L - \bar{U}_C)^2}$
$\bar{U}_R = \sqrt{\bar{E}^2 - \bar{U}_L^2}$	$\bar{U}_R = \sqrt{\bar{E}^2 - \bar{U}_C^2}$	$\bar{U}_R = \sqrt{\bar{E} - (\bar{U}_L - \bar{U}_C)^2}$
$\bar{U}_L = \sqrt{\bar{E}^2 - \bar{U}_R^2}$	$\bar{U}_C = \sqrt{\bar{E}^2 - \bar{U}_R^2}$	$(\bar{U}_L - \bar{U}_C) = \sqrt{\bar{E} - \bar{U}_R^2}$

We can also determine the phase angle for each circuit, as follows:

$\phi = \cos^{-1}\left(\dfrac{\bar{U}_R}{\bar{E}}\right)$	$\phi = \cos^{-1}\left(\dfrac{\bar{U}_R}{\bar{E}}\right)$	$\phi = \cos^{-1}\left(\dfrac{\bar{U}_R}{\bar{E}}\right)$

Impedance diagrams

We can convert each of the above *voltage* phasor diagrams into a corresponding **impedance diagram**, by simply *dividing throughout by the reference phasor,* $\bar{I}$, as follows:

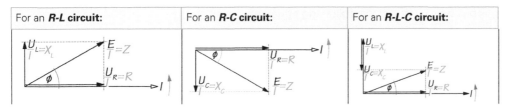

| For an **R-L circuit:** | For an **R-C circuit:** | For an **R-L-C circuit:** |

| **Figure 2.19** | **Figure 2.20** | **Figure 2.21** |

From which we get the following equations:

$Z = \dfrac{\bar{E}}{\bar{I}}$	$Z = \dfrac{\bar{E}}{\bar{I}}$	$Z = \dfrac{\bar{E}}{\bar{I}}$
$R = \dfrac{\bar{U}_R}{\bar{I}}$	$R = \dfrac{\bar{U}_R}{\bar{I}}$	$R = \dfrac{\bar{U}_R}{\bar{I}}$
$X_L = \dfrac{\bar{U}_L}{\bar{I}}$	$X_C = \dfrac{\bar{U}_C}{\bar{I}}$	$X_L = \dfrac{\bar{U}_L}{\bar{I}}$
		$X_C = \dfrac{\bar{U}_C}{\bar{I}}$
		$X = \dfrac{(\bar{U}_L - \bar{U}_C)}{\bar{I}}$

Applying **Pythagoras's Theorem** to each impedance diagram:

$Z^2 = R^2 + X_L^2$	$Z^2 = R^2 + X_C^2$	$Z^2 = R^2 + (X_L - X_C)^2$

From which we can find the values of Z, X_L, and X_C:

$Z = \sqrt{R^2 + X_L^2}$	$Z = \sqrt{R^2 + X_C^2}$	$Z = \sqrt{R^2 + (X_L - X_C)^2}$
$R = \sqrt{Z^2 - X_L^2}$	$R = \sqrt{Z^2 - X_C^2}$	$R = \sqrt{Z^2 + (X_L - X_C)^2}$
$X_L = \sqrt{Z^2 - R^2}$	$X_C = \sqrt{Z^2 - R^2}$	$(X_L - X_C) = \sqrt{Z^2 - R^2}$

We can also determine the phase angle for each circuit, as follows:

$\phi = \cos^{-1}\left(\dfrac{R}{Z}\right)$	$\phi = \cos^{-1}\left(\dfrac{R}{Z}\right)$	$\phi = \cos^{-1}\left(\dfrac{R}{Z}\right)$

Power diagrams

We can convert any of the above *voltage* phasor diagrams into a corresponding **power diagram**, simply by *multiplying throughout by the reference phasor*, $\bar{I}$, as follows.

Remember, '**apparent power**' *(S)* is expressed in volt amperes, '**true power**' *(P)* is expressed in watts, and reactive power *(Q)* is expressed in reactive volt amperes.

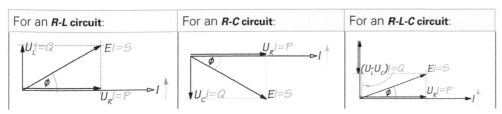

| For an **R-L circuit:** | For an **R-C circuit:** | For an **R-L-C circuit:** |

| **Figure 2.22** | **Figure 2.23** | **Figure 2.24** |

From which we get the following equations:

apparent power, $S = \bar{E}\,\bar{I}$	apparent power, $S = \bar{E}\,\bar{I}$	apparent power, $S = \bar{E}\,\bar{I}$
true power, $P = \bar{U}_R\,\bar{I}$	true power, $P = \bar{U}_R\,\bar{I}$	true power, $P = \bar{U}_R\,\bar{I}$
reactive power, $Q = \bar{U}_L\,\bar{I}$	reactive power, $Q = \bar{U}_C\,\bar{I}$	$Q = (\bar{U}_L - \bar{U}_C)\,\bar{I}$

Applying **Pythagoras's Theorem** to each power diagram:

$S^2 = P^2 + Q^2$	$S^2 = P^2 + Q^2$	$S^2 = P^2 + Q^2$

From which, we can find the values of apparent power *(S)*, true power *(P)*, and reactive power *(Q)*:

$S = \sqrt{P^2 + Q^2}$	$S = \sqrt{P^2 + Q^2}$	$S = \sqrt{P^2 + Q^2}$
$P = \sqrt{S^2 - Q^2}$	$P = \sqrt{S^2 - Q^2}$	$P = \sqrt{S^2 - Q^2}$
$Q = \sqrt{S^2 - P^2}$	$Q = \sqrt{S^2 - P^2}$	$Q = \sqrt{S^2 - P^2}$

We can also convert any *impedance* diagram into an **power diagram**, by *multiplying throughout by the square of the current ($\bar{I}^2$)*, as follows.

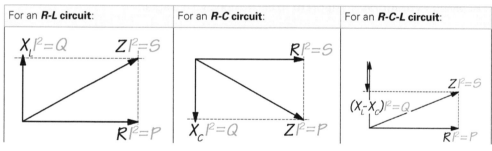

For an **R-L circuit**:	For an **R-C circuit**:	For an **R-C-L circuit**:
Figure 2.25	**Figure 2.26**	**Figure 2.27**

From which we get the following equations:

apparent power, $S = \bar{I}^2 Z$	apparent power, $S = \bar{I}^2 Z$	apparent power, $S = \bar{I}^2 Z$
true power, $P = \bar{I}^2 R$	true power, $P = \bar{I}^2 R$	true power, $P = \bar{I}^2 R$
reactive power, $Q = \bar{I}^2 X_L$	reactive power, $Q = \bar{I}^2 X_C$	$Q = \bar{I}^2 (X_L - X_C)$

R-L, R-C, and R-L-C parallel a.c. circuits

Again, whenever we draw a schematic diagram for **R-L**, **R-C**, or **R-L-C** circuits, we must remember that the circuit symbols we use are representing the *quantities*, resistance, inductance, and capacitors; they do *not* represent resistors, inductors, or capacitors!

Once again, the key to understanding and solving any a.c. circuit, series or parallel, is *the ability to construct a phasor diagram, and to apply basic geometry and trigonometry to it.*

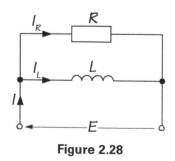

Figure 2.28

R-L parallel circuit

To draw the phasor diagram for a parallel **R-L** circuit as shown in Figure 2.8, we *always* start by drawing the reference phasor along the positive, horizontal, axis. Because the **supply voltage** is common to each branch of a parallel circuit, the supply voltage is $(\bar{E})$ *always* used as the reference phasor (Figure 2.29a).

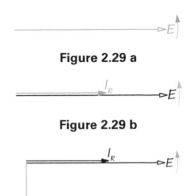

Figure 2.29 a

Figure 2.29 b

Next, we draw the phasor representing the current through the resistive branch $(\bar{I}_R)$ *in phase* with the reference phasor (Figure 2.29b). Then we draw the phasor representing the current through the inductive branch $(\bar{I}_L)$, *lagging* the reference current by 90° (Figure 2.29c).

Finally, we vectorially add $\bar{I}_R$ and $\bar{I}_L$ to show the supply current, $\bar{I}$ (Figure 2.29d).

The mnemonic, **CIVIL**, will help us remember whether a branch current lags or leads the supply voltage.

So, for a parallel **R-L circuit**, the supply current will *lag* the supply voltage by some angle, ϕ, which we call the '**phase angle**'.

Figure 2.29 c

R-C parallel circuit

The phasor diagram for an **R-C** parallel circuit (Figure 2.30) is drawn in just the same way as it was for an R-L parallel circuit, *except* that the current through the capacitive branch $(\bar{I}_C)$ is drawn *leading* the reference phasor (Figure 2.31).

So, for a parallel **R-C circuit**, the supply current will *lead* the supply voltage by some angle, φ, which we call the '**phase angle**'.

Figure 2.29 d

R-L-C parallel circuit

To draw the phasor diagram for an **R-L-C** parallel circuit (Figure 2.32), we *always* start by drawing the **reference phasor** along the positive, horizontal, axis. Because the supply voltage is common to each branch of a parallel circuit, the reference phasor is always represented by the supply voltage.

Next, we draw the phasor representing the current through the resistive branch $(\bar{I}_R)$ *in phase* with the reference phasor. Then we draw the phasor representing the current through the inductive branch $(\bar{I}_L)$, *lagging* the reference phasor by 90° (Figure 2.29c). Next, we drawn the phasor representing the current through the capacitive branch $(\bar{I}_C)$, *leading* the reference phasor by 90°.

Once again, in order to avoid representing a condition called parallel resonance, *we should always make $\bar{I}_L$ bigger than $\bar{I}_C$, or vice versa.*

The completed current phasor diagram should then look like Figure 2.33.

So, for a parallel **R-L-C circuit**, the load current will either lag or lead the supply voltage by some angle, ϕ, which we call the '**phase angle**'. If current, $\bar{I}_L$, is *larger* than current, $\bar{I}_C$, then the supply current will *lag* the supply voltage; if current, $\bar{I}_C$, is *larger* than current, $\bar{I}_L$, then the supply current will *lead* the supply voltage.

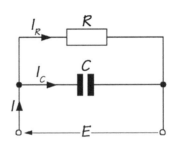

Figure 2.30

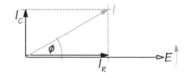

Figure 2.31

Solving parallel a.c. circuits by generating equations

Now, solving an 'electrical' problem becomes a simple matter of applying either *basic geometry* (Pythagoras's Theorem) or *basic trigonometry* (sine, cosine, or tangent ratios), where the phasor diagrams are simple, right-angled, triangles!

Applying **Pythagoras's Theorem** to each of these phasor diagrams will give us the following relationships:

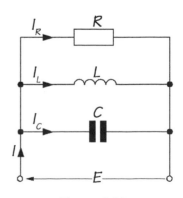

Figure 2.32

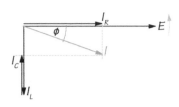

Figure 2.33

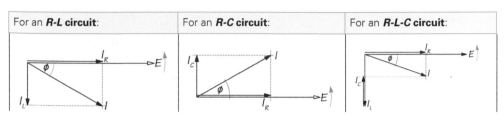

For an **R-L** circuit:	For an **R-C** circuit:	For an **R-L-C** circuit:
Figure 2.34	**Figure 2.35**	**Figure 2.36**
$\overline{I}^2 = \overline{I}_R^2 + \overline{I}_L^2$	$\overline{I}^2 = \overline{I}_R^2 + \overline{I}_C^2$	$\overline{I}^2 = \overline{I}_R^2 + (\overline{I}_L - \overline{I}_C)^2$

From which we can derive the following equations:

$\overline{I} = \sqrt{\overline{I}_R^2 + \overline{I}_L^2}$	$\overline{I} = \sqrt{\overline{I}_R^2 + \overline{I}_C^2}$	$\overline{I} = \sqrt{\overline{I}_R^2 + (\overline{I}_L - \overline{I}_C)^2}$
$\overline{I}_R = \sqrt{\overline{I}^2 - \overline{I}_L^2}$	$\overline{I}_R = \sqrt{\overline{I}^2 - \overline{I}_C^2}$	$\overline{I}_R = \sqrt{\overline{I} - (\overline{I}_L - \overline{I}_C)^2}$
$\overline{I}_L = \sqrt{\overline{I}^2 - \overline{I}_R^2}$	$\overline{I}_C = \sqrt{\overline{I}^2 - \overline{I}_R^2}$	$(\overline{I}_L - \overline{I}_C) = \sqrt{\overline{I} - \overline{I}_R^2}$

We can also determine the phase angle for each circuit, as follows:

$\phi = \cos^{-1}\left(\dfrac{\overline{I}_R}{\overline{I}}\right)$	$\phi = \cos^{-1}\left(\dfrac{\overline{I}_R}{\overline{I}}\right)$	$\phi = \cos^{-1}\left(\dfrac{\overline{I}_R}{\overline{I}}\right)$

Admittance diagrams

You will recall that for a **parallel d.c. circuit** the total resistance is found as follows:

$$\frac{1}{R} = \frac{1}{R_1} + \frac{1}{R_2} + \frac{1}{R3} + \text{etc.}$$

The *reciprocal* of resistance is termed '**conductance**' (symbol: **G**), expressed in **siemens** (symbol: **S**). So, we could rewrite the above equation in terms of conductances, that is:

$$G = G_1 + G_2 + G_3 + \text{etc.}$$

In a.c. circuits, the reciprocal of impedance is called '**admittance**' (symbol: **Y**), and the reciprocal of reactance is called '**susceptance**' (symbol: **B**). Specifically, the reciprocal of inductive reactance is called '**inductive susceptance**' (symbol: **B_L**) and the reciprocal of capacitive reactance is called '**capacitive susceptance**' (symbol: **B_C**).

Generally, in a.c. parallel circuits, it's usually *much* easier to work with admittance, conductance, and susceptance, than it is to work with impedance, resistance, and reactance.

So, to create an **admittance diagram** (equivalent to an impedance diagram in a series circuit) we *divide the current phasors by the reference phasor* – i.e. by the supply voltage:

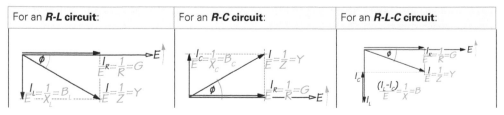

For an **R-L** circuit:	For an **R-C** circuit:	For an **R-L-C** circuit:
Figure 2.37	**Figure 2.38**	**Figure 2.39**

From which we get the following equations:

$$Y = \frac{1}{Z} = \frac{\overline{I}}{\overline{E}}$$

$$G = \frac{1}{R} = \frac{\overline{I_R}}{\overline{E}}$$

$$B_L = \frac{1}{X_L} = \frac{\overline{I_L}}{\overline{E}}$$

$$Y = \frac{1}{Z} = \frac{\overline{I}}{\overline{E}}$$

$$G = \frac{1}{R} = \frac{\overline{I_R}}{\overline{E}}$$

$$B_C = \frac{1}{X_C} = \frac{\overline{I_C}}{\overline{E}}$$

$$Y = \frac{1}{Z} = \frac{\overline{I}}{\overline{E}}$$

$$G = \frac{1}{R} = \frac{\overline{I_R}}{\overline{E}}$$

$$B = \frac{1}{X} = \frac{(\overline{I_L} - \overline{I_C})}{\overline{E}}$$

Applying **Pythagoras's Theorem** to each impedance diagram:

$$Y^2 = G^2 + B_L^2$$

or, if you prefer:

$$\left(\frac{1}{Z}\right)^2 = \left(\frac{1}{R}\right)^2 + \left(\frac{1}{X_L}\right)^2$$

$$Y^2 = G^2 + B_C^2$$

or, if you prefer:

$$\left(\frac{1}{Z}\right)^2 = \left(\frac{1}{R}\right)^2 + \left(\frac{1}{X_C}\right)^2$$

$$Y^2 = G^2 + (B_L - B_C)^2$$

or, if you prefer:

$$\left(\frac{1}{Z}\right)^2 = \left(\frac{1}{R}\right)^2 + \left(\frac{1}{(X_L - X_C)}\right)^2$$

From which we can find the values of Y, G, B_L, and B_C:

$$Y = \sqrt{G^2 + B_L^2}$$

$$G = \sqrt{Y^2 - B_L^2}$$

$$B_L = \sqrt{Y^2 - G^2}$$

$$Y = \sqrt{G^2 + B_C^2}$$

$$G = \sqrt{Y^2 - B_C^2}$$

$$B_C = \sqrt{Y^2 - G^2}$$

$$Y = \sqrt{G^2 + (B_L - B_C)^2}$$

$$G = \sqrt{Y^2 + (B_L - B_C)^2}$$

$$(B_L - B_C) = \sqrt{Y^2 - G^2}$$

If you prefer, you could *still* work in terms of impedance, resistance, and reactance:

$$\frac{1}{Z} = \sqrt{\left(\frac{1}{R}\right)^2 + \left(\frac{1}{X_L}\right)^2}$$

$$\frac{1}{R} = \sqrt{\left(\frac{1}{Z}\right)^2 - \left(\frac{1}{X_L}\right)^2}$$

$$\frac{1}{X_L} = \sqrt{\left(\frac{1}{Z}\right)^2 - \left(\frac{1}{R}\right)^2}$$

$$\frac{1}{R} = \sqrt{\left(\frac{1}{Z}\right)^2 - \left(\frac{1}{X_L}\right)^2}$$

$$\frac{1}{R} = \sqrt{\left(\frac{1}{Z}\right)^2 - \left(\frac{1}{X_C}\right)^2}$$

$$\frac{1}{X_C} = \sqrt{\left(\frac{1}{Z}\right)^2 - \left(\frac{1}{R}\right)^2}$$

$$\frac{1}{Z} = \sqrt{\left(\frac{1}{R}\right)^2 + \left(\frac{1}{X_L - X_C}\right)^2}$$

$$\frac{1}{R} = \sqrt{\left(\frac{1}{Z}\right)^2 - \left(\frac{1}{X_L - X_C}\right)^2}$$

$$\left(\frac{1}{X_L - X_C}\right) = \sqrt{\left(\frac{1}{Z}\right)^2 - \left(\frac{1}{R}\right)^2}$$

We can also determine the phase angle for each circuit, as follows:

$$\phi = \cos^{-1}\left(\frac{G}{Y}\right)$$

$$\phi = \cos^{-1}\left(\frac{G}{Y}\right)$$

$$\phi = \cos^{-1}\left(\frac{G}{Y}\right)$$

Or again, if you prefer: in terms of impedance and resistance:

$$\phi = \cos^{-1}\left(\frac{(1/R)}{(1/Z)}\right) = \frac{Z}{R}$$

$$\phi = \cos^{-1}\left(\frac{(1/R)}{(1/Z)}\right) = \frac{Z}{R}$$

$$\phi = \cos^{-1}\left(\frac{(1/R)}{(1/Z)}\right) = \frac{Z}{R}$$

Power diagrams

We can convert any of the above *current* phasor diagrams into a **power diagram**, by *multiplying throughout by the reference phasor*, $\overline{E}$, as follows. Remember, '**apparent power**' *(S)* is expressed in volt amperes, '**true power**' *(P)* is expressed in watts, and reactive power *(Q)* is expressed in reactive volt amperes.

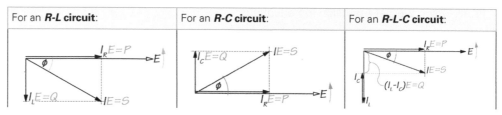

For an **R-L circuit**:	For an **R-C circuit**:	For an **R-L-C circuit**:
Figure 2.40	**Figure 2.41**	**Figure 2.42**

From which we get the following equations:

$S = \bar{E}\bar{I}$	$S = \bar{E}\bar{I}$	$S = \bar{E}\bar{I}$
$P = \bar{E}\bar{I}_R$	$P = \bar{E}\bar{I}_R$	$P = \bar{E}\bar{I}_R$
$P = \bar{E}\bar{I}_R$	$Q = \bar{E}\bar{I}_C$	$Q = \bar{E}(\bar{I}_L - \bar{I}_C)$

Applying **Pythagoras's Theorem** to each power diagram:

$S^2 = P^2 + Q^2$	$S^2 = P^2 + Q^2$	$S^2 = P^2 + Q^2$

From which, we can find the values of apparent power *(S)*, true power *(P)*, and reactive power *(Q)*:

$S = \sqrt{P^2 + Q^2}$	$S = \sqrt{P^2 + Q^2}$	$S = \sqrt{P^2 + Q^2}$
$P = \sqrt{S^2 - Q^2}$	$P = \sqrt{S^2 - Q^2}$	$P = \sqrt{S^2 - Q^2}$
$Q = \sqrt{S^2 - P^2}$	$Q = \sqrt{S^2 - P^2}$	$Q = \sqrt{S^2 - P^2}$

Hopefully, what should have been made abundantly clear from the above is that *all* the equations are derived by applying Pythagoras's Theorem, or the sine, cosine, or tangent ratios to phasor diagrams, *which are nothing more than right-angled triangles!*

Accordingly, *there is absolutely no need whatsoever to commit any of these equations to memory!*

> You will learn **far** more about the behaviour of a.c. circuits by **learning how to derive all these equations** than you will ever learn by trying to commit them to memory!

Worked example

A series circuit comprises a resistance of 1.5 Ω in series with an inductance of 160 mH and a capacitance of 50 μF, connected across a 12 V, 50 Hz, supply. Calculate:

a inductive reactance
b capacitive reactance
c impedance
d current
e voltage drop across the resistive component
f voltage drop across the inductive component
g voltage drop across the capacitive component
h phase angle
i apparent power
j true power
k reactive power
l power factor.

Solution

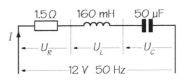

Figure 2.43

As always, the first step is to draw a labelled schematic diagram of the circuit (Figure 2.43).

The next step is to sketch a phasor diagram representing the circuit (Figure 2.44). Remember, we must always sketch $\bar{U}_L$ longer than $\bar{U}_C$ (or *vice versa* – it won't affect the answers), so that we avoid representing the circuit at resonance.

We need to convert it into an *impedance diagram*, by dividing throughout by the reference phasor, $\bar{I}$ (Figure 2.45).

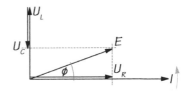

Figure 2.44

There's nothing in the phasor diagram to help us determine the *inductive reactance* or the *capacitive reactance*, using the information supplied in the question, so we must fall back on their equations – which we have committed to memory:

a $X_L = 2\pi f L = 2\pi \times 50 \times (160 \times 10^{-3}) = 50.27\ \Omega$ **(Answer a.)**

b $X_C = \dfrac{1}{2\pi f C} = \dfrac{1}{2\pi \times 50 \times (50 \times 10^{-6})} = 63.66\ \Omega$ **(Answer b.)**

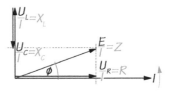

Figure 2.45

c From the impedance diagram, the value of **Z** must be the vectorial-sum of **R** and the difference between **XL** and **XC**, that is:

$$Z = \sqrt{R^2 + (X_L - X_C)^2} = \sqrt{1.5^2 + (50.27 - 63.66)^2}$$
$$= \sqrt{1.5^2 + (-13.39)^2} = \sqrt{2.25 + 179.29}$$
$$= 13.47\ \Omega \text{ (Answer c.)}$$

d $\bar{I} = \dfrac{\bar{E}}{Z} = \dfrac{12}{13.47} = \mathbf{0.89\ A}$ **(Answer d.)**

e $\bar{U}_R = IR = 0.89 \times 1.5 = \mathbf{1.34\ V}$ **(Answer e.)**

f $\bar{U}_L = \bar{I}X_L = 0.89 \times 50.27 = \mathbf{44.74\ V}$ **(Answer f.)**

g $\bar{U}_C = \bar{I}X_C = 0.89 \times 63.66 = \mathbf{56.66\ V}$ **(Answer g.)**

h phase angle, $\phi = \cos^{-1}\dfrac{R}{Z} = \cos^{-1}\dfrac{1.5}{13.47} = \mathbf{83.6°}$ **(Answer h.)**

For the remainder of this problem, we need to draw a power diagram. This can be achieved simply by redrawing the voltage phasor diagram, and *multiplying* throughout by the reference phasor, $\bar{I}$, in order to create a power diagram (Figure 2.46).

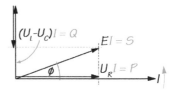

Figure 2.46

i Apparent Power, $S = \bar{E}\bar{I} = 100 \times 0.89 = \mathbf{89\ V \cdot A}$ **(Answer i.)**

j True Power, $P = \bar{I}\bar{U}_R = 0.89 \times 1.34 = \mathbf{1.19\ W}$ **(Answer j.)**

k Reactive Power, $Q = \bar{I}(\bar{U}_L - \bar{U}_C)$
$$= 0.89 \times (11.92) = \mathbf{10.6\ var} \text{ (Answer k.)}$$

We can double-check this answer by using the equation:

$$Q = \sqrt{S^2 - P^2} = \sqrt{89^2 - 1.19^2} \approx 11\,\text{var (Answer)}$$

Conclusion

Now that we've completed this chapter, we need to examine its **objectives** listed at its start. Placing a question mark at the end of each objective turns that objective into a **test item**. If we can answer those test items, then we've met the objectives of this chapter.

Resolving phasors

Objectives

On completion of this chapter you should be able to:

1. resolve any phasor into its horizontal and vertical components.

2. add two phasors, using their horizontal and vertical components.

Introduction

In this chapter, we are going to learn a specific, and very useful, tool, which will enable us to *add* or *subtract* two or more phasors *that lie at various angles, other than right-angles, relative to each other*.

Until now, we have only dealt with **series** and **parallel** a.c. circuits and, so, we've only ever needed to add or subtract phasors *that lie at right-angles relative to each other*, as shown in Figure 3.1 (left). And, as these form simple right-angled triangles, that means we've only ever needed to apply either **Pythagoras's Theorem** or one or other of the **sine**, **cosine**, or **tangent** ratios in order to solve them.

But, unfortunately, as we move on to deal with **series-parallel a.c. circuits**, we will find that we must be able to add or subtract phasors which do *not* lie at right-angles to each other (Figure 3.1 (right)). And, because they *don't* form right-angled triangles, we *cannot* use Pythagoras's Theorem or apply those trigonometric ratios with which we are already so familiar.

Well, not directly anyway!

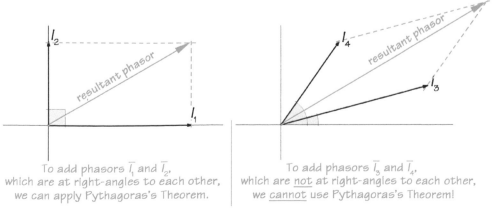

To add phasors $\overline{I_1}$ and $\overline{I_2}$, which are at right-angles to each other, we can apply Pythagoras's Theorem.

To add phasors $\overline{I_3}$ and $\overline{I_4}$, which are <u>not</u> at right-angles to each other, we <u>cannot</u> use Pythagoras's Theorem!

Figure 3.1

Electrical Science for Technicians. 978-1-138-84926-6 © Adrian Waygood.
Published by Taylor & Francis. All rights reserved.

So *how* can we add or subtract phasors that are *not* at right-angles to each other?

Well, of course, there *are* advanced mathematical techniques we *could* use to do this, but it would require us to learn these new techniques before we can apply them. Alternatively, we could simply use the techniques with which we are already so familiar, by using a relatively straightforward technique that involves first *resolving* those phasors into their *horizontal* and *vertical components*.

Horizontal and vertical components of a phasor

Figure 3.2 shows how a current phasor, $\bar{I}$, is *resolved* into its **horizontal** and **vertical components**, using the *cosine* and *sine* ratios.

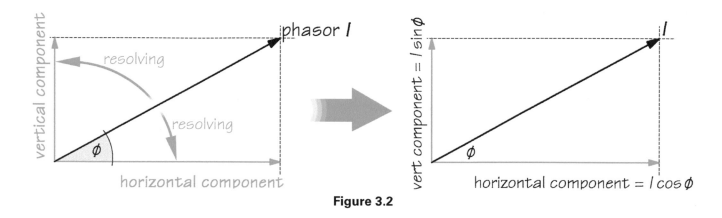

Figure 3.2

$$\text{horizontal component} = \left(\bar{I} \cos\phi \right) \text{ and vertical component} = \left(\bar{I} \sin\phi \right)$$

So what exactly do we *mean* when we talk about a phasor's '**horizontal**' and '**vertical components**'?

Well, the horizontal and vertical components of, say, a current phasor don't actually exist as *real*, *separate*, *currents*. They are just a mathematical model. But if they *did* exist, then the circuit through which they flow would be completely unaware of any difference whatsoever between these *component* currents or the *actual* current itself.

Worked example 1

Find the horizontal and vertical components of a phasor representing a current of 10 A that lies at an angle of +30° relative to the horizontal positive axis, as illustrated in Figure 3.3.

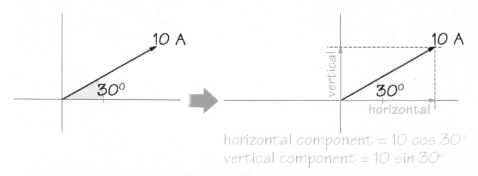

horizontal component = 10 cos 30°
vertical component = 10 sin 30°

Figure 3.3

Solution

$$\text{horizontal component} = 10\cos 30°$$
$$= 10 \times 0.866$$
$$= \textbf{8.66 A (Answer)}$$

$$\text{vertical component} = 10\sin 30°$$
$$= 10 \times 0.5$$
$$= \textbf{5 A (Answer)}$$

As evidence that the horizontal and vertical components in the above example are *exactly equivalent* to the *actual* current from which they were resolved, let's find out the amount of power those separate *components* would produce if they *were* to flow through a circuit of, say, resistance 10 Ω, and compare that with the power produced by the *actual* current itself.

[Remember, these are the horizontal and vertical *components* of a current passing through a resistor, they are *not* separate in-phase (resistive) and quadrature (reactive) currents flowing through a resistive-reactive circuit! Accordingly, *both* components result in the expenditure of true power.]

Let's start by determining the power produced by the 10 A current itself:

$$P = I^2 R = 10^2 \times 10 = 1000 \text{ W}$$

Now, let's find the *total* power produced by the two resolved components of this current:

$$P_{\text{horizontal component}} = I^2_{\text{horizontal}} R = 8.66^2 \times 10 \approx 750 \text{ W}$$
$$P_{\text{vertical component}} = I^2_{\text{vertical}} R = 5^2 \times 10 = 250 \text{ W}$$
$$P_{\text{total}} = P_{\text{horizontal component}} + P_{\text{vertical component}}$$
$$\approx 750 + 250$$
$$\approx 1000 \text{ W}$$

As you can see, the resolved horizontal and vertical *components* produce *exactly* the same amount of power as the *actual* current passing through the circuit, proving that, when taken together, those 'components' *are exactly equivalent to the 'actual' current itself.*

Of course, this technique applies to voltage phasors as well as to current phasors, as demonstrated in the next worked example.

Worked example 2

Find the horizontal and vertical components of a phasor representing a voltage of 100 V at +210° relative to the horizontal positive axis, as shown in Figure 3.4.

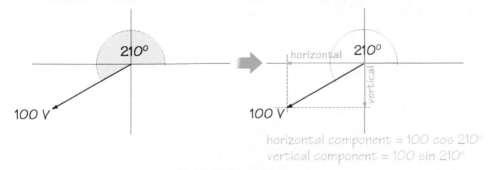

horizontal component = 100 cos 210°
vertical component = 100 sin 210°

Figure 3.4

Solution

Important – remember 'positive' angles, by convention, are *always* measured **counterclockwise** from the horizontal positive axis, so:

$$\text{horizontal component} = 100 \cos 210°$$
$$= 10 \times (-0.866)$$
$$= \textbf{-86.6 V (Answer)}$$
$$\text{vertical component} = 100 \sin 210°$$
$$= 10 \times (-0.5)$$
$$= \textbf{-50 V (Answer)}$$

So what, then, is the *point* of resolving a phasor into its horizontal and vertical components?

Well, it simply provides us with a convenient method of adding two or more phasors that are not perpendicular to each other – something we usually have to do whenever we solve series-parallel circuits. And it allows us to do so using the mathematics with which we are already familiar, rather than having to introducing more complicated mathematics.

We can also use this technique to *subtract* two or more phasors but, for our purposes, this is *not* required either in this chapter or in what follows.

Adding phasors using their horizontal and vertical components

Two or more phasors can be **added**, then, by

▶ first, resolving each of the phasors into their horizontal and vertical components, then

▶ adding their corresponding horizontal components together, and adding their vertical components together and, finally

▶ finding the resultant phasor by vectorially adding the new horizontal and vertical components.

These steps should be made clear in the following worked examples.

Worked example 3

Add two current phasors, one which is 25 A at 30° and the other which is 30 A at 60°, as illustrated in Figure 3.5, specifying the value of the resultant phasor together with its angle relative to the positive real axis.

Solution

Again, remember *positive* angles are always measured *counterclockwise* from the real positive axis.

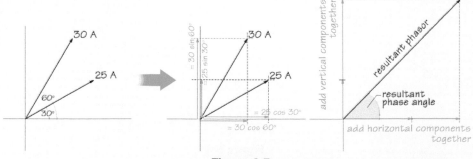

Figure 3.5

First, we start by resolving the two individual phasors into their horizontal and vertical components.

For the 25 A phasor:

horizontal component $= 25 \cos 30° = 25 \times 0.866 =$ **21.65 A**

vertical component $= 25 \sin 30° = 25 \times 0.5 =$ **12.5 A**

For the 30 A phasor:

horizontal component $= 30 \cos 60° = 30 \times 0.5 =$ **15 A**

vertical component $= 30 \sin 60° = 30 \times 0.866 =$ **25.98 A**

Next, we **add the two horizontal components** together, and we **add the two vertical components** together:

Sum of the horizontal components $= 21.65 + 15 =$ **36.65 A**

Sum of the vertical components $= 12.5 + 25.98 =$ **38.48 A**

Finally, we vectorially add these two components, by applying **Pythagoras's Theorem** to find their resultant:

magnitude of resultant $= \sqrt{\text{(horizontal component)}^2 + \text{(vertical component)}^2}$

$$= \sqrt{36.65^2 + 38.48^2}$$

$$= \textbf{53.14 A (Answer)}$$

To find the **angle** of the resultant current, relative to the positive real axis, we can then simply apply the *tangent* ratio, as follows:

$$\angle\theta = \tan^{-1}\frac{\text{opposite}}{\text{adjacent}}$$

$$= \tan^{-1}\frac{\text{vertical component}}{\text{horizontal component}}$$

$$= \tan^{-1}\frac{38.48}{36.65}$$

$$= \tan^{-1}1.05 = 46.39° \quad \textbf{(Answer)}$$

Worked example 4

Find the phasor-sum of the two phasors shown in Figure 3.6: 100 V at an angle of 120°, and 200 V at an angle of 210°.

Solution

Remember positive angles are always measured *counterclockwise* from the real positive axis.

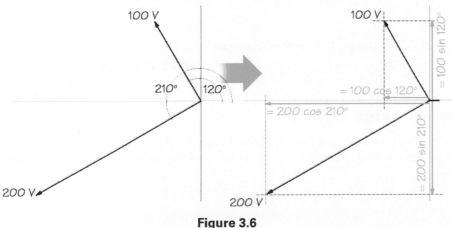

Figure 3.6

Again, we start by resolving the two individual phasors into their horizontal and vertical components.

For the 100 V phasor:

$$\text{horizontal component} = 100 \cos 120° = 100 \times (-0.5) = \textbf{-50 V}$$
$$\text{vertical component} = 100 \sin 120° = 100 \times 0.866 = \textbf{86.6 A}$$

For the 200 V phasor:

$$\text{horizontal component} = 200 \cos 210° = 200 \times (-0.866) = \textbf{-173.2 V}$$
$$\text{vertical component} = 200 \sin 210° = 200 \times (-0.5) = \textbf{-100 V}$$

Next, we **add the two horizontal components** together, and we **add the two vertical components** together:

$$\text{Sum of the horizontal components} = (-50) + (-173.2) = \textbf{-223.2 V}$$
$$\text{Sum of the vertical components} = 86.6 + (-100) = \textbf{-13.4 V}$$

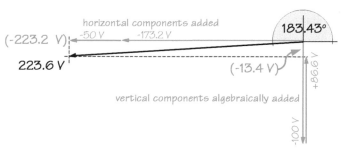

Figure 3.7

Finally, we vectorially add these two components, by applying **Pythagoras's Theorem** to find their resultant:

$$\text{magnitude of resultant} = \sqrt{(\text{horizontal component})^2 + (\text{vertical component})^2}$$
$$= \sqrt{(-223.2)^2 + (-13.4)^2}$$
$$= \textbf{223.6 V (Answer)}$$

To find the **angle** of the resultant current, relative to the positive real axis, we can then simply apply the *tangent* ratio, as follows:

$$\text{angle,} \quad \angle\theta = \tan^{-1}\frac{\text{opposite}}{\text{adjacent}}$$
$$= \tan^{-1}\frac{\text{vertical component}}{\text{horizontal component}}$$
$$= \tan^{-1}\frac{(-13.4)}{(-223.2)}$$
$$= \tan^{-1} 0.06 = 3.43°$$
$$\text{angle from horizontal positive axis} = 3.43° + 180°$$
$$= \textbf{183.43° (Answer)}$$

The next step . . .

Now that we have learned this simple technique for adding phasors, we are ready to proceed to the next chapter, which deals with solving **series-parallel a.c. circuits**.

CHAPTER 4

Series-parallel a.c. circuits

Objectives

On completion of this chapter you should be able to:

1. explain why constructing phasor diagrams for series-parallel a.c. circuits can become so difficult.

2. solve simple series-parallel a.c. circuits, using the technique of resolving phasors.

Introduction

In the companion book, **An Introduction to Electrical Science**, we learnt how to solve **a.c. series** and **a.c. parallel** circuits.

In this chapter, we are going to move on and learn how to solve **series-parallel a.c.** circuits.

Up to this point, it has been continually emphasised that, after sketching a fully labelled circuit diagram, the single most important step in solving *any* a.c. circuit is always to *construct its phasor diagram*, from which practically *all* the equations we are ever likely to need can then be derived – as summarised in an earlier chapter – without having to make any serious effort to commit them to memory.

Constructing a phasor diagram for a **series-parallel** circuit follows *exactly* the same principles we've already learned for series and for parallel circuits. However, for series-parallel circuits *they can be significantly more difficult to construct*.

In fact, even for relatively straightforward examples of series-parallel circuits, we have now reached the point where attempting to construct a phasor diagram will start to *impede*, rather than to support, our progress in this topic!

So we must look for an alternative way of solving series-parallel a.c. circuits; *one that eliminates the need to construct complicated phasor diagrams*. And we shall learn more about this in the following chapter.

But, for now, let's start by solving a relatively straightforward series-parallel problem using the methods we've already learnt: i.e. by first constructing a phasor diagram. This will help us fully appreciate *why* its construction is more complicated than for a series or a parallel circuit, and will leave us wondering *whether there is an easier way of solving series-parallel a.c. circuits?*

In the series-parallel circuit shown in Figure 4.1, the *upper* branch has a resistance of 50 Ω in series with an inductive reactance of 20 Ω, and the *lower* branch has a resistance of 30 Ω, in series with a capacitive reactance of 40 Ω. The two branches

Electrical Science for Technicians. 978-1-138-84926-6 © Adrian Waygood. Published by Taylor & Francis. All rights reserved.

are connected across a 230 V, 50 Hz, a.c. supply. The question is, *how do we determine the values of the supply current and the phase angle for this circuit?*

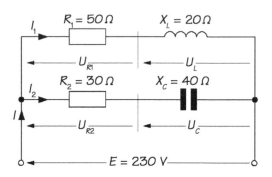

Figure 4.1

Let's start by learning how we construct its phasor diagram.

Initially, we need to treat the branches *as though they were two, completely **independent***, *series circuits*. So we start by drawing *two, separate, phasor diagrams:* one for the *upper* branch, and the other for the *lower* branch. And, of course, we know that the reference phasor for any *series* circuit is *always* **current** because it's common to each component within that series circuit.

So, for the *upper* branch:	**And for the *lower* branch:**
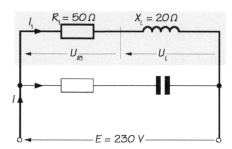	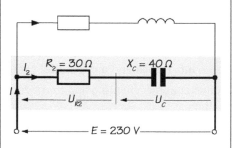
Figure 4.2a	**Figure 4.3a**
To draw the phasor diagram for the upper branch, we use current $\overline{I_1}$ as the reference phasor:	To draw the phasor diagram for the lower branch, we use current $\overline{I_2}$ as the reference phasor:
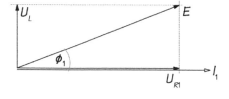	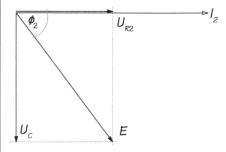
Figure 4.2b	**Figure 4.3b**
We then convert this voltage phasor diagram into an **impedance diagram** by *dividing* throughout by the reference phasor, $\overline{I_1}$:	We then convert this voltage phasor diagram into an **impedance diagram** by *dividing* throughout by the reference phasor, $\overline{I_2}$:

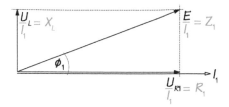

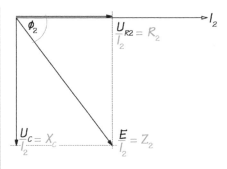

Figure 4.2c

Figure 4.3c

Now, we can determine the impedance of the upper branch, Z_1, as follows:

$$Z_1 = \sqrt{R_1^2 + X_L^2} = \sqrt{50^2 + 20^2}$$
$$= \sqrt{2900} = 53.85\ \Omega$$

. . . enabling us to calculate the current in the upper branch, $\bar{I}_1$:

$$\bar{I}_1 = \frac{\bar{E}}{Z_1} = \frac{230}{53.85} = 4.27\ A$$

. . . and its phase angle:

$$\phi_1 = \cos^{-1}\frac{R_1}{Z_1} = \cos^{-1}\frac{50}{53.85} = 21.8°\ \text{lagging}$$

Now, we can determine the impedance of the lower branch, Z_2, as follows:

$$Z_2 = \sqrt{R_2^2 + X_C^2} = \sqrt{30^2 + 40^2}$$
$$= \sqrt{2500} = 50\ \Omega$$

. . . enabling us to calculate the current in the upper branch, $\bar{I}_2$:

$$\bar{I}_2 = \frac{\bar{E}}{Z_2} = \frac{230}{50} = 4.6\ A$$

. . . and its phase angle:

$$\phi_2 = \cos^{-1}\frac{R_2}{Z_2} = \cos^{-1}\frac{30}{50} = 53.13°\ \text{leading}$$

So, we now have *two* individual phasor and impedance diagrams: one set representing the *upper* branch, and the other set representing the *lower* branch of our series-parallel circuit.

The question, now, is *how do these two phasor diagrams relate to each other?* And, more importantly, *how do we combine them to create a* single *phasor diagram that represents the complete circuit?*

We already know that we *can't* simply add the two branch currents together algebraically; we must add them *vectorially*.

As the two branches are in *parallel* with each other, we should realize that *the supply voltage, $\bar{E}$, must be common to each branch.* So we must choose $\bar{E}$ as the reference phasor for the *complete* circuit and, therefore, *for the combined phasor diagram.*

So what we must do, then, is to *merge* the two phasor diagrams we've already drawn, *so that they each share the same common supply voltage* – i.e. with $\bar{E}$, the new reference phasor, being drawn, as usual, along the horizontal, positive, axis.

We do this by rotating the phasor diagram for the *upper* branch in a *clockwise* direction, through **21.8°** (its phase angle), and rotating the phasor diagram for the *lower* branch in a *counterclockwise* direction, through **51.13°** (its phase angle), *until each of their supply voltage phasors lie along the horizontal axis and are superimposed over each other to become the new reference phasor (E) for the complete circuit.*

This process is illustrated in Figure 4.4.

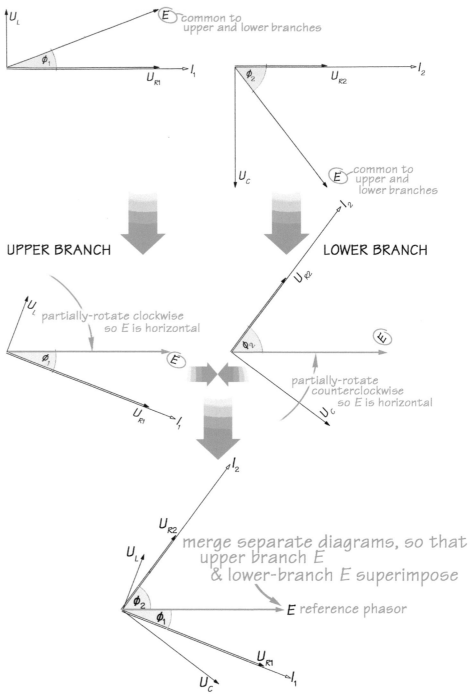

UPPER BRANCH

LOWER BRANCH

MERGED PHASOR DIAGRAMS

Figure 4.4

So we can now use this *combined* phasor diagram to determine the value of the
supply current $(\overline{I})$ which, of course, must be the **phasor-sum** of the two branch
currents, $\overline{I}_1$ and $\overline{I}_2$.

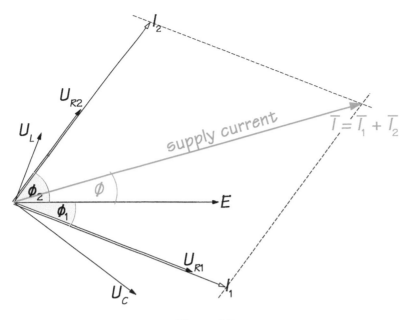

Figure 4.5

Now, if we had drawn this phasor diagram *accurately*, and to scale, then we could easily determine the supply current by completing a parallelogram as shown in Figure 4.5 and, then, simply *measuring* the length of the resultant phasor that represents the supply current:

But this is much easier said than done, as drawing the phasor diagram to scale, from scratch, is impractical! Far better, then, to determine the supply current *mathematically*, using Pythagoras's Theorem and basic trigonometry.

So, let's move on!

In the last chapter, we learnt that we can add two or more phasors that are not at right-angles to each other, by **resolving** each of them into their **horizontal** and **vertical** components and, then, by adding the horizontal components together and the vertical components together, we can reverse the process, to find the resultant phasor.

So let's return to our problem, and let's start by looking at the two branch currents, $\overline{I}_1$ and $\overline{I}_2$, separately, as shown in Figures 4.6 and 4.7:

For current $\overline{I}_1$:

Figure 4.6

For current $\overline{I}_2$:

Figure 4.7

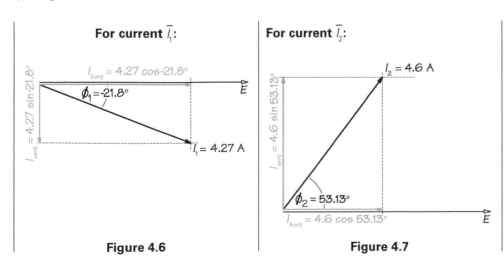

horizontal component $= \bar{I_1} \cos \phi_1$	horizontal component $= \bar{I_2} \cos \phi_2$
$= 4.27 \cos(-21.8°)$	$= 4.6 \cos(53.13°)$
$= 4.27 \times 0.928$	$= 4.6 \times 0.60$
$= 3.96$ A	$= 2.76$ A
vertical component $= \bar{I_1} \sin \phi_1$	vertical component $= \bar{I_2} \sin \phi_2$
$= 4.27 \sin(-21.8°)$	$= 4.6 \sin(53.13°)$
$= 4.27 \times (-0.371)$	$= 4.6 \times 0.8$
$= -1.58$ A	$= 3.68$ A

Now we can vectorially *add* the horizontal components together, and the vertical components together – as in Figure 4.8:

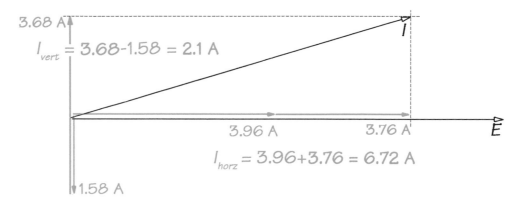

Figure 4.8

total of horizontal components $= 3.96 + 2.76 = 6.72$ A

total of vertical components $= (-1.58) + 3.68 = 2.1$ A

So, now, we have a simple right-angled triangle, from which we can use Pythagoras's Theorem to determine the hypotenuse (that is, the supply current):

$$\bar{I}, \text{ supply current} = \sqrt{(\text{horizontal component})^2 + (\text{vertical component})^2}$$

$$= \sqrt{6.72^2 + 2.10^2}$$

$$= \sqrt{45.16 + 4.41}$$

$$= \sqrt{49.57} = \textbf{7.04 A (Answer)}$$

. . . and we can use the tangent ratio to determine the phase angle:

$$\angle \phi = \tan^{-1} \frac{\text{vertical component}}{\text{horizontal component}} = \tan^{-1} \frac{2.10}{7.04} = \tan^{-1} 0.298$$

$$= \textbf{16.59° leading (Answer)}$$

In this chapter, hopefully, we have learnt *two* things.

Firstly, it is much more difficult to construct a phasor diagram for a **series-parallel** circuit, than it is for either a series circuit or for a parallel circuit. And, for a more-complicated series-parallel circuit than the one in our worked example (e.g. one with, say, three branches!), constructing a phasor diagram can become *really* difficult indeed!

Secondly, solving a series-parallel circuit is quite time consuming and prone to error because, as we have seen in the above worked example, we will usually need to

(a) resolve the individual branch-current phasors into their horizontal and vertical components and, then, (b) add those horizontal and vertical components in order to determine the resultant phasor that represents the supply current. And that's only the beginning, because we may be asked not only to determine the supply current (as in our worked example), but also to go on to find the equivalent resistance, reactance, voltage drops, true power, reactive power, apparent power, and power factor!

We appear, therefore, to have reached a point where relying on a phasor diagram to help us solve an a.c. circuit has started become both difficult and time consuming.

'Surely,' we might ask ourselves, 'there *must* be an easier way forward?'

Well, fortunately, the answer is *'Yes, there is!'* But, to be able to do so requires us to learn a completely different approach to solving a.c. circuits; one that builds on *everything* we've learnt so far – so, don't worry, *none* of that will be wasted!

This new technique uses something we call the '**j-operator**': a purely mathematical approach we shall be learning to use in the following chapter.

Do not be put off by the expression, *'a purely mathematical approach'*! Once we've mastered the basic rules (and they are *very* straightforward), we will actually find it very much *easier* and *faster* than the techniques we've been using until now – *regardless of how complicated the circuit itself is!*

So it's a technique well worth learning!

So why, we may ask, have we even bothered with the techniques we have learnt thus far? The answer is that the techniques we are *about* to learn won't really make a great deal of sense to us *unless we are already thoroughly familiar with phasor diagrams and the equations that they generate*.

Equations, incidentally, that we have probably been unconsciously committing to memory *despite* having been told not to do so!

For these reasons, then, we will *not* spend any more time solving series-parallel a.c. circuits using the techniques demonstrated in this chapter. Instead, we shall return to series-parallel circuits *after* we've mastered these new techniques for solving a.c. circuits.

Conclusion

Now that we've completed this chapter, we need to examine its **objectives** listed at its start. Placing a question mark at the end of each objective turns that objective into a **test item**. If we can answer those test items, then we've met the objectives of this chapter.

Solving a.c. circuits using symbolic notation

Objectives

On completion of this chapter you should be able to:

1. define a phasor quantity using (a) polar notation, and (b) rectangular notation.

2. convert a phasor expressed in polar notation to rectangular notation, and *vice versa.*

3. explain the symbol, 'j', in terms of (a) an 'operator', and (b) an 'imaginary number'.

4. explain what is meant by the mathematical term, 'imaginary number'.

5. explain what is meant by the term, 'complex number'.

6. add complex numbers longhand, and using a scientific calculator.

7. subtract complex numbers longhand, and using a scientific calculator.

8. multiply complex numbers longhand, and using a scientific calculator.

9. divide complex numbers longhand, and using a scientific calculator.

10. invert complex numbers longhand, and using a scientific calculator.

11. define the voltage drops in series *R-L*, *R-C*, and *R-L-C* circuits using symbolic notation.

12. define impedance using symbolic notation.

13. define the currents in a parallel *R-L*, *R-C*, and *R-L-C* circuits using symbolic notation.

14. use symbolic notation to solve series, parallel, and series-parallel a.c. circuits.

Introduction

In the last chapter, we learnt how difficult it can become to construct a phasor diagram for an **series-parallel a.c.** circuit.

The worked example we solved in that particular chapter was, in fact, a relatively straightforward example of a series-parallel circuit. For more complicated series-parallel circuits, constructing their phasor diagrams can be *very* complicated indeed!

Based on the phasor diagram we constructed in the previous chapter consider, for example, how would we construct a complete phasor diagram for each of the series-parallel circuits shown in Figure 5.1!

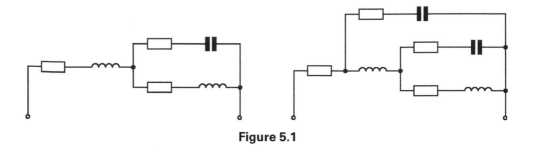

Figure 5.1

You may wish to try doing this in your own time, but we're *not* going to spend time doing it here, when – as we shall learn in this chapter – there is an alternative and *easier* way of solving these circuits! A way that *significantly reduces the number of calculations we must perform.*

The whole purpose of constructing a phasor diagram, don't forget, is simply to help us *generate appropriate equations* that we can then use to solve the circuit it represents by, essentially, reducing the 'electrical' problem to what is really a simple exercise in applying *basic geometry* (i.e. Pythagoras's Theorem) or *basic trigonometry* (i.e. the cosine, sine, and/or tangent ratios) to triangles.

In the companion book, **An Introduction to Electrical Science**, we are urged *not* to deliberately commit the various equations generated by phasor diagrams to memory but, instead, to learn how to *derive* them! But, by now, we've probably found that we've subconsciously done so; if not *all* of them, certainly the more important ones!

So, now, we need to move on and learn a completely new way of solving a.c. circuits; *one that does not rely on us first having to construct a phasor diagram.*

The new method we will be learning in this chapter *builds* on everything we've already learnt about solving a.c. circuits, *so that nothing we've learnt will be wasted!*

In fact, *without understanding what we have already learnt about constructing phasor diagrams, very little of what follows will make a great deal of sense!*

Methods of representing phasor quantities

There are *two* methods of representing phasor quantities. We call these methods '**polar notation**' and '**rectangular notation**' respectively and, in fact, we have *already* been using these methods without necessarily knowing them by these names.

Polar notation

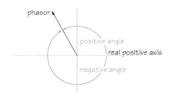

Figure 5.2

Whenever we define a phasor in terms of its *magnitude* (length) and its *phase angle* (e.g. '20 A lagging by 30°'), we are expressing it in '**polar notation**'.

As we already know, by common agreement, the angle specified in polar notation, is *always* considered to be 'positive' when measured in a *counterclockwise direction* from the real (horizontal) positive axis (Figure 5.2).

So, for example, using polar notation we can define the phasors in Figures 5.3 and 5.4 . . .

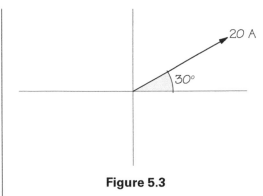

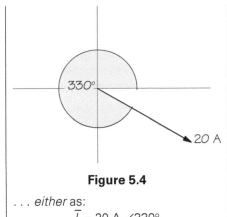

Figure 5.3

. . . *either* as:

$$\overline{I} = 20 \text{ A} \angle 30°$$

. . . *or* as:

$$\overline{I} = 20 \text{ A} \angle -330°$$

Figure 5.4

. . . *either* as:

$$\overline{I} = 20 \text{ A} \angle 330°$$

. . . *or* as:

$$\overline{I} = 20 \text{ A} \angle -30°$$

Rectangular notation

When, in earlier chapters, we determined the *horizontal* and *vertical components* of a phasor, we were expressing that phasor in '**rectangular notation**'.

By common agreement, the **horizontal** component is *always* written in front of the **vertical** component (*horizontal*, *vertical*). This follows *exactly* the same conventions with which we are already familiar for specifying the co-ordinates on graphs (*x*, *y*), and for locations on maps ('*eastings*', '*northings*').

For example, using 'rectangular notation', we can define each of the phasors shown in Figure 5.5 . . .

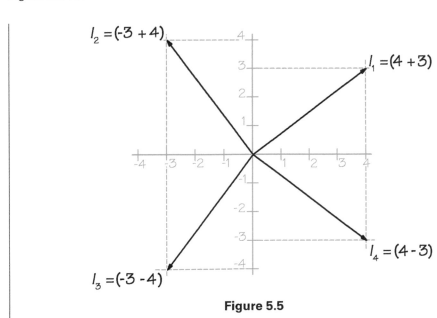

Figure 5.5

. . . as:

$$\overline{I}_1 = (4+3) \qquad \overline{I}_2 = (-4+3) \qquad \overline{I}_3 = (-3-4) \qquad \overline{I}_4 = (4-3)$$

Adding, subtracting, multiplying, and dividing phasors using rectangular and polar notation

We can *add* or *subtract* phasors using **rectangular notation** but not using polar notation and it's very much easier to *multiply* or *divide* phasors using **polar notation** than it is using rectangular notation.

The significance of this will be revealed in a later chapter, when we deal with unbalanced three-phase loads.

We'll learn how to add and subtract (as well as how to multiply and divide) phasors in *rectangular* notation later in this chapter. But, in the meantime . . .

Multiplying and dividing phasors in *polar* notation is very simple, as follows:

▶ **Multiplication in polar notation**. To *multiply* two phasors expressed in polar notation, we simple *multiply* their magnitudes and *add* their angles. For example:

$$(12 \angle 30°) \times (15 \angle 20°) = (12 \times 15) \angle (30° + 20°) = \mathbf{180 \angle 50°} \text{ (Answer)}$$

▶ **Division in polar notation**. To *divide* two phasors expressed in polar notation, we simply *divide* their magnitudes and subtract the denominator angle from the nominator angle. For example:

$$\frac{(12 \angle 30°)}{(15 \angle 20°)} = \left(\frac{12}{15}\right) \angle (30° - 20°) = \mathbf{0.8 \angle 10°} \text{ (Answer)}$$

It's worth pointing out that it's *impossible* to add or to subtract phasors using polar notation. Addition and subtraction can only be done using rectangular notation.

Converting between notations

It's both useful and important to be able to convert between polar notation and rectangular notation, and *vice versa*.

For example, in the series-parallel worked example in the previous chapter, when we found the **horizontal** and **vertical components** of the two branch-current phasors, we were *converting from polar notation into rectangular notation*. Then, after adding the resulting horizontal and vertical components, we found the final current by *converting from rectangular notation back to polar notation*.

▶ **Polar to rectangular notation**. To convert *from* polar notation *to* rectangular notation, we simply apply the cosine and sine ratios, as shown in the following worked example.

Worked example 1

Convert $\overline{I} = 20$ A $\angle 30°$ into rectangular notation.

Solution

$$\text{horizontal component} = \overline{I} \cos \phi = 20 \cos 30° = \mathbf{17.32 \ A}$$
$$\text{vertical component} = \overline{I} \sin \phi = 20 \sin 30° = \mathbf{10 \ A}$$

This can be written as: (**17.32**, **10.00**) **A (Answer)**

▶ **Rectangular to polar notation**. To convert *from* rectangular notation *to* polar notation, we simply apply **Pythagoras's Theorem** to determine the *magnitude*, and the **arctan (tan⁻¹)** ratio to determine the *angle* measured from the horizontal positive axis, as shown in the following worked example.

Worked example 2

Convert $\bar{I} = (3, 4)$ A into polar notation.

Solution

Applying Pythagoras's Theorem:

$$I = \sqrt{(\text{horizontal component})^2 + (\text{vertical component})^2}$$

$$= \sqrt{3^2 + 4^2} = \sqrt{25} = \textbf{5A}$$

Angle, relative to the horizontal positive axis:

$$\angle\phi = \tan^{-1}\left(\frac{\text{opposite}}{\text{adjacent}}\right) = \tan^{-1}\left(\frac{4}{3}\right) = \tan^{-1}1.33 = \textbf{53.06°}$$

This can be written as: **5 A ∠53.06° A (Answer)**

Using symbolic notation: the 'j-operator'

Now that we have absorbed the necessary background information, the time has come to move on to learn *a completely different approach to solving a.c. circuits*, but one that builds on what we have already learnt – this time, through the use of a purely mathematical approach using what is termed the '**j-operator**' (some textbooks call this the '**operator, j**'). Generically, this method is termed '**symbolic notation**'.

In this case, the symbol, '**j**', *doesn't* represent a variable (and, so, it is *never* italicized in the way that we should *always* italicize variables) but, instead, it indicates an 'instruction', or '**operation**', performed *on* any variable that it precedes. In this context, by 'variable', we mean a **phasor** quantity, such as current ($\bar{I}$) or voltage ($\bar{E}$ or $\bar{U}$).

We are, probably without realizing it, already familiar with various mathematical symbols that indicate *instructions* or *operations* on what to do with the variable that they precede; these include the symbols: **+, −, ×**, and **÷**.

We know, for example, that whenever we apply a **negative (−)** symbol to a variable such as a current phasor, it's telling us that we must *reverse* the direction of that phasor so, in effect, that negative symbol identifies an 'operation' that has *'rotated that phasor through 180°'* (Figure 5.6).

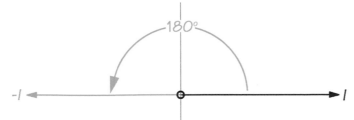

Applying a negative (-) symbol to phasor I is the same thing as rotating that phasor through 180°.

Figure 5.6

So now we are now going to meet a *new* mathematical symbol, '**j**', which *also* instructs us to *perform an operation*; but, in this case, it tells us to '**rotate a phasor counterclockwise through 90°**' or, more accurately, it *identifies a phasor that **has** been rotated, counterclockwise, by 90°* (Figure 5.7). Remember, by convention, positive angles are *always* measured in the *counterclockwise* direction.

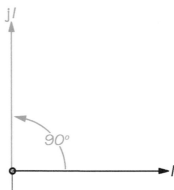

Applying the symbol 'j' to phasor *I* is the same thing as rotating that phasor, counterclockwise, through 90°.

Figure 5.7

If we were to *combine* the '**negative**' operation with the '**j**' operation (i.e. '**−j**'), then it identifies an operation that has '***rotated the phasor through −90° (i.e. clockwise)'***, as shown in Figure 5.8.

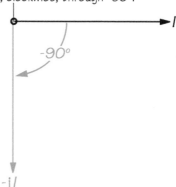

Applying the symbol '-j' to phasor *I* is the same thing as rotating that phasor, clockwise, through -90°.

Figure 5.8

So, just like the symbols with which we are already familiar (**+**, **−**, **×**, and **÷**), we call the symbol '**j**' an '**operator**', i.e. it describes an operation that has been performed on the variable that it precedes.

> The symbol, '**j**', identifies *an operation that has been performed on the variable that follows it*. For this reason, we refer to it as the '**j-operator**'.

But the symbol, '**j**', is *more* than just an *operation*; it actually represents a *number* as well!

But this is no ordinary number; it's what mathematicians call an '**imaginary number**'!

So what do we mean by an 'imaginary number'? An imaginary number is the name mathematicians give to *the square root of a negative number*.

Until now, if you've ever given any thought to it at all, you've probably assumed that it's *impossible* to find the square root of a *negative* number.

After all, if *every* number, whether positive or negative, becomes *positive* once it's been 'squared' then it must be impossible to *reverse the process* and find the square root of a negative number because, surely, they cannot exist?

For example, how can we find the square root of, say, −16? It *can't* be **−4**, of course, because (−4)² = **+16**. And, obviously, it *can't* be **+4** either, because (+4)² = **+16**, too!

So it would be reasonable to assume that it *isn't* possible to find the square root of −16? Or any other negative number, come to that!

Or is it?

Well, during the eighteenth century, a Swiss mathematician, by the name of Leonhard Euler (1707–1783), gave a great deal of thought to this question, and decided that it *would* be possible to find the square root of a negative number, *provided a 'new' symbol was used to represent it.*

Euler called this new symbol, '**i**', which stands for '**imaginary**'. And this 'imaginary number' was allocated a value of $\sqrt{-1}$.

What mathematicians mean by *'imaginary'*, we'll discuss later! And it's actually not quite as daft as it sounds!

So, using his new 'imaginary number', '**i**', let's see how Euler would have represented the square root of **−16**. He would simply have applied the following logical steps:

$$\sqrt{-16} \text{ is exactly the same thing as}: \quad \sqrt{16 \times (-1)}$$
$$\text{...which is exactly the same thing as}: \quad \sqrt{16} \times \sqrt{-1}$$
$$\text{...which is exactly the same thing as}: \quad \sqrt{16} \times i$$
$$\text{...which is exactly the same thing as}: \quad 4i$$

So Euler was able to write the 'square root of **−16**' simply as '**4i**', which his fellow mathematicians would then be able to immediately recognize the meaning of (even if the rest of us wouldn't).

We'll meet a far more logical explanation for *why* $i = \sqrt{-1}$, shortly.

Now because, in electrical engineering, we normally use the lower-case letter, *i*, to represent an *instantaneous value of a.c. current*, to avoid any confusion electrical engineers always use a lower-case, '**j**', *instead* of '**i**', to represent an imaginary number (mathematicians stubbornly cling to 'i', of course!).

Furthermore, in order to constantly remind ourselves that '**j**' is not *just* an imaginary number, but also an *instruction to perform an operation* on whatever variable follows it, we *always* place it *in front* of a variable. For example, engineers always write '**j4**', never '4j'!

And, again, it's worth reminding ourselves that 'j' is *not* a variable, so it is *never* italicised.

To summarise, then:

> The symbol, '**j**', represents an 'instruction' or 'operation' performed on the variable it precedes: it tells us the variable (in this context, a phasor) *has been rotated through an angle of 90° in a counterclockwise direction from its previous position.*

> By definition, the symbol '**j**' also represents $\sqrt{-1}$.

Applying the j-operator to a phasor

So let's consider a phasor of magnitude, $\overline{I}$, lying along the horizontal positive axis, as shown in Figure 5.9:

Figure 5.9

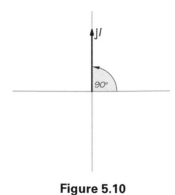

Figure 5.10

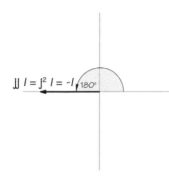

Figure 5.11

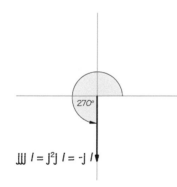

Figure 5.12

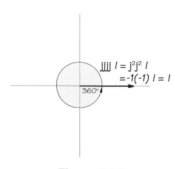

Figure 5.13

In Figure 5.10, we see that an *operation* by '**j**' upon the phasor, $\overline{I}$, has *rotated that phasor through an angle of 90° in a counterclockwise direction* from the horizontal positive axis. In its new position, the phasor is written as: $j\overline{I}$.

So $j\overline{I}$ simply represents *a phasor, $\overline{I}$, that's lying along the positive vertical axis* – as shown above. It's as easy as that!

In Figure 5.11, we see that another operation by '**j**' has rotated the phasor through a *further* 90° in the counterclockwise direction – or 180° from its original position along the horizontal positive axis, where it is expressed as $jj\overline{I}$, or $j^2\overline{I}$.

However, we would normally represent a current phasor that is lying along the negative horizontal axis simply as $-\overline{I}$, so the expression ($j^2\,\overline{I}$) **must therefore be exactly equivalent to** $-\overline{I}$ (that is : $-1 \times \overline{I}$).

So, a *value* can now be assigned to the operator, '**j**':

$$\text{since:} \quad j^2 = -\overline{I}$$
$$\text{then:} \quad j^2\,\overline{I} = (-1) \times \overline{I}$$
$$\text{so if:} \quad j^2 = -1, \quad \text{then:} \quad j = \sqrt{-1}$$

This is an alternative and, perhaps, a more understandable way of explaining *why* the symbol, '**j**', is also used to represent the square root of minus one.

Returning, now, to our rotating phasor, in Figure 5.12 another operation by the operator '**j**' has moved the phasor through yet *another* 90°counterclockwise – i.e. 270° (or –90°) from its original position along the horizontal positive axis.

The term $jjj\,\overline{I}$ is *exactly* the same as $j^2j\,\overline{I}$:

$$\text{i.e:} \quad jjj\,\overline{I} = j^2j\,\overline{I}$$
$$\text{(but since } j^2 = -1\text{)}$$
$$\text{then:} \quad j^2j\,\overline{I} = (-1)j\,\overline{I} = -j\,\overline{I}$$

So $-j\overline{I}$ simply represents *a phasor lying along the negative vertical axis.*

In Figure 5.13, a further, and final, operation by '**j**' has moved the phasor through yet another 90° counterclockwise, *and back to its original position along the horizontal positive axis.*

Because the phasor is now back to its original position, ($jjjj\,\overline{I}$) must be equivalent to $\overline{I}$, because*:*

$$(jjjj\,\overline{I}) = (j^2j^2)\overline{I}$$
$$= [(-1)\times(-1)]\,\overline{I}$$
$$= \overline{I}$$

So, as illustrated in Figure 5.14, in very simple terms, the **j-operator** *defines the direction in which a phasor is pointing!* That is:

► any phasor that is lying along the **positive vertical axis** is identified by the symbol '**+j**' (although we don't need to show the '+' sign);
► any phasor that is lying along the **negative vertical axis** is identified by the symbol '**–j**'.

As already explained, mathematicians call $\sqrt{-1}$ an 'imaginary number', whereas *all* other numbers they call 'real numbers'. So we *could* say, therefore, that any phasor that lies along the positive or negative *horizontal axis* exists in the '**real dimension**',

Figure 5.14

whereas any phasor that lies along the positive or negative *vertical axis* exists in the '**imaginary dimension**'.

A mathematician's choice of the word, 'imaginary', is unfortunate. Used in this context, 'imaginary' *doesn't* really mean 'existing in one's mind' but, instead, as simply *'lying along the vertical axis'*!

> Another way of explaining this is that a '**real number**' represents the '**in-phase**' component of a phasor, while an '**imaginary number**' simply represents its '**quadrature**' component.

So '**imaginary**' is *nothing* more than a simple **technical term** used by mathematicians, and has *nothing* whatsoever to do with a number being 'fictional', 'made-up', 'unreal', 'nonexistent', or 'fanciful'!

To summarise what we have learnt so far:

> You can think of '**j**' being a 'tag' or an 'instruction' which, when attached to a variable quantity representing a current or voltage phasor, simply tells us that the quantity is lying along the *vertical* axis.
>
> The operator '**+j**' tells us that a phasor is lying along the *positive vertical axis*, whereas the operator '**–j**' tells us that a phasor is lying along the *negative vertical axis*.

> But '**j**' is *also* a **symbol**, representing the square root of minus one ($\sqrt{-1}$).

Complex numbers

We can define each of the current phasors in Figure 5.15 by simply combining their 'real' (i.e. horizontal) components with their 'imaginary' (i.e. vertical) components, as follows:

$$\overline{I_1} = (4+j3) \qquad \overline{I_2} = (-3+j4) \qquad \overline{I_3} = (-3-j4) \qquad \overline{I_4} = (4-j3)$$

From the above, it should also be clear that combinations of 'real' and 'imaginary' numbers are examples of *rectangular notation*.

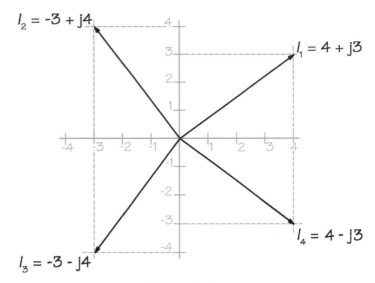

Figure 5.15

Whenever an expression, like any of those listed above, combines both a 'real' *and* an 'imaginary' number, we describe it as being a '**complex number**' (Figure 5.16).

Figure 5.16

In this context, the term 'complex' is, again, another unfortunate choice of words, as it is *not* being used in the sense of meaning 'complicated'!

No doubt, by now, you must be wondering where the heck all of this is leading, and how the j-operator can possibly help us solve a.c. circuits?

But we must wait a little longer because, *before* we move on to find this out, we must learn the *basic rules of complex number mathematics* which are quite straightforward. In fact, 'complex number mathematics' *isn't complicated at all*, so there is absolutely no reason to be wary of it!

It's made even simpler if you have a scientific calculator with a 'complex numbers' function, and most do! This function will enable our scientific calculations to do *all* the following 'complex number mathematics' for us.

Complex number mathematics

So, before we proceed to learn how to *apply* complex numbers to a.c. circuits, then, we must learn the **basic rules** for **adding**, **subtracting**, **multiplying**, and **dividing** complex numbers.

Before proceeding, however, you will no doubt be pleased to hear that practically *all* modern scientific calculators have a 'complex numbers' function, allowing them to perform all the calculations that follow. But it is essential that you should at least *understand* the basics of complex number mathematics even if you then only use a calculator when you come to *use* these techniques to solve a.c. circuits.

Adding complex numbers

Suppose we want to add two phasors, $\overline{I}_1$ and $\overline{I}_2$, where:

$$\overline{I}_1 = (a+jb) \quad \text{and} \quad \overline{I}_2 = (c+jd)$$

We know that **a** and **c** are the *horizontal* components of $\overline{I}_1$ and $\overline{I}_2$, and that **jb** and **jd** are the *vertical* components of $\overline{I}_1$ and $\overline{I}_2$ so, to find the phasor-sum of $\overline{I}_1$ and $\overline{I}_2$, we simply add their horizontal components together (**a + c**) and add their vertical components together (**jb + jd**). So the normal rules of addition apply:

$$\overline{I_1} = \quad (a \quad + \quad jb)$$
$$\overline{I_2} = \quad (c \quad + \quad jd) \; +$$
$$= \quad (a+c)+ \quad j(b+d)$$

i.e: $\overline{I_1}+\overline{I_2} = (a+jb)+(c+jd) = (a+c)+(jb+jd) = (a+c)+j(b+d)$

So, when two or more phasors are added, their *real parts must be added together* to form the 'real' part of the resultant, and their *imaginary parts must be added together* to form the 'imaginary' part of the resultant.

Worked example 3

Add the following two voltage phasors together:

$$\overline{E_1} = (100+j\,30) \quad \text{and} \quad \overline{E_2} = (20+j\,120)$$

Solution

$$\overline{E_1}+\overline{E_2} = (100+j\,30)+(20+j\,120)$$
$$= (100+20)+(j\,30+j\,120)$$
$$= (100+20)+j(30+120)$$
$$= 120+j\,150 \quad \text{(Answer)}$$

Worked example 4

Add the following two voltage phasors together:

$$\overline{E_1} = (150+j\,100) \quad \text{and} \quad \overline{E_2} = (130-j\,60)$$

Solution

$$\overline{E_1}+\overline{E_2} = (150+j\,100)+(130-j\,60)$$
$$= (150+130)+(j\,100-j\,60)$$
$$= (150+130)+j(100-60)$$
$$= 280+j\,40 \quad \text{(Answer)}$$

Subtracting complex numbers

Suppose that you wish to subtract from phasor $\overline{I_1}$, a second phasor, $\overline{I_2}$, where:

$$\text{from} \quad \overline{I_1} = (a+jb) \quad \text{subtract} \quad \overline{I_2} = (c+jd)$$

The normal rules of subtraction apply:

$$\overline{I_1} = \quad (a \quad + \quad jb)$$
$$\overline{I_2} = \quad (c \quad + \quad jd) \; -$$
$$= \quad (a-c)+ \quad j(b-d)$$

ie: $\overline{I_1}-\overline{I_2} = (a-c)+(jb-jd) = (a-c)+j(b-d)$

So, when one phasor is *subtracted* from another, the real component of the resultant is the *difference between the 'real' parts of the two phasors*, and the imaginary part of the resultant is the *difference between the 'imaginary' parts of the two phasors*.

Worked example 5

From $\bar{E}_1 = (100 + j\,30)$ subtract $\bar{E}_2 = (20 - j\,120)$

Solution

$$
\begin{aligned}
\bar{E}_1 - \bar{E}_2 &= (100 + j\,30) - (20 - j\,120) \\
&= 100 + j\,30 - 20 + j\,120 \text{ (expanding bracket)} \\
&= (100 - 20) + (j\,30 + j\,120) \\
&= (100 - 20) + j(30 + 120) \\
&= \mathbf{80 + j\,150 \ (Answer)}
\end{aligned}
$$

Worked example 6

From $\bar{E}_1 = (60 + j\,40)$ subtract $\bar{E}_2 = (100 - j\,25)$

Solution

$$
\begin{aligned}
\bar{E}_1 - \bar{E}_2 &= (60 + j\,40) - (100 - j\,25) \\
&= 60 + j\,40 - 100 + j\,25 \text{ (expanding bracket)} \\
&= (60 - 100) + (j\,40 + j25) \\
&= (60 - 100) + j\,(40 + 25) \\
&= \mathbf{-40 + j\,65 \ (Answer)}
\end{aligned}
$$

Multiplying complex numbers

Suppose that you wish to multiply two phasors, $\bar{I}_1$ and $\bar{I}_2$, where:

$$\bar{I}_1 = (a + jb) \quad \text{and} \quad \bar{I}_2 = (c + jd)$$

The normal rules of algebraic multiplication apply, that is:

$$
\begin{array}{r}
(a + \quad jb) \\
(c + \quad jd)\times \\
\hline
ac + \quad jbc + \\
\quad jad + \quad j^2\,bd \\
\hline
ac + \quad jbc + \quad jad + \quad j^2\,bd \\
\hline
\end{array}
$$

$$
\begin{aligned}
(\bar{I}_1 \times \bar{I}_2) &= ac + jbc + jad + [(-1) \times bd] \text{ replacing } j^2 \text{ with } -1 \\
&= ac + jbc + jad - bd \\
&= ac - bd + jbc + jad \\
\text{rearranging} \quad &= (ac - bd) + j\,(bc + ad)
\end{aligned}
$$

Worked example 7

Multiply the phasors $\bar{E}_1$ and $\bar{E}_2$ together, where:

$$\bar{E}_1 = (8 + j\,5) \quad \text{and} \quad \bar{E}_2 = (10 + j\,9)$$

Solution

$$\begin{array}{r} (8 \ + \ j5) \\ (10 \ + \ j9) \times \\ \hline 80 \ + \ j50 \\ + \ j72 + \ j^2 45 \\ \hline 80 + \ j122 + \ j^2 45 \end{array}$$

$$\left(\bar{E}_1 \times \bar{E}_2\right) = 80 + j\,122 + j^2 45$$

$$= 80 + j\,122 + [(-1) \times 45] \text{ replacing } j^2 \text{ with } \sqrt{-1}$$

$$= 80 + j122 - 45$$

$$= \mathbf{35 + j\ 122\ V\ (Answer)}$$

> So, when one phasor is *multiplied* by another, the usual rules of multiplication apply. The resulting expression can be simplified by *replacing any j2 operator* by −1.

Dividing complex numbers

Suppose you wish to divide phasor $\bar{I}_1$ by phasor $\bar{I}_2$, where:

$$\bar{I}_1 = (a + jb) \quad \text{and} \quad \bar{I}_2 = (c + jd)$$

$$\text{i.e: } \quad \frac{\bar{I}_1}{\bar{I}_2} = \frac{(a + jb)}{(c + jd)}$$

Unfortunately, *dividing* by a complex number is simply *not possible*. So how can we get around it? Well, mathematicians (being mathematicians!) have come up with a really clever way of solving the problem, called '**rationalisation**'!

You will recall that, if we multiply *both* the numerator *and* the denominator of a fraction by exactly the *same amount*, then it makes absolutely no difference whatsoever to the ratio of that fraction. For example:

$$\frac{3}{4} \times \left(\frac{2}{2}\right) = \frac{6}{8} \equiv \frac{3}{4} \qquad \frac{3}{4} \times \left(\frac{5}{5}\right) = \frac{15}{20} \equiv \frac{3}{4} \qquad \frac{3}{4} \times \left(\frac{7}{7}\right) = \frac{21}{28} \equiv \frac{3}{4}$$

To **rationalise**, you simply take the complex denominator, *change the sign between its real and imaginary parts* and, then, multiply both the top and bottom of the expression by this new complex number, as shown below:

$$\frac{\bar{I}_1}{\bar{I}_2} = \frac{(a + jb)}{(c + jd)} = \quad \leftarrow \text{(change the centre sign, and multiply both top and bottom)}$$

$$= \frac{(a + jb)}{(c + jd)} \times \left(\frac{c - jd}{c - jd}\right) \text{'rationalisation'} - \text{causes no change in ratio}$$

$$= \frac{(a + jb)(c - jd)}{(c + jd)(c - jd)}$$

$$= \frac{ac + j\,bc - j\,ad - \mathbf{j^2}\,bd}{c^2 - \mathbf{j^2}\,d^2}$$

$$= \frac{ac + j\,bc - j\,ad - (-\mathbf{1} \times bd)}{c^2 - (-\mathbf{1} \times d^2)} \qquad \text{(replace } j^2 \text{ with} -1)$$

$$= \frac{ac + j\,bc - j\,ad + bd}{c^2 + d^2}$$

$$\text{then} \quad = \frac{(ac + bd) + j\,(bc - ad)}{c^2 + d^2}$$

So the process of **rationalising**, *changes the complex denominator into a real number* – and, of course, it is quite easy to divide a complex number by a real number!

Incidentally, you don't *really* have to bother to multiply-out the denominator, as *it is always* the **sum** *of the squares of the two variables* – in the above example, it doesn't matter whether the denominator was $(c + jd)$ or $(c - jd)$, after rationalisation the new denominator *always* becomes $(c^2 + d^2)$.

> So, when one phasor is *divided* by another, we must follow the rules of 'rationalisation'. The resulting expression can be simplified by replacing any j2 operator by −1.

Worked example 8

Divide the phasor $\bar{E}_1$ by the phasor $\bar{E}_2$, where:

$$\bar{E}_1 = (100 + j\,30) \quad \text{and} \quad \bar{E}_2 = (20 - j\,120)$$

Solution

$$\frac{\bar{E}_1}{\bar{E}_2} = \frac{(100 + j\,30)}{(20 + j\,120)}$$

$$= \frac{(100 + j\,30)}{(20 + j\,120)} \times \left(\frac{20 - j\,120}{20 - j\,120}\right) \quad \text{(rationalising)}$$

$$= \frac{2000 - j11400 - j^2 3600}{20^2 + 120^2}$$

$$= \frac{2000 - j11400 - (-3600)}{20^2 + 120^2} \quad \text{(replacing } j^2 \text{ with } -1\text{)}$$

$$= \frac{5600 - j\,11400}{20^2 + 120^2}$$

$$= \frac{5600 - j\,11400}{14\,800}$$

$$= \mathbf{0.378 - j0.770} \quad \textbf{(Answer)}$$

Inverting complex numbers

As we shall learn, being able to *invert* a complex numbers is very useful; particularly when we solve parallel circuits, and is simply a variation on dividing complex numbers, described above:

Worked example 9

Invert the following complex number: $\bar{E}_1 = (100 + j\,30)$.

Solution

$$= \frac{1}{100 + j30}$$

$$= \frac{1}{100 + j30} \times \left(\frac{100 - j30}{100 - j30}\right) \text{(rationalising)}$$

$$= \frac{100 - j30}{100^2 + 30^2}$$

$$= \frac{100 - j30}{10\,900} = \mathbf{0.009 - j\,0.003 \ (Answer)}$$

Complex number mathematics, using scientific calculators

Complex number mathematics has become *much* less time consuming, and less prone to error, thanks to the functions available with the modern scientific pocket calculator. Complex numbers can be *added*, *subtracted*, *multiplied*, *divided*, and *inverted* quickly by performing some very simple keystrokes – thus avoiding the lengthy number-crunching shown in these worked examples.

Unfortunately, different calculator manufacturers have different ways of enabling their complex mathematics mode, and use different approaches for entering data, so it is simply not possible to offer any guidance in this chapter. Unfortunately, too, the method of entering/reading data often changes according to the age of a particular calculator.

It is essential to carefully read the instruction manual supplied with your particular calculator. Once you have familiarised yourself with its instructions, you can practise, by using your calculator to double-check the above examples, as well as the answers to each of the worked examples shown throughout the rest of this chapter.

> **Warning**: *Most* scientific calculators (e.g. *Casio*, *Sharp*) output the answer to a complex-mathematics calculation in the following (logical) order: **real part**, followed by the **imaginary (j) part**. However, *some* calculators (e.g. *Hewlett Packard*), according to their age, unfortunately supply the answers in the *reverse* order – **beware of this, and examine the manufacturer's instruction manual *very carefully* to find out how it outputs its answers!**

Applying complex numbers to a.c. circuits

By now, you are probably wondering *where* what we've learned, so far, about the **j-operator** is actually leading? Well, consider the simple series circuits, below:

Series *R-L* circuit

In a **series circuit** (Figure 5.17), the common quantity is current so, as usual, the supply current is taken as the reference phasor, with $\bar{E}$ being the phasor-sum of the voltage drops, $\bar{U}_R$ and $\bar{U}_L$, where $\bar{U}_R$ is the in-phase, or 'real' component, and $\bar{U}_L$ is the quadrature, or 'imaginary' component. The resulting phasor diagram will look like Figure 5.18.

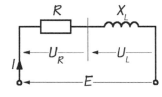

Figure 5.17

So, instead of drawing a phasor diagram, this situation could, instead, be expressed mathematically using the 'j-operator', as follows:

$$\bar{E} = \bar{U}_R + j\bar{U}_L$$

Just as we divided throughout this expression by the reference phasor, $\bar{I}$, to obtain an **impedance diagram**, we can divide this complex expression by $\bar{I}$ to obtain a complex expression for the impedance *(Z)* of the circuit:

$$\frac{\bar{E}}{\bar{I}} = \frac{\bar{U}_R}{\bar{I}} + \frac{j\bar{U}_L}{\bar{I}}$$
$$Z = R + jX_L$$

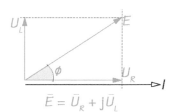

$$\bar{E} = \bar{U}_R + j\bar{U}_L$$

Figure 5.18

From this expression, we can see that **inductive reactance** (X_L) is *always* expressed as a **positive** imaginary number. For example, in the following expression, the '**+j**' defines a circuit as having an impedance made up of 30 Ω **resistance** in series with an *inductive* **reactance** of 40 Ω:

$$Z = (30 + \text{j}\,40)\ \Omega$$

As we will see shortly, when working with parallel circuits, it's often very useful to be able to determine the **admittance** *(Y)* of a circuit, where admittance is the *reciprocal* of impedance, expressed in **siemens (S)**.

So, to determine the **admittance**, *Y*, of this series *R-L* circuit:

$$Y = \frac{1}{Z}$$

$$= \frac{1}{(30 + \text{j}\,40)}$$

$$= \frac{1}{(30 + \text{j}\,40)} \times \left(\frac{30 - \text{j}\,40}{30 - \text{j}\,40}\right) \text{(rationalising)}$$

$$= \frac{(30 - \text{j}\,40)}{(30^2 + 40^2)}$$

$$= \frac{(30 - \text{j}\,40)}{2500}$$

$$= \textbf{0.012} - \textbf{j}\,\textbf{0.016 S (Answer)}$$

Or you can use your scientific calculator by inputting the complex impedance, and finding its reciprocal.

We will see *why* it's useful to change from impedance to admittance when we move on to look at parallel circuits in a moment.

Series *R-C* circuit

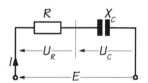

R   *X*_c

Now, let's consider a series **R-C** circuit shown in Figure 5.19.

Again, as it's a series circuit, current is used as the reference phasor, and $\bar{E}$ is the phasor-sum of the voltage drops, $\bar{U}_R$ and $\bar{U}_C$, with the resulting phasor diagram looking like Figure 5.20.

This phasor diagram can be expressed, mathematically, using the operator 'j', as follows:

$$\bar{E} = \bar{U}_R - \text{j}\,\bar{U}_C$$

Figure 5.19

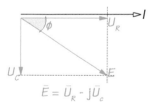

$$\bar{E} = \bar{U}_R - \text{j}\,\bar{U}_C$$

Figure 5.20

Once again, just as we divided throughout by the reference phasor, $\bar{I}$, to obtain an **impedance diagram**, we can divide this complex expression by $\bar{I}$ to obtain a complex expression for the impedance *(Z)* of the circuit:

$$\frac{\bar{E}}{\bar{I}} = \frac{\bar{U}_R}{\bar{I}} - \text{j}\frac{\bar{U}_C}{\bar{I}}$$

$$Z = R - \text{j}\,X_c$$

So, **capacitive reactance** (X_c) is *always* expressed as a **negative** imaginary number. For example, in the following expression, the '**–j**' defines a circuit having an impedance made up of 4 Ω **resistance** in series with an *capacitive* **reactance** of 3 Ω:

$$Z = (4 - \text{j}\,3)\ \Omega$$

Again, as we shall see later, it's very useful to be able to convert from **impedance** *(Z)* to **admittance** *(Y)*, where admittance is the reciprocal of impedance:

$$Y = \frac{1}{Z}$$
$$= \frac{1}{(4 - j3)}$$
$$= \frac{1}{(4 - j3)} \times \left(\frac{4 + j3}{4 + j3} \right) \text{(rationalising)}$$
$$= \frac{(4 + j3)}{(4^2 + 3^2)}$$
$$= \frac{(4 + j3)}{25}$$
$$= \mathbf{0.16 + j\,0.12\ S\ (Answer)}$$

Or you can use your scientific calculator, by inputting the complex impedance, and finding its reciprocal.

Series *R-L-C* circuit

Now, let's consider the series *R-L-C* circuit shown in Figure 5.21:

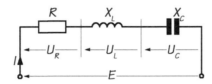

Figure 5.21

Again, as it's a series circuit, current is used as the reference phasor, and $\bar{E}$ is the phasor-sum of the voltage drops, $\bar{U}_R$, $\bar{U}_L$, and $\bar{U}_C$, with the resulting phasor diagram looking like this (Figure 5.22):

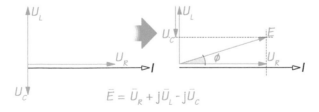

Figure 5.22

From the above, a series circuit comprising a resistance of *R* ohms, in series with an inductive reactance of X_L ohms and a capacitive reactance of X_C ohms, would be expressed as:

$$Z = R + jX_L - jX_C$$

As explained, above, *inductive* reactance is always represented by '**+j**', whereas *capacitive* reactance is always represented by '**−j**'. For example, the following expression defines an impedance made up of 20 Ω resistance in series with an **inductive reactance** of 15 Ω, and a **capacitive reactance** of 5 Ω:

$$Z = 20 + j15 - j5$$

Worked example 10

Express the effective impedance of a series circuit of resistance 12 Ω, inductive reactance 20 Ω, and capacitive reactance 50 Ω, in complex notation.

Solution

$$Z = (12 + j20 - j50)$$
$$= \mathbf{(12 - j30)}\ \Omega$$

Again, it's always useful to be able to work out its **admittance**, by using the techniques already described. You can do this longhand, or using your scientific calculator.

The illustrations in Figure 5.23 summarise how we can express combinations of resistance, inductive reactance, and capacitive reactance, as complex numbers.

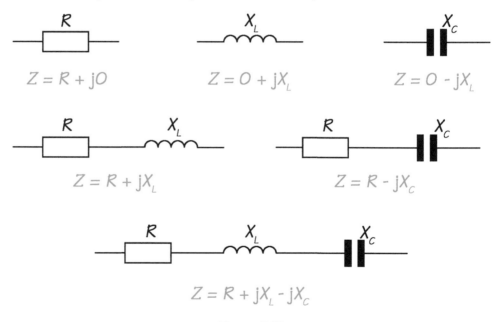

Impedance, expressed in Complex Notation

$$Z = R + j0 \qquad Z = 0 + jX_L \qquad Z = 0 - jX_L$$

$$Z = R + jX_L \qquad Z = R - jX_C$$

$$Z = R + jX_L - jX_C$$

Figure 5.23

The **j-operator**, then, also provides us with a useful way of describing an a.c. circuit. For example, instead of saying, '*A circuit comprises a resistance of 10 Ω in series with an inductive reactance of 15 Ω . . .*', we can instead say, '*A circuit has an impedance of* $(10 + j15)\ \Omega$ . . .'.

Parallel *R-L* circuit

In the **parallel *R-L*** circuit shown in Figure 5.24, the *supply voltage* is common to each branch and so is taken as the reference phasor. The current through the resistive branch, $\overline{I_R}$, is *in phase* with the supply voltage, while the current through the inductive branch, $\overline{I_L}$, *lags* (remember the mnemonic, '**C**I**VIL**') the supply voltage by 90°, and the supply current, $\overline{I}$, is the phasor-sum of the two branch currents. The resulting phasor diagram will look like Figure 5.25.

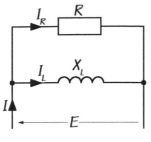

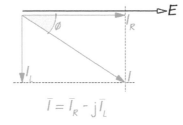

| **Figure 5.24** | **Figure 5.25** |

Instead of drawing a phasor diagram, this situation could, instead, be expressed mathematically using the 'j-operator', as follows:

$$\overline{I} = \overline{I}_R - j\overline{I}_L$$

Note that we use **–j** in front of $\overline{I}_L$, because this current is *lagging* the supply voltage by right angles.

Parallel *R-C* circuit

Now, let's consider the parallel ***R-C*** circuit in Figure 5.26:

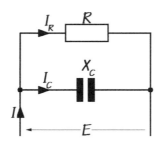

Figure 5.26

In a **parallel *R-C*** circuit, the supply voltage is, again, common to each branch and so is taken as the reference phasor. The current through the resistive branch, $\overline{I}_R$, is *in phase* with the supply voltage, while the current through the capacitive branch, $\overline{I}_C$, *leads* (remember the mnemonic, '**CIV**IL') the supply voltage by 90°, and the supply current, $\overline{I}$, is the phasor-sum of the two branch currents. The resulting phasor diagram will look like this (Figure 5.27):

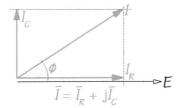

Figure 5.27

This phasor diagram can be expressed, mathematically, using the operator 'j', as follows:

$$\overline{I} = \overline{I}_R + j\overline{I}_C$$

Note, this time, we use a **+j** in front of $\overline{I}_C$, because this current is *leading* the supply voltage by right angles.

Parallel R-L-C circuit

Now, let's consider the parallel **R-L-C** circuit shown in Figure 5.28:

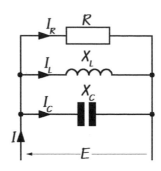

Figure 5.28

Again, the supply voltage, $\bar{E}$, is used as the reference phasor, and is the supply current is the phasor-sum of the branch currents, $\bar{I}_R$, $\bar{I}_L$, and $\bar{I}_C$, with the resulting phasor diagram looking like this (Figure 5.29):

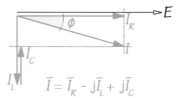

$$\bar{I} = \bar{I}_R - j\bar{I}_L + j\bar{I}_C$$

Figure 5.29

This phasor diagram can be expressed, mathematically, using the operator 'j', as follows:

$$\bar{I} = \bar{I}_R - j\bar{I}_L + j\bar{I}_C$$

Again, we use a **–j** in front of $\bar{I}_L$, because this current is *lagging* the supply voltage by right angles, and we use a **+j** in front of $\bar{I}_C$, because this current is *leading* the supply voltage by right angles.

Using the j-operator to solve a.c. circuits

This is where we begin to learn how very much easier it is to solve series-parallel circuits using the j-operator, rather than by constructing a phasor diagram and then proceeding in the way that we are already familiar with.

But, before that, let's use the j-operator to solve a couple of relatively straightforward a.c. circuits. In practice, it's hardly worthwhile bothering to use the j-operator to solve these particular examples; you could say that we are 'using a sledgehammer to crack a nut'! And we are! The only purpose is to learn how to use the j-operator using a couple of straightforward examples, before proceeding on to more difficult circuits.

Again, you should follow this example but using your scientific calculator in its complex mathematics mode.

Worked example 11: series circuit

A circuit consists of a resistance of 30 Ω in series with an inductive reactance of 40 Ω, and is connected across a 100 V a.c. supply. Calculate (a) the resulting current, and (b) the voltage drop across the resistive and reactive components.

Solution

As always, we start by sketching the circuit diagram and inserting all the information supplied in the question (Figure 5.30). This saves us from having to constantly refer back to the question to obtain any information.

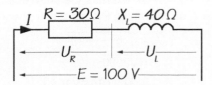

Figure 5.30

Next, we express the impedance of the circuit, in complex form:

$$Z = 30 + j40$$

(a) To find the current, we simply divide the supply voltage by the circuit's complex impedance (note that if the supply voltage is stated as being '100 V', by default we can *always* assume that its complex form must, therefore, be '100 ± j0', although it's unnecessary to write it in that way).

We'll be solving this problem longhand, but you can always use your scientific calculator to confirm each answer.

$$\overline{I} = \frac{\overline{E}}{Z} = \frac{100 + j0}{30 + j40} \text{ which is the same as } \frac{100}{30 + j40}$$

$$= \frac{100}{30 + j40} \times \frac{30 - j40}{30 - j40} \text{(rationalising)}$$

$$= \frac{3000 - j4000}{30^2 + 40^2}$$

$$= \frac{3000 - j4000}{2500}$$

$$= \textbf{1.2} - \textbf{j1.6 A (Answer a.)}$$

The '−j' indicates that the current is *lagging* the supply voltage, which is *exactly* what it should be doing in a series *R-L* circuit.

We could leave the answer in complex (rectangular) form but, often, we are often asked to express the answer in **polar form**, as follows:

By Pythagoras's Theorem,

$$\overline{I} = \sqrt{1.2^2 + (-1.6)^2} = \sqrt{4} = 2 \text{ A (Answer a.)}$$

To find the (phase) angle, we use the arctangent function ($\tan^{-1}$), as follows:

$$\angle\phi = \tan^{-1}\frac{opposite}{adjacent} = \tan^{-1}\frac{-1.6}{1.2} = \tan^{-1}-1.33 = -53.05° \textbf{ (Answer b.)}$$

Again, the negative sign confirms that the current is *lagging* the supply voltage, which is what you would expect in a series *R-L* circuit.

The −j1.6 A in our answer tells us that the phase angle is *lagging*, which is what we'd expect for an *R-L* circuit and, this is confirmed from the polar form (−53.05°).

(b) To find the voltage drop across the *resistive* component, we simply multiply the resistance by the current through it:

$$\bar{U}_R = \bar{I}R = (1.2 - j1.6) \times (30 + j0) = 36 - j48 \text{ V}$$

Again, it's often expected that we express the answer as the resulting magnitude of its polar form, so, using Pythagoras's Theorem:

$$\bar{U}_R = \sqrt{36^2 + (-48)^2}$$
$$= \sqrt{3600}$$
$$= 60 \text{ V } \textbf{(Answer c.)}$$

To find the voltage drop across the *inductive* component, we simply multiply the inductive reactance by the current through it:

$$\bar{U}_L = \bar{I} X_L = (1.2 - j1.6) \times (0 + j40)$$
$$= j48 \times (-j^2 64)$$

(since $j^2 = -1$) :

$$\bar{U}_L = 64 + j48 \text{ V}$$

Again, it's often expected that we provide the answer as the magnitude of its polar form, so, applying Pythagoras's Theorem:

$$\bar{U}_L = \sqrt{64^2 + 48^2}$$
$$= \sqrt{6400}$$
$$= 80 \text{ V } \textbf{(Answer c.)}$$

To confirm the two answers in part (c), if we were to *add* the two voltage drops together, in complex form, then they should equal the supply voltage:

$$\bar{U}_R + \bar{U}_L = (36 - j48) + (64 + j48) = (100 + j0) \text{ V}$$

. . . which they do!

Worked example 12: parallel circuit

A circuit consists of a resistance of 20 Ω in parallel with a capacitive reactance of 115 Ω, and is connected across a 100 V a.c. supply. Calculate (a) current through each branch, and (b) the supply current.

Solution

As always, we start by sketching the circuit diagram, and inserting all the information supplied in the question (Figure 5.31), which will save us from having to constantly refer back to the question to obtain any information.

Before we start, we should realise that the complex impedance of the upper branch should, strictly speaking, be written as $(20 + j0)$ Ω, although we can *also* write it simply as '20 Ω', and the complex impedance of the lower branch should, strictly speaking, be written as $(0 - j15)$ Ω, although we can *also* write it simply as '−j15 Ω'.

Also, when the supply voltage is given, simply, as '100 V', we can always assume that, by default, this means: $(100 + j0)$ V .

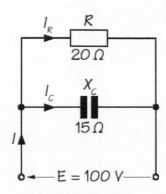

Figure 5.31

(a) To find the current through the *upper branch* ($\bar{I}_{upper}$), we simply divide the supply voltage by the impedance of the upper branch.

$$\bar{I}_{upper} = \frac{\bar{E}}{Z_{upper}}$$

$$= \frac{100 + j0}{20 + j0}$$

$$= \frac{100 + j0}{20 + j0} \times \left(\frac{20 - j0}{20 - j0}\right) \text{(rationalising)}$$

$$= \frac{2000 - j0}{20^2 + 0^2}$$

$$= \frac{2000 - j0}{400}$$

$$= 5 - j0 \text{ A}$$

Again, it's not uncommon to be asked to present the answer in polar form, but we won't bother doing that here:

To find the current through the *lower branch* ($\bar{I}_{lower}$), we simply divide the supply voltage by the complex impedance of the lower branch:

$$\bar{I}_{lower} = \frac{\bar{E}}{Z_{lower}}$$

$$= \frac{100 + j0}{0 - j15}$$

$$= \frac{100 + j0}{0 - j15} \times \left(\frac{0 + j15}{0 + j15}\right) \text{(rationalising)}$$

$$= \frac{0 + j1500}{0^2 + 15^2}$$

$$= \frac{0 + j1500}{225}$$

$$= 0 + j6.67 \text{ A}$$

(The '**+j**' makes sense, because the current through the capacitive branch must be *leading* the supply voltage.)

(b) The supply current, of course, is the phasor-sum of the two branch currents, so:

$$\bar{I}_{supply} = \bar{I}_{upper} + \bar{I}_{lower}$$
$$= (5+j0)+(0+j6.65)$$
$$= (5+j6.65)\ A$$

We can also express the answer in *polar form*, if requested, as follows:

Applying Pythagoras's Theorem:	Applying arctan:
$\bar{I}_{supply} = \sqrt{5^2+6.67^2}$ $= \sqrt{69.5}$ $= 8.34\ A$	$\angle\phi = \tan^{-1}\left(\dfrac{opposite}{adjacent}\right)$ $= \tan^{-1}\left(\dfrac{6.67}{5}\right)$ $= \tan^{-1}1.33$ $= 53.06°$

$$\bar{I} = \textbf{8.34}\ \angle\textbf{53.06°\ (Answer)}$$

The real power of the j-operator

Now, to demonstrate the **real power** of using the j-operator, *let's solve exactly the same series-parallel worked example, that we solved in the previous chapter.* As you will see, using the j-operator will reduce the amount of calculations significantly!

Worked example 13

In the series-parallel circuit shown in Figure 5.32, the *upper* branch has a resistance of 50 Ω in series with an inductive reactance of 20 Ω, and the *lower* branch has a resistance of 30 Ω, in series with a capacitive reactance of 40 Ω. The two branches are connected across a 230 V, 50 Hz, a.c. supply. The question is, *how do we determine the supply current and phase angle for this circuit?*

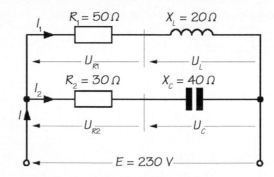

Figure 5.32

Solution

The first step is to express the impedance of each branch using j-notation:

$$Z_{upper} = (50+j20)\ \Omega \quad and \quad Z_{lower} = (30-j40)\ \Omega$$

For a series-parallel circuit, it's *always* easier to work with **admittances** (expressed in **siemens**), rather than impedances. This is because you simply *add* admittances, whereas you must add *reciprocals* and, then, *invert* the answer when adding impedances.

$$Z_{upper} = (50 + j20)\,\Omega$$

$$Y_{upper} = \frac{1}{Z_{upper}} = \frac{1}{50 + j20}$$

$$= \frac{1}{50 + j20} \times \left(\frac{50 - j20}{50 - j20}\right) \text{rationalising}$$

$$= \frac{50 - j20}{50^2 + 20^2}$$

$$= \frac{50 - j20}{2900}$$

$$= 0.017 - j0.007\ \text{S}$$

$$Z_{lower} = (30 - j40)\,\Omega$$

$$Y_{lower} = \frac{1}{Z_{lower}} = \frac{1}{30 - j40}$$

$$= \frac{1}{30 - j40} \times \left(\frac{30 + j40}{30 + j40}\right) \text{rationalising}$$

$$= \frac{30 + j40}{30^2 + 40^2}$$

$$= \frac{30 + j40}{2500}$$

$$= 0.012 + j0.016\ \text{S}$$

Next, we can find the *total* admittance, Y_{total}, by *adding* the two branch complex admittances together:

$$Y_{total} = Y_{upper} + Y_{lower} = (0.017 - j\,0.007) + (0.012 + j0.016)$$

$$= 0.017 + 0.012 - j0.007 + j0.016$$

$$= 0.029 + j0.009\ \text{S}$$

Finally, the supply current is found by *multiplying* the supply voltage by the total admittance of the circuit (which, of course, is equivalent to *dividing* the supply voltage by the total *impedance* of the circuit):

$$\overline{I} = \overline{E}\,Y_{total} = 230 \times (0.029 + j0.009) = 6.67 + j2.07\ \text{A}$$

As the quadrature current value is preceded by '**+j**', we know that the current must be *leading* the supply voltage – in other words, the circuit is predominantly capacitive. As it's more usual to express current in polar form, we need to convert from rectangular form to polar form:

$$\text{By Pythagoras's Theorem, } I = \sqrt{6.67^2 + 2.07^2}$$

$$= \sqrt{48.77} = \textbf{6.98A (Answer)}$$

To find its phase angle, we use the arctan ($\tan^{-1}$) function:

$$\angle\phi = \tan^{-1}\frac{\text{opposite}}{\text{adjacent}} = \tan^{-1}\frac{2.07}{6.67} = \tan^{-1}0.31 = \textbf{17.22° leading (Answer)}$$

These answers differ slightly from those in the original worked example from the previous chapter due to the way we rounded-off figures as we worked through each solution.

If you compare *this* solution with the solution for the identical question in the previous chapter, it should be clear *why* the **j-operator** is such a useful technique.

Although we have solved the above worked example longhand, bear in mind that it can be solved very much quicker, and with less chance of error, if you use a scientific calculator set to its 'complex mathematics' mode. You are urged to check the above calculations using your scientific calculator set to 'complex numbers', in order to get used to using your calculator for this purpose.

What you *must* do, however, is to *take great care with the positive and negative signs, as assigning an incorrect sign will completely change your answers.*

Let's now look at some more complicated series-parallel circuits to reinforce our understanding of this new technique.

Worked example 14

For the circuit shown in Figure 5.33, calculate (a) the total impedance of the circuit, (b) the value of the supply current in polar form, (c) the voltage drop across the series component of the circuit in polar form, (d) the voltage drop across the parallel component of the circuit in polar form, (e) the currents through each branch of the parallel component in polar form.

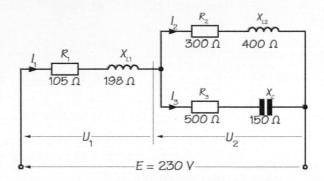

Figure 5.33

Solution

(a) To find the **total impedance** of the circuit, we start by determining the equivalent impedance of the parallel component of the circuit. Remember, for parallel branches, it's always easier to work with *admittances*, rather than impedances:

Let the impedance of upper branch be Z_2:

$$Z_2 = (300 + j400)\,\Omega$$

(**+j**, because it's an *inductive* component.) So, it's admittance, Y_2 is given by:

$$Y_2 = \frac{1}{Z_2} = \frac{1}{300 + j400}$$

$$= \frac{1}{300 + j400} \times \left(\frac{300 - j400}{300 - j400}\right) \text{(rationalising)}$$

$$= \frac{300 - j400}{300^2 + 400^2}$$

$$= \frac{300 - j400}{250\,000}$$

$$= 0.0012 - j0.0016\,\text{S}$$

Let the impedance of upper branch be Z_3:

$$Z_3 = (500 - j150)\,\Omega$$

(**−j** because it's a *capacitive* component.) So, it's admittance, Y_3, is given by:

$$Y_3 = \frac{1}{Z_3} = \frac{1}{500 - j150}$$

$$= \frac{1}{500 - j150} \times \left(\frac{500 + j150}{500 + j150}\right) \text{(rationalising)}$$

$$= \frac{500 + j150}{500^2 + 150^2}$$

$$= \frac{500 + j150}{272\,500}$$

$$= 0.0018 + j0.0005\,\text{S}$$

The equivalent admittance (Y_A) of the parallel part of the circuit will be the *sum* of the two branch admittances:

$$Y_A = Y_2 + Y_3$$

$$= (0.0012 - j0.0016) + (0.0018 + j0.0005)$$

$$= 0.0030 - j0.0011\,\text{S}$$

To find the equivalent impedance (Z_A) of the parallel part of the circuit, we need to find the *reciprocal* of its equivalent admittance:

$$Z_A = \frac{1}{Y_A} = \frac{1}{0.0030 - j0.0011}$$

$$= \frac{1}{0.0030 - j0.0011} \times \left(\frac{0.0030 + j0.0011}{0.0030 + j0.0011} \right) \text{(rationalising)}$$

$$= \frac{0.0030 + j0.0011}{0.0030^2 + 0.0011^2}$$

$$= \frac{0.0030 + j0.0011}{(0.0102 \times 10^{-3})}$$

$$\approx 248 + j108 \ \Omega$$

So the total impedance of the circuit will be:

$$Z = Z_1 + Z_A$$

$$= (105 + j198) + (248 + j108)$$

$$= \mathbf{353 + j306} \ \Omega \ \textbf{(Answer a.)}$$

(b) The supply current, I_1, is given as follows. Remember, the supply voltage is by default, $(230 \pm j0)$ V :

$$\overline{I_1} = \frac{\overline{E}}{Z} = \frac{230 + j0}{353 + j306}$$

$$= \frac{230}{353 + j306} \times \left(\frac{353 - j306}{353 - j306} \right) \text{(rationalising)}$$

$$= \frac{81190 - j70\,380}{353^2 + 306^2}$$

$$= \frac{81190 - j70\,380}{218\,245}$$

$$= 0.372 - j0.322 \ \text{A}$$

Again, it's usual to express an answer in polar form, so:

$$I_1 = \sqrt{0.372^2 + (-0.322^2)} = \sqrt{0.242} = 0.492 \ \text{A}$$

and the phase angle is given by:

$$\angle \phi = \tan^{-1} \frac{opposite}{adjacent} = \tan^{-1} \frac{-0.322}{0.372} = \tan^{-1} -0.8656 = -40.88°$$

So, the current, in polar form:

$$I_1 = \mathbf{0.492 \ A} \ \angle -\mathbf{40.88°} \ \textbf{(Answer b.)}$$

[The '**−j**' in the complex expression for the supply current, indicates that the current is *lagging* the supply voltage and, so, the circuit is predominantly **inductive**.]

(c) The voltage drop, $\overline{U_1}$, across the series part of the circuit is the product of the supply current, $\overline{I_1}$, and the impedance, Z_1, of that part of the circuit:

$$\overline{U_1} = \overline{I_1} Z_1 = (0.372 - j0.322) \times (105 + j198)$$

$$= 102.82 + j39.85 \ \text{V}$$

Expressed in polar form:

$$\overline{U_1} = \sqrt{102.82^2 + 39.85^2} = \sqrt{12\,160} = 110.27 \ \text{V}$$

$$\angle\phi_1 = \tan^{-1}\frac{\text{opposite}}{\text{adjacent}} = \tan^{-1}\frac{39.85}{102.82} = \tan^{-1}0.3876 = 21.19°$$

So the voltage drop across the series part of the circuit, expressed in polar form is:

$$\bar{U}_1 = \mathbf{110.27\ V\ \angle21.19°\ \ (Answer\ c.)}$$

(d) The voltage drop, $\bar{U}_2$, across the parallel part of the circuit will be the phasor *difference* between the voltage across the entire circuit, $\bar{E}$, and the voltage drop across the series part of the circuit, $\bar{U}_1$:

$$\bar{U}_2 = \bar{E} - \bar{U}_1 = (230 + j0) - (102.82 + j39.85) = 127.18 - j39.85\ V$$

Again, it's usual to express values in polar form, so:

$$\bar{U}_2 = \sqrt{127.18^2 + (-39.852)^2} = \sqrt{17\,763} = 133.28\ V$$

$$\angle\phi = \tan^{-1}\frac{\text{opposite}}{\text{adjacent}} = \tan^{-1}\frac{-39.85}{127.18} = \tan^{-1}-0.3133 = -117.4°$$

So, $\bar{U}_2 = \mathbf{133.28\ V\ \angle-117.4°\ \ (Answer\ d.)}$

[We can check whether this answer is correct by multiplying the supply current, $\bar{I}_1$, by the equivalent impedance of the parallel circuit, ZA, which should give us the same answer.]

(e) Finally, we can now calculate the value of the currents through the upper branch and lower branch of the parallel part of the circuit:

$$\bar{I}_2 = \frac{\bar{U}_2}{Z_2} = \frac{127.18 - j39.85}{300 + j400}$$
$$= \frac{127.18 - j39.85}{300 + j400} \times \left(\frac{300 - j400}{300 - j400}\right)$$
$$= \frac{22\,214 - j62\,827}{300^2 + 400^2}$$
$$= \frac{22\,214 - j62\,827}{250\,000}$$
$$= 0.089 - j0.251\ A$$

$$\bar{I}_2 = \frac{\bar{U}_2}{Z_3} = \frac{127.18 - j39.85}{500 - j150}$$
$$= \frac{127.18 - j39.85}{500 - j150} \times \left(\frac{500 + j150}{500 + j150}\right)$$
$$= \frac{69\,568 - j848}{500^2 + 150^2}$$
$$= \frac{69\,568 - j848}{272\,500}$$
$$= 0.255 - j0.003\ A$$

These can be expressed in polar form, as follows:

Applying Pythagoras's Theorem:

$$\bar{I}_2 = \sqrt{0.089^2 + (-0.251)^2}$$
$$= \sqrt{0.0079 + 0.063}$$
$$= \sqrt{0.0709}$$
$$= 0.266\ A$$

Applying Pythagoras's Theorem:

$$\bar{I}_2 = \sqrt{0.255^2 + (-0.003)^2}$$
$$= \sqrt{0.065 + 0.009}$$
$$= \sqrt{0.074}$$
$$= 0.272\ A$$

[We could also have calculated currents $\bar{I}_2$ and $\bar{I}_3$ by *multiplying* the voltage drop across the parallel branches by the admittance of each of the two branches.]

[We can also confirm that our answers are correct, by vectorially adding the two branch currents, to confirm that they equal the value of the supply voltage.]

Worked example 15

Calculate the total impedance of a circuit comprising two parallel branches, A and B, having impedances of $(20+j0)\,\Omega$ and $(5-j31.4)\,\Omega$ respectively, which are connected in series with a third impedance of $(0-j21.2)\,\Omega$. If the supply voltage is 230 V, what value of load current will the circuit draw, expressed in complex form?

Solution

Again, for parallel branches, it's always much easier to work with admittances, rather than impedances, so:

$$Y_A = \frac{1}{Z_A}$$

$$= \frac{1}{20+j0}$$

$$= \frac{1}{20+j0} \times \left(\frac{20-j0}{20-j0}\right)$$

$$= \frac{20-j0}{20^2+0^2}$$

$$= \frac{20-j0}{400}$$

$$= 0.05 - j0\ \text{S}$$

$$Y_B = \frac{1}{Z_B}$$

$$= \frac{1}{5+j31.4}$$

$$= \frac{1}{5+j31.4} \times \left(\frac{5-j31.4}{5-j31.4}\right)$$

$$= \frac{5-j31.4}{5^2+31.4^2}$$

$$= \frac{5-j31.4}{1011}$$

$$= 0.004\,95 - j0.0311\ \text{S}$$

The total admittance for the parallel part of the circuit will be the sum of the admittances for the upper and lower branches:

$$Y = Y_{upper} + Y_{lower} = (0.05-j0) + (0.004\,95 - j0.0311)$$

$$= 0.054\,95 - j0.0311\ \text{S}$$

Next, we need to determine the *total impedance* of the complete circuit, so we must first find the impedance of the parallel part of the circuit which, of course, is the reciprocal of its admittance:

$$Z_{parallel} = \frac{1}{Y_{parallel}} = \frac{1}{0.054\,95 - j0.0311}$$

$$= \frac{1}{0.054\,95 - j0.0311} \times \frac{0.054\,95 + j0.0311}{0.054\,95 + j0.0311}\ \text{(rationalising)}$$

$$= \frac{0.054\,95 + j0.0311}{0.054\,95^2 + 0.0311^2}$$

$$= \frac{0.054\,95 + j0.0311}{0.00399}$$

$$= 13.77 + j7.79\ \Omega$$

So the total impedance, Z, for the complete circuit must be:

$$Z = Z_{parallel} + Z_{series} = (13.77+j7.79) + (0-j21.2)$$

$$= \left(13.77 - j13.41\right)\Omega$$

To find the value of the load current, we must divide the supply voltage by the total impedance:

$$\bar{I} = \frac{\bar{E}}{Z} = \frac{230}{(13.77 - j13.41)}$$

$$= \frac{230}{(13.77 - j13.41)} \times \left(\frac{13.77 + j13.41}{13.77 + j13.41}\right)$$

$$= \frac{3167 + j3084}{13.77^2 + 13.41^2}$$

$$= \frac{3167 + j3084}{369.44}$$

$$= \left(8.57 + j8.35\right) \text{ A (Answer)}$$

Worked example 16

Two coils, A and B, of resistance 10 Ω and 20 Ω and inductance 0.02 H and 0.03 H respectively, are connected in parallel across a 200 V, 50 Hz, supply. Calculate the total current drawn from the supply, expressed in polar notation.

Solution

We start by determining the inductive reactance of each coil:

For coil A:

$$X_L = 2\pi f L_A$$
$$= 2\pi \times 50 \times 0.02 = 6.28 \text{ Ω}$$

For coil B:

$$X_L = 2\pi f L_B$$
$$= 2\pi \times 50 \times 0.03 = 9.42 \text{ Ω}$$

As this is a parallel circuit, it's always easier to work in terms of **admittance** *(Y)*:

For coil A:

$$Y_A = \frac{1}{Z_A} = \frac{1}{10 + j6.28}$$

$$= \frac{1}{10 + j6.28} \times \left(\frac{10 - j6.28}{10 - j6.28}\right)$$

$$= \frac{10 - j6.28}{10^2 + 6.28^2}$$

$$= \frac{10 - j6.28}{139.44}$$

$$= \left(0.072 - j0.045\right) \text{ S}$$

For coil B:

$$Y_B = \frac{1}{Z_B} = \frac{1}{20 + j9.42}$$

$$= \frac{1}{20 + j9.42} \times \left(\frac{20 - j9.42}{20 - j9.42}\right)$$

$$= \frac{20 - j9.42}{20^2 + 9.42^2}$$

$$= \frac{20 - j9.42}{488.74}$$

$$= \left(0.041 - j0.019\right) \text{ S}$$

Now we can determine the admittance of the complete circuit:

$$Y = Y_A + Y_B = (0.072 - j0.045) + (0.041 - j0.019) = \left(0.113 - j0.064\right) \text{ S}$$

. . . from which we can determine the total supply current:

$$\bar{I} = \bar{E}Y = 200 \times (0.113 - j0.064) = (22.6 - j12.8) \text{ A}$$

Expressed in polar notation:

Applying Pythagoras's Theorem:

$$I = \sqrt{22.6^2 + (-12.8)^2}$$
$$= \sqrt{511 + 164}$$
$$= 25.98 \text{ A}$$

$$\angle\phi = \tan^{-1}\left(\frac{\text{opposite}}{\text{adjacent}}\right)$$
$$= \tan^{-1}\left(\frac{-12.8}{22.6}\right)$$
$$= \tan^{-1} -0.566$$
$$= -29.51°$$

So the supply current, expressed in polar form:

$$\overline{I} = \mathbf{25.98} \angle(\mathbf{-29.51°}) \textbf{ A (Answer)}$$

Conclusion

Constructing a phasor diagram for **series-parallel a.c. circuits** is complicated – *very complicated* in the case of circuits such as the one in the worked example, above!

The use of the **j-operator** allows us to solve series-parallel circuits without the need to construct complicated phasor diagrams. Although the two worked examples shown in this chapter have been solved long-handed, it is much easier and quicker to solve them using a scientific calculator in its 'complex mathematics' mode.

You are urged, therefore, to follow each of the two worked examples, step by step, using your scientific calculator.

Now that we've completed this chapter, we need to examine its **objectives** listed at its start. Placing a question mark at the end of each objective turns that objective into a **test item**. If we can answer those test items, then we've met the objectives of this chapter.

CHAPTER 6

Solving balanced and unbalanced three-phase circuits, using symbolic notation

Objectives

On completion of this chapter you should be able to:

1. explain the result of applying the operator, '**a**', to a phasor-quantity.

2. express each of the following operators in polar, and rectangular, form:

   a. a

   b. a²

   c. −a

   d. −a².

3. use the operator, 'a', to solve three-phase problems for both balanced and unbalanced loads

4. use Millman's Theorem to solve unbalanced, three-wire, star-connected loads.

Introduction

In an earlier chapter, we learned that there's a limit to how far we can go in constructing phasor diagrams to help us solve single-phase a.c. circuit problems, because of the increasing complexity of those diagrams. Beyond that point (usually reached when we try to solve series-parallel circuits), phasor diagrams become very difficult to construct, and it is very much easier to use **symbolic notation**, in that case the '**j-operator**', to solve such circuits.

Surprisingly, perhaps, the construction of phasor diagrams for three-phase circuits doesn't really get as complicated as those for single-phase circuits. However, solving *unbalanced* three-phase circuits can get very complicated indeed if we try to extend the techniques we learnt for solving *balanced* three-phase circuits in the previous chapter.

So the time has come to meet *another* 'operator' – this time, the '**a-operator**'! This operator, which is also called the '**120° operator**', makes solving three-phase unbalanced-load problems *very* much easier than by constructing and analysing three-phase phasor diagrams.

Electrical Science for Technicians. 978-1-138-84926-6 © Adrian Waygood.
Published by Taylor & Francis. All rights reserved.

In an earlier chapter, we learnt how the symbol, '**j**', represents the 'operation' or 'instruction' telling us to *'rotate a phasor through an angle of 90° in a counterclockwise direction'*.

Well, the symbol '**a**' is also considered to represent an 'operation' or 'instruction', telling us to *'rotate a phasor'* but, this time, ***'through an angle of 120° in a counterclockwise direction'***.

Once again, it's important to remember that, by common agreement, angles that are measured in a *counterclockwise* direction are considered to be *positive* angles.

Let's apply this, below:

(a) Consider a phasor, of magnitude, $\bar{E}$, lying along the positive real axis (Figure 6.1).

Figure 6.1

(b) An operation by '**a**' rotates that phasor through an angle of 120° in a *counterclockwise* direction measured from the positive real axis (Figure 6.2). Of course, '$a\bar{E}$' may be expressed either in *polar* or in *complex (rectangular)* notation, i.e.:

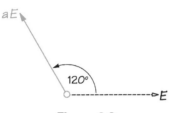

$$\text{In } \textbf{polar} \text{ form: } a\bar{E} = \bar{E} \angle 120°$$
$$\text{therefore: } a = 1 \angle 120°$$
$$\text{And, in } \textbf{rectangular} \text{ form: } a = (-0.5 + j0.866)$$

Figure 6.2

We can *multiply* or *divide* in **polar notation** easier than in rectangular notation. But, on the other hand, it's much easier to *add* or *subtract* in **rectangular notation** than in polar notation. So it's useful to be able to express the 'a-operator' in *each* of these notations.

To summarise:

$$\boxed{a = 1 \angle 120° \quad \text{or} \quad a = (-0.5 + j0.866)}$$

(c) A further operation by '**a**' rotates the phasor through yet another 120° counterclockwise – or 240° from the horizontal positive axis (Figure 6.3). Two consecutive operations by '**a**' is, of course, equivalent to '**a²**'.

Again, '**a²**' may be expressed in *polar* or in *complex (rectangular)* form, as follows:

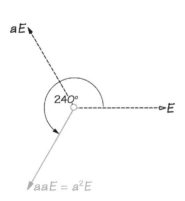

$$\text{In } \textbf{polar form}: \quad a^2 E = aaE$$
$$\text{therefore, } a^2 = (1 \angle 120°) \times (1 \angle 120°)$$
$$= 1 \angle (120° + 120°)$$
$$= 1 \angle 240°$$
$$\text{And in } \textbf{rectangular} \text{ form: } a^2 = (-0.5 - j0.866)$$

To summarise:

Figure 6.3

$$\boxed{a^2 = 1 \angle 240° \quad \text{or} \quad a^2 = (-0.5 - j0.866)}$$

(d) Yet a further operation by '**a**' rotates the phasor through yet another 120°, counterclockwise, and *back to its original position* – i.e. along the positive real axis (Figure 6.4).

Expressing '**a³**' in polar form confirms that the phasor has, indeed, arrived back at the positive real axis:

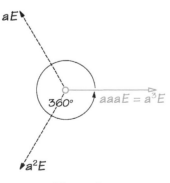

$$\text{In } \textbf{polar} \text{ form:}$$
$$a^3 \bar{E} = aaa\bar{E}$$
$$\text{therefore, } a^3 = (1 \angle 120° \times 1 \angle 120° \times 1 \angle 120°)$$
$$= (1 \angle (120° + 120° + 120°))$$
$$= (1 \angle 360°)V$$
$$\equiv (1 \angle 0°)$$

Figure 6.4

$$\text{And in } \textbf{rectangular} \text{ form: } a^3 = (1 + j0)$$

To summarise:

$$a^3 = 1\angle 0° \quad \text{or}: \quad a^3 = (1+j0)$$

Negative values of 'a'

Earlier, we learnt that '**–j**' represents a phasor that has been rotated 90° in a ***clockwise*** direction from the horizontal positive axis. This does *not* apply to the a-operator!

We must, however, be aware that '**–a**' does **not** *represent a phasor that has been rotated **clockwise** through an angle of 120°!*

Explanation:

$$\text{in } \textbf{rectangular} \text{ form}, -a = -1(a)$$
$$= -1(-0.5 + j0.866)$$
$$= (0.5 - j0.866)$$
$$\text{in } \textbf{polar} \text{ form}, -a = 1\angle 300°$$
$$\text{or} \quad -a = 1\angle -60°$$

So, '**–a**' does *not* rotate a phasor through 120° clockwise, as we might expect but, instead, *rotates it just 60° clockwise!*

To summarise:

$$-a = 1\angle -60° \quad \text{or}: \quad -a = (0.5 - j0866)$$

Let's apply the same logic to '**–a²**'.

$$\text{in } \textbf{rectangular} \text{ form}, -a^2 = -1(a^2)$$
$$= -1(-0.5 - j0.866)$$
$$= (0.5 + j0.866)$$
$$\text{in } \textbf{polar} \text{ form}, -a^2 = 1\angle 60°$$

So, '**–a²**' does *not* rotate a phasor through 240° clockwise, as we might expect but, instead, *rotates it just 60° counterclockwise!*

$$-a^2 = 1\angle 60° \quad \text{or}: \quad -a^2 = (0.5 + j0866)$$

The significance of this will become clearer a little further into this chapter.

Relating the 'a-operator' to a three-phase supply

Figure 6.5, below, represents a **phasor diagram** showing the *phase voltages* (i.e. line-to-neutral voltages) for a four-wire three-phase star-connected supply. These supply voltages (both *phase* and *line*) are carefully maintained by the electricity network companies and, except in very exceptional circumstances, *are always symmetrical* – that is, identical in magnitude and displaced from each other by 120°.

Supply voltages generally only become 'unsymmetrical' as a result of an electrical fault or when, for example, the star point of a three-phase distribution transformer becomes disconnected from earth as a result of, say, the activities of copper thieves!

We'll be using the letters '**A-B-C**' to identify individual *line conductors* or their corresponding *terminals*, together with the letter '**N**' to identify a *neutral conductor* and its corresponding *terminal*. The *sequence* of these letters represents the normal *phase sequence* of the supply voltage: i.e. the sequence or order in which the line potentials reach their peak values.

The normal phase sequence, therefore, is in the order: **A-B-C**.

So, for the following star-connected load, we can identify the **phase voltages** as $\bar{E}_{AN}$, $\bar{E}_{BN}$, and $\bar{E}_{CN}$ respectively. When constructing a three-phase phasor diagram, it's usual to use $\boldsymbol{\bar{E}_{AN}}$ as the *reference phasor*, and to draw it along the positive real axis.

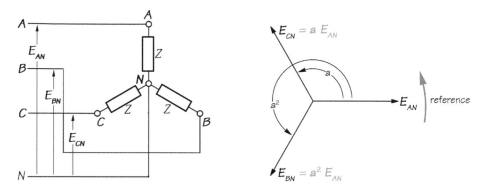

Figure 6.5

The grey-coloured curved arrow at the extreme right, in Figure 6.5, is simply there to remind us that phasors *always* 'rotate' in a counterclockwise direction. So we can imagine these three phase voltages 'rotating' past the positive real axis in the sequence $\bar{E}_{AN}$, followed by $\bar{E}_{BN}$, followed by $\bar{E}_{CN}$.

If we now apply the a-operator to the phasor diagram in Figure 6.5, we should realise that:

$$\bar{E}_{BN} \quad \text{is \textit{exactly} equivalent to} \quad a^2 \bar{E}_{AN}$$

Similarly,

$$\bar{E}_{CN} \quad \text{is \textit{exactly} equivalent to} \quad a\bar{E}_{AN}$$

So, if we wish, we could actually relabel the left-hand phasor diagram, in Figure 6.6, to that shown in the right-hand phasor diagram. Both phasor diagrams are *exactly equivalent* to each other.

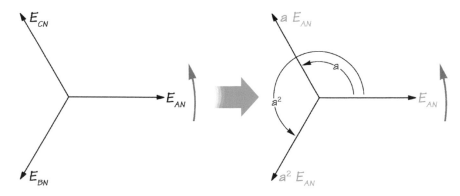

This phasor diagram can be....   ...relabelled as shown above.

Figure 6.6

It's *really* important to commit to memory the relationships shown in Figure 6.6, above. The vectorial-sum of three identical phasors, which are displaced from each other by 120°, is *zero*. So, in the above case, we can say that:

$$\bar{E}_{AN} + a\bar{E}_{AN} + a^2\bar{E}_{AN} = 0$$
$$\bar{E}_{AN}(1 + a + a^2) = 0$$

From which we can obtain another very useful relationship:

$$(1 + a + a^2) = 0$$

Summary

Earlier, we learnt that **−a** instructed us *to rotate a phasor through −60°* (*not −120°*, as we might have supposed!) so, as illustrated below, $-a\bar{E}_{AN}$ is equivalent to $-\bar{E}_{CN}$ or $\bar{E}_{NC}$. Similarly, we learnt that **−a²** instructed us *to rotate a phasor through +60°* (*not −240°*, as we might have supposed!) so, as illustrated below, $-a^2\bar{E}_{AN}$ is equivalent to $-\bar{E}_{BN}$ or $\bar{E}_{NB}$.

By sketching *all* possible combination of phase voltages as a phasor diagram, and creating the corresponding table shown below, we can very easily refer to them in order to quickly convert between the position of the '**a-operator**', and its corresponding **polar** and **rectangular** equivalents. We can then refer to this table as we work through the examples that follow throughout the rest of this chapter.

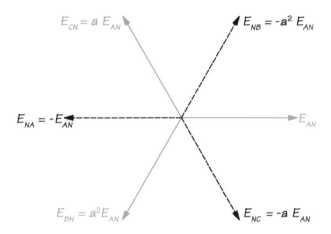

Figure 6.7

It's now possible to create a table that summarises the phase voltages for the star-connected system, shown in Figure 6.7, expressed in terms of the **a-operator**, together with its **polar** and **rectangular** equivalents:

Phasor*	a-operator	Polar equivalent	Rectangular equivalent
$\bar{E}_{AN}$	$\bar{E}_{AN}$	$1\angle 0°$	$1 + j0$
$\bar{E}_{BN}$	$a^2\bar{E}_{AN}$	$1\angle 240°$	$-0.5 - j0.866$
$\bar{E}_{CN}$	$a\bar{E}_{AN}$	$1\angle 120°$	$-0.5 + j0.866$
$\bar{E}_{NA}$	$-\bar{E}_{AN}$	$\angle -180°$	$-1 + j0$
$\bar{E}_{NB}$	$-a^2\bar{E}_{AN}$	$1\angle 60°$	$0.5 + j0.866$
$\bar{E}_{NC}$	$-a\bar{V}_{AN}$	$1\angle -60°$	$0.5 - j0.866$

Table 6.1

Columns 1 and 2 of Table 6.1 are based on the **phase voltages** of the *star*-connected load illustrated in Figure 6.7. But we can *replace* these with the phase voltages $(\bar{E}_{AB}\text{-}\bar{E}_{BC}\text{-}\bar{E}_{CA})$ for a *delta*-connected system, as illustrated in Figure 6.8 or, indeed, for the phase currents $(\bar{I}_{AN}\text{-}\bar{I}_{BN}\text{-}\bar{I}_{CN})$ for a *balanced* star-connected load, or the phase currents $(\bar{I}_{AB}\text{-}\bar{I}_{BC}\text{-}\bar{I}_{CA})$ for a *balanced* delta-connected load. Remember, the system voltages are *always* 'balanced', whereas the currents *depend on the load being balanced*.

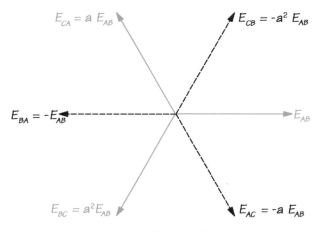

Figure 6.8

We will need to refer back to this very useful table regularly throughout the rest of this chapter. Also, it's important to re-emphasise that *it's much easier to multiply and divide using* **polar** *notation, we can only add and subtract in* **rectangular** *notation*, so this table will save us from having to perform the conversion between polar and rectangular, and *vice versa*, repeatedly.

Balanced versus unbalanced three-phase loads

In the previous chapter, we only dealt with **balanced loads**.

Let's remind ourselves what we *mean* by a three-phase 'balanced load'. This exists when the impedance of each phase of a three-phase load *is identical in all respects*. And, if each phase has an identical impedance, *then each of the resulting phase currents will be identical in magnitude and phase angle*.

In the case of a *balanced* three-phase star-connected load, we learnt that the phasor-sum of the three phase currents is *zero* and, therefore, *no current flows through the neutral conductor*.

But for an *unbalanced* three-phase star-connected load, each phase current will be *different*, and their phasor-sum – i.e. the neutral current – could be significant and *will need to return through the neutral conductor*.

Similarly, an *unbalanced* three-phase delta-connected load will result in *different* phase and line currents.

Applying the 'a-operator' to balanced loads

Using the 'a-operator' to confirm the relationship between line and phase voltages (star-connected system)

In the companion book, **An Introduction to Electrical Science**, we applied simple geometry to a phasor diagram for a three-phase, four-wire, (star- or wye-connected) system in order to determine the following relationship between the line voltages ($\bar{E}_L$) and the phase voltages ($\bar{E}_P$):

$$\bar{E}_L = \sqrt{3}\,\bar{E}_P$$

As an example of how to use the a-operator, let's use it to *confirm* this relationship, as illustrated in Figure 6.9.

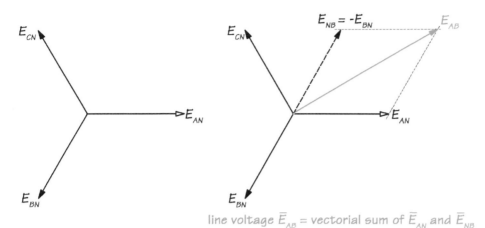

line voltage $\bar{E}_{AB}$ = vectorial sum of $\bar{E}_{AN}$ and $\bar{E}_{NB}$

Figure 6.9

The line voltage, $\bar{E}_{AB}$, is the phasor-sum of phase voltages $\bar{E}_{AN}$ and $\bar{E}_{NB}$, where:

$$\bar{E}_{NB} = -\bar{E}_{BN}$$

Using the 'a-operator', we can express phase voltages $-\bar{E}_{BN}$ as follows:

$$-\bar{E}_{BN} = -(a^2\,\bar{E}_{AN})$$

(Refer to Table 6.1 to convert between polar and rectangular notation.)

$$\text{Line voltage,} \quad \bar{E}_{AB} = \bar{E}_{AN} + \bar{E}_{NB}$$
$$= \bar{E}_{AN} - \bar{E}_{BN}$$

(where $\bar{E}_{BN}$ is equivalent to: $a^2\bar{E}_{AN}$) $=$

$$= \bar{E}_{AN} - \left(a^2\,\bar{E}_{AN}\right)$$
$$= \bar{E}_{AN}(1\angle 0° - 1\angle 240°)$$

We subtract using rectangular notation. *We can use the table, above, to quickly convert between polar and rectangular notation:*

$$= \bar{E}_{AN}[(1+j0) - (-0.5 - j0.866)]$$
$$= \bar{E}_{AN}[(1+j0+0.5+j0.866)]$$
$$= \bar{E}_{AN}(1.5+j0.866)$$
$$= \bar{E}_{AN}(1.732\angle 30°)$$
$$\text{or} \quad = \sqrt{3}\,\bar{E}_{AN}\,\angle 30°$$

This confirms that the line voltage ($\bar{E}_{AB}$) is $\sqrt{3}$ *larger* than its corresponding phase voltage ($\bar{E}_{AN}$), and also that it *leads* that phase voltage by 30°.

As the supply voltages are symmetrical, this relationship should also be true for the other two line voltages, $\bar{E}_{BC}$ and $\bar{E}_{CA}$.

Just for practice, let's confirm this using the same technique:

Line voltage $\bar{E}_{BC} = \bar{E}_{BN} + \bar{E}_{NC}$

$$= \bar{E}_{BN} - \bar{E}_{CN}$$

$$= a^2 \bar{E}_{AN} - a\bar{E}_{AN}$$

$$= \bar{E}_{AN}(1\angle 240° - 1\angle 120°)$$

(remember, it's easier to subtract using rectangular form)

$$= \bar{E}_{AN}[(-0.5 - j0.866) - (-0.5 + j0.866)]$$

$$= \bar{E}_{AN}[(-0.5 - j0.866 + 0.5 - j0.866)]$$

$$= \bar{E}_{AN}[(-0.5 + 0.5) + j(-0.866 - 0.866)]$$

$$= \bar{E}_{AN}(0 - j1.732)$$

$$= \bar{E}_{AN}(1.732 \angle -90°)$$

or $= \sqrt{3}\, \bar{E}_{AN} \angle -90°$

Line voltage $\bar{E}_{CA} = \bar{E}_{CN} + \bar{E}_{NA}$

$$= \bar{E}_{CN} - \bar{E}_{AN}$$

$$= a\bar{E}_{AN} - \bar{E}_{AN}$$

$$= \bar{E}_{AN}(1\angle 120° - 1\angle 0°)$$

(remember, it's easier to subtract using rectangular form)

$$= \bar{E}_{AN}[(-0.5 + j0.866) - (1 + j0)]$$

$$= \bar{E}_{AN}[(-0.5 + j0.866 - 1 - j0)]$$

$$= \bar{E}_{AN}[(-0.5 - 1) + j(0.866 - 0)]$$

$$= \bar{E}_{AN}(-1.5 + j0.866)$$

$$= \bar{E}_{AN}(1.732 \angle 150°)$$

or $= \sqrt{3}\, \bar{E}_{AN} \angle 150°$

Having obtained an expression for line voltage, $\bar{E}_{AB}$, an alternative (and *much* quicker!) way of determining the line voltages $\bar{E}_{BC}$ and $\bar{E}_{CA}$, is as follows:

$$\bar{E}_{BC} = a^2 \bar{E}_{AB}$$

$$= 1\angle 240° \times \sqrt{3}\, \bar{E}_{AN} \angle 30°$$

$$= \sqrt{3}\, \bar{E}_{AN} \angle(240° + 30°)$$

$$= \sqrt{3}\, \bar{E}_{AN} \angle 270°$$

$$\text{or} = \sqrt{3}\, \bar{V}_{AN} \angle -90°$$

. . . and:

$$\bar{E}_{CA} = a\bar{E}_{AB}$$

$$= 1\angle 120° \times \sqrt{3}\, \bar{E}_{AN} \angle 30°$$

$$= \sqrt{3}\, \bar{E}_{AN} \angle(20° + 30°)$$

$$= \sqrt{3}\, \bar{E}_{AN} \angle 150°$$

Using the 'a-operator' to determine the neutral current for a balanced star-connected system

Another thing we learnt about balanced loads is that, for a three-phase, four-wire, system, the phasor-sum of the phase currents is zero – which means that no current returns via the neutral conductor.

As shown in Figure 6.10, to vectorially add the three phase currents we first add $\overline{I}_{AN}$ and $\overline{I}_{BN}$, the resultant of which is equal and opposite phase voltage $\overline{I}_{CN}$.

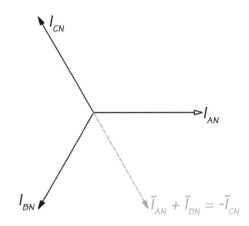

Figure 6.10

So, let's use the a-operator to confirm that the phasor-sum of the three phase currents is zero.

(Refer to Table 6.1 to convert between polar and rectangular notation.)
Applying the a-operator to phasors $\overline{I}_{BN}$ and $\overline{I}_{CN}$, using the phase current, $\overline{I}_{AN}$, as the reference phasor:

$$\overline{I}_{BN} = a^2\,\overline{I}_{AN} \quad \text{and} \quad \overline{I}_{CN} = a\,\overline{I}_{AN}.$$

Let $\overline{I}_{N}$ be the neutral current (i.e. the phasor-sum of the three phase voltages):

$$\overline{I}_{N} = \overline{I}_{AN} + \overline{I}_{BN} + \overline{I}_{CN}$$
$$= \overline{I}_{AN} + a^2\,\overline{I}_{AN} + a\,\overline{I}_{AN}$$
$$= \overline{I}_{AN}(1\angle 0° + 1\angle 240° + 1\angle 120°)$$

(remember, it's easier to add using rectangular form)

$$= \overline{I}_{AN}[(1+j0) + (-0.5 - j0.866) + (-0.5 + j0.866)]$$
$$= \overline{I}_{AN}[(1 + j0 - 0.5 - j0.866 - 0.5 + j0.866)]$$
$$= \overline{I}_{AN}[(1 - 0.5 - 0.5) + j(0 - 0.866 + 0.866)]$$
$$= \overline{I}_{AN}(0 + j0)$$
$$= 0 \text{ ampere}$$

Using the 'a-operator' to confirm the relationship between line and phase currents (delta-connected system) for a balanced load

In the companion book, ***An Introduction to Electrical Science***, we applied simple geometry to a phasor diagram for a three-phase, three-wire, (delta-connected) system in order to determine the following relationship between the line currents ($\overline{I}_L$) and a phase currents ($\overline{I}_P$) for a balanced load:

$$\overline{I}_L = \sqrt{3}\,\overline{I}_P$$

Let's now demonstrate how we can use the 'a-operator' to confirm this relationship.

But first, a word of explanation. If we apply Kirchhoff's Current Law to junction 'A' of a delta-connected load (Figure 6.11), then the line current (in this case, $\overline{I}_A$) will be the phasor-sum of the two phase currents, in this case:

$$\overline{I}_A + \overline{I}_{CA} = \overline{I}_{AB}$$
$$\overline{I}_A = \overline{I}_{AB} - \overline{I}_{CA}$$

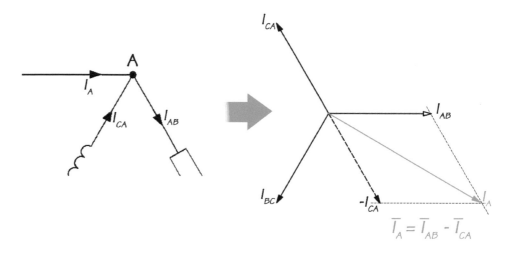

Figure 6.11

Let the phase current, $\overline{I}_{AB}$, be the reference phasor:

then: $\overline{I}_{BC} = a^2\,\overline{I}_{AB}$ and $\overline{I}_{CA} = a\,\overline{I}_{AB}$.

(Refer to Table 6.1 to convert between polar and rectangular notation.)

Solution

Line current, $\overline{I}_A = \overline{I}_{AB} - \overline{I}_{CA}$
$$= \overline{I}_{AB} + \overline{I}_{AC}$$
$$= \overline{I}_{AB} - a\,\overline{I}_{AB}$$
$$= \overline{I}_{AB}(1\angle 0° - 1\angle 120°)$$

(remember, it's easier to subtract using rectangular form)
$$= \overline{I}_{AB}[(1+j0) - (-0.5 + j0.866)]$$
$$= \overline{I}_{AB}[(1+j0+0.5 - j0.866)]$$
$$= \overline{I}_{AB}[(1+0.5) + j(0 - 0.866)]$$
$$= \overline{I}_{AB}(1.5 - j0.866)$$
$$= \overline{I}_{AB}(1.732\angle - 30°)$$

or $= \sqrt{3}\,\overline{I}_{AB}\angle - 30°\ \text{A}$

This matches the phasor diagram, shown in Figure 6.11.

Let's repeat this for the other two line currents, $\overline{I}_B$ and $\overline{I}_C$:

Line current, $\quad \overline{I}_B = \overline{I}_{BC} + \overline{I}_{BA}$

$\qquad = \overline{I}_{BC} - \overline{I}_{AB}$

$\qquad = a^2\,\overline{I}_{AB} - \overline{I}_{AB}$

$\qquad = \overline{I}_{AB}(1\angle 240° - 1\angle 0°)$

$\qquad$ (remember, it's easier to subtract using rectangular form)

$\qquad = \overline{I}_{AB}[(-0.5 - j0.866) - (1 + j0)]$

$\qquad = \overline{I}_{AB}[(-0.5 - j0.866 - 1 - j0)]$

$\qquad = \overline{I}_{AB}[(-0.5 - 1) + j(-0.866 - 0)]$

$\qquad = \overline{I}_{AB}(-1.5 - j0.866)$

$\qquad = \overline{I}_{AB}(1.732\angle -150°)$

$\quad$or$\quad = \sqrt{3}\,\overline{I}_{AB}\angle -150°$ ampere

Line current, $\quad \overline{I}_C = \overline{I}_{CA} + \overline{I}_{CB}$

$\qquad = \overline{I}_{CA} - \overline{I}_{BC}$

$\qquad = a\,\overline{I}_{AB} - a^2\,\overline{I}_{AB}$

$\qquad = \overline{I}_{AB}(1\angle 120° - 1\angle 240°)$

$\qquad$ (remember, it's easier to subtract using rectangular form)

$\qquad = \overline{I}_{AB}[(-0.5 + j0.866) - (-0.5 - j0.866)]$

$\qquad = \overline{I}_{AB}[(-0.5 + j0.866 + 0.5 + j0.866)]$

$\qquad = \overline{I}_{AB}[(-0.5 + 0.5) + j(0.866 + 0.866)]$

$\qquad = \overline{I}_{AB}(0 - j1.732)$

$\qquad = \overline{I}_{AB}(1.732\angle 90°)$

$\quad$or$\quad = \sqrt{3}\,\overline{I}_{AB}\angle 90°$ ampere

An alternative (and *much* quicker!) way of determining the line currents $\overline{I}_B$ and $\overline{I}_C$: since the load is **symmetrical**, then the line currents must *also* be symmetrical, and 120° apart, so:

$$\overline{I}_B = a^2\,\overline{I}_A$$

$$= 1\angle 240° \times \sqrt{3}\,\overline{I}_{AB}\angle -30°$$

$$= \sqrt{3}\,\overline{I}_{AB}\angle 210°$$

$$\text{or} = \sqrt{3}\,\overline{I}_{AB}\angle -150° \text{ ampere}$$

$$\overline{I}_C = a\,\overline{I}_A$$

$$= 1\angle 120° \times \sqrt{3}\,\overline{I}_{AB}\angle -30°$$

$$= \sqrt{3}\,\overline{I}_{AB}\angle 90° \text{ ampere}$$

Applying the 'a-operator' to unbalanced loads

This is where we start to use *the real power of the 'a-operator'*: solving *unbalanced* three-phase loads!

An **unbalanced load** is one in which the phase currents are *different* in both magnitude and phase angle. Even though the *load* is unbalanced, it is important to emphasise that, under normal circumstances, the **supply voltages are symmetrical** because they are determined by the supply system and *not* by the load – regardless of whether that load is balanced or unbalanced.

This means that, although the phase and line *currents* may vary, the line and phase *voltages* are *always* symmetrical.

The *exception* to this general rule regarding symmetrical voltages is the *three-wire, star-connected, load* – but more on this, later.

Worked example 1: unbalanced four-wire star-connected load

For the four-wire star-connected load impedances (expressed in ohms) given in Figure 6.12, calculate each phase current, and the resulting neutral current. The phase voltage is 254 V.

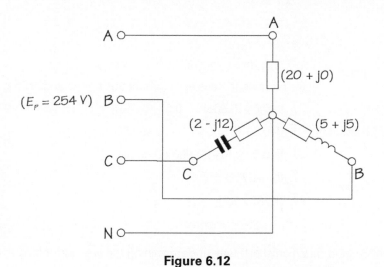

Figure 6.12

(Refer to Table 6.1 to convert between polar and rectangular notation.)

Solution

Let $\bar{E}_{AN}$ be the reference phasor. In which case, we can express that voltage as: $\bar{E}_{AN} \angle 0°$.

Applying the a-operator, we can write:

$$\bar{E}_{BN} = a^2 \bar{E}_{AN} \quad \text{and} \quad \bar{E}_{CN} = a \bar{E}_{AN}.$$

Next, we express the impedance of each phase in *polar form* (because we are going to *divide* these into their phase voltages to determine their phase currents):

$$Z_{AN} = (20 + j0) = 20 \angle 0° \text{ ohm}$$
$$Z_{BN} = (5 + j5) = 7.97 \angle 45° \text{ ohm}$$
$$Z_{CN} = (2 - j12) = 12.17 \angle -80.54° \text{ ohm}$$

Now, we can determine the value of each phase current:

$$\bar{I}_{AN} = \frac{\bar{E}_{AN}}{Z_{AN}} = \frac{254\angle 0°}{20\angle 0°} = 12.7\angle 0° = \mathbf{(12.7 + j\,0)\ A}$$

$$\bar{I}_{BN} = \frac{\bar{E}_{BN}}{Z_{BN}} = \frac{a^2\,\bar{E}_{AN}}{Z_{BN}} = \frac{1\angle 240°\times 254\angle 0°}{7.07\angle 45°} = (35.93\ \angle 195°) = \mathbf{(-34.71 - j\,9.3)\ A}$$

$$\bar{I}_{CN} = \frac{\bar{E}_{CN}}{Z_{CN}} = \frac{a\,\bar{E}_{AN}}{Z_{CN}} = \frac{1\angle 120°\times 254\angle 0°}{12.71\angle -80.54°} = (20.87\ \angle 200.54°) = \mathbf{(-19.54 - j\,7.32)\ A}$$

The **neutral current** is, of course, *the phasor-sum of the three phase currents:*

$$\bar{I}_N = \bar{I}_{AN} + \bar{I}_{BN} + \bar{I}_{CN}$$
$$= (12.7 + j\,0) + (-34.71 - j\,9.3) + (-19.54 - j\,7.32)$$
$$= (12.7 + j\,0 - 34.71 - j\,9.3 - 19.54 - j\,7.32)$$
$$= (12.7 - 34.71 - 19.54) + j(0 - 9.3 - 7.32)$$
$$= \mathbf{(-41.55 - j\,16.62)\ A\ (Answer)}$$
$$\text{or} = \mathbf{44.75\ \angle -158.2°\ A\ (Answer)}$$

That's it! It's as simple as that!

Worked example 2: unbalanced delta-connected load

For the delta-connected load impedances (expressed in ohms) given in Figure 6.13, calculate each phase current, and each line current. The phase voltage is 230 V.

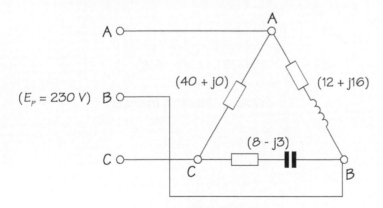

Figure 6.13

(refer to Table 1 to convert between polar and rectangular notation)

Solution

Before we start, we mustn't forget that, for delta load, the phase voltages are numerically equal to the line voltages and, both are symmetrical.

Let $\bar{E}_{AB}$ be the reference phasor, in which case it can be written as $\bar{E}_{AB}\angle 0°$.

And, applying the a-operator, we can write:

$$\bar{E}_{BC} = a^2\,\bar{E}_{AB} \quad \text{and} \quad \bar{E}_{CA} = a\,\bar{E}_{AB}$$

Express the impedance of each phase in *polar form* (because we are going to *divide* these into their corresponding phase voltages to determine the phase currents):

$$Z_{AB} = 12 + j16 = (20 \angle 53.13°) \text{ ohms}$$
$$Z_{BC} = 8 - j3 = (8.54 \angle -20.56°) \text{ ohms}$$
$$Z_{CA} = 40 + j0 = (40 \angle 0°) \text{ ohms}$$

Determine the phase currents:

$$\bar{I}_{AB} = \frac{\bar{E}_{AB}}{Z_{AB}} = \frac{230 \angle 0°}{20 \angle 53.13°}$$
$$= 11.5 \angle -53.13° = \textbf{(6.9 - j9.2) A (Answer)}$$

$$\bar{I}_{BC} = \frac{\bar{E}_{BC}}{Z_{BC}} = \frac{a^2 \bar{E}_{AB}}{Z_{BC}} = \frac{1 \angle 240° \times 230 \angle 0°}{8.54 \angle -20.56°}$$
$$= 26.93 \angle 260.56° = \textbf{(-4.42 - j26.57) A (Answer)}$$

$$\bar{I}_{CA} = \frac{\bar{E}_{CA}}{Z_{CA}} = \frac{a \bar{E}_{AB}}{Z_{CA}} = \frac{1 \angle 120° \times 230 \angle 0°}{40 \angle 0°}$$
$$= 5.75 \angle 120° = \textbf{(-2.88 - j4.98) A (Answer)}$$

Next, we determine the line currents:

$$\bar{I}_A = \bar{I}_{AB} + \bar{I}_{AC}$$
$$= \bar{I}_{AB} - \bar{I}_{CA}$$
$$= (6.9 - j9.2) - (-2.88 + j4.98)$$
$$= (6.9 - j9.2 + 2.88 - j4.98)$$
$$= (9.78 - j14.98)$$
$$= \textbf{17.23} \angle \textbf{-55.41° A (Answer)}$$

$$\bar{I}_B = \bar{I}_{BC} + \bar{I}_{BA}$$
$$= \bar{I}_{BC} - \bar{I}_{AB}$$
$$= (-4.42 - j26.57) - (6.9 - j9.2)$$
$$= (-4.42 - j26.57 - 6.9 + j9.2)$$
$$= (-11.32 - j17.37)$$
$$= \textbf{20.73} \angle \textbf{-123.09° A (Answer)}$$

$$\bar{I}_C = \bar{I}_{CA} + \bar{I}_{CB}$$
$$= \bar{I}_{CA} - \bar{I}_{BC}$$
$$= (-2.88 + j4.98) - (-4.42 - j26.57)$$
$$= (-2.88 + j4.98 + 4.42 + j26.57)$$
$$= (1.62 - j31.55)$$
$$= \textbf{31.59} \angle \textbf{87.06° A (Answer)}$$

Worked example 3: power of an unbalanced four-wire star-connected load

For the four-wire star-connected load impedances (expressed in ohms) given in Figure 6.14, calculate the total power developed by the load, if the circuit's phase voltage is 254 V.

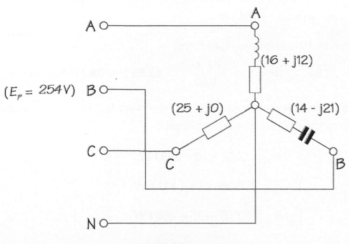

Figure 6.14

(Refer to Table 6.1 to convert between polar and rectangular notation.)

Solution

Let the phase voltage, $\bar{E}_{AN}$ be the reference phasor, in which case:

$$\bar{E}_{BN} = a^2 \bar{E}_{AN} \quad \text{and} \quad \bar{E}_{CN} = a\bar{E}_{AN}$$

Express the impedance of each phase in *polar form* (because we are going to *divide* these into their corresponding phase voltages to determine the phase currents):

$$Z_{AN} = 16 + j12 = (20\angle36.87°) \text{ ohms}$$
$$Z_{BN} = 14 - j21 = (25.24\angle-56.31°) \text{ ohms}$$
$$Z_{CA} = 25 + j0 = (25\angle0°) \text{ ohms}$$

Now, we can determine the value of each phase current:

$$\bar{I}_{AN} = \frac{\bar{E}_{AN}}{Z_{AN}} = \frac{254\angle0°}{20\angle36.87°} = \frac{254}{20}\angle(0° - 36.87°) = \mathbf{12.7\angle-36.87°\ A\ (Answer)}$$

$$\bar{I}_{BN} = \frac{\bar{E}_{BN}}{Z_{BN}} = \frac{a^2\bar{E}_{AN}}{Z_{BN}} = \frac{1\angle240° \times 254\angle0°}{25.24\angle-56.31°} = \frac{1 \times 254}{25.24}\angle(240° + 0° - (-56.31°))$$
$$= \mathbf{10.06\angle296.3°\ A\ (Answer)}$$

$$\bar{I}_{CN} = \frac{\bar{E}_{CN}}{Z_{CN}} = \frac{a\bar{E}_{AN}}{Z_{CN}} = \frac{1\angle120° \times 254\angle0°}{25\angle0°} = \frac{254}{25}\angle(120° + 0° - 0°)$$
$$= \mathbf{10.16\angle120.0°\ A\ (Answer)}$$

To find the neutral current, we need to vectorially add the three phase currents:

$$\bar{I}_N = \bar{I}_{AN} + \bar{I}_{BN} + \bar{I}_{CN} = (12.7\angle-36.87°) + (10.06\angle296.3°) + (10.16\angle120.0°)$$

111

In order to *add* these currents, we need to convert from polar to rectangular:

$$\overline{I}_N = \overline{I}_{AN} + \overline{I}_{BN} + \overline{I}_{CN}$$
$$= (10.16 - j7.62) + (4.46 - j9.02) + (-5.08 + j8.80)$$
$$= (10.16 + 4.46 - 5.08) + j(-7.62 - 9.02 + 8.80)$$
$$= (9.54 - j7.84) \text{ A}$$

Expressing the neutral current in polar form,

$$\overline{I}_N = \sqrt{9.54^2 - 7.84^2} \left(\tan{-1}\frac{-7.84}{9.54} \right)° = \textbf{12.35} \angle -\textbf{39.41°A (Answer)}$$

To find the *power* developed by each phase, we need to multiply the square of the phase current by the resistance of that phase.

$$P_{AN} = \left(\overline{I}_{AN} \right)^2 \times R_{AN} = 12.7^2 \times 16 = \textbf{2581 W}$$

$$P_{BN} = \left(\overline{I}_{BN} \right)^2 \times R_{BN} = 10.06^2 \times 14 = \textbf{1417 W}$$

$$P_{CN} = \left(\overline{I}_{CN} \right)^2 \times R_{CN} = 10.16^2 \times 25 = \textbf{2581 W}$$

$$P_{total} = P_{AN} + P_{BN} + P_{CN} = 2581 + 1417 + 2581 = \textbf{6579 W (Answer)}$$

Unbalanced three-wire star-connected loads

Throughout this chapter, it's been emphasised that, because **supply voltages** are carefully controlled by the electricity network company, they are always *symmetrical* – i.e. identical in magnitude and displaced from each other by 120 electrical degrees. Loss of symmetry is rare and usually results from an electrical fault on the supply system. In many cases, it results from what is known as a 'floating neutral' following the removal of the supply's star point earthing conductor by copper thieves!

With a *four-wire*, star-connected, load the **neutral point** *(N')* of the *load* is at the same potential as the **star point** *(N)* of the *alternator* or of the supply transformer's *secondary windings*. This is because the two neutral points are connected together by the neutral conductor. So the load's phase voltages *must* be symmetrical, *regardless* of whether that load is balanced or unbalanced (Figure 6.15).

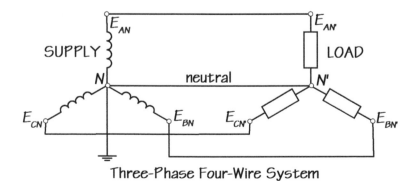

Three-Phase Four-Wire System

Figure 6.15

However, in the case of a *three-wire* star-connected load, there is *no* neutral conductor and, if the load is *unbalanced*, then its neutral point *is not necessarily at the same potential as the neutral point of the alternator or transformer*. This means that the load's phase voltages *cannot be symmetrical*, even though the supply voltages themselves remain symmetrical!

This problem explains why a neutral conductor is *normally* supplied and why, in practice, there are very few three-wire star-connected loads with the exception of three-phase motors. Another way of looking at a 'three-wire' star-connected system is to consider it to be really a four-wire system, but with a *break in the neutral conductor* – as illustrated in Figure 6.16.

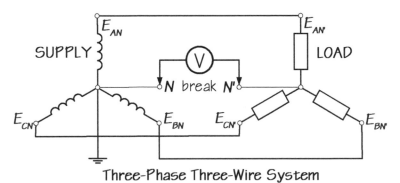

Three-Phase Three-Wire System

Figure 6.16

This 'displaced' load neutral makes solving unbalanced three-wire, star-connected, loads very difficult or, at least, time consuming, to solve using the techniques used so far in this chapter.

Fortunately, such problems can be simplified to some extent, by using what's known as '**Millman's Theorem**', named in honour of the Ukranian-born American academic Jacob Millman (1911–1991). The proof of this theorem is beyond the scope of this book, but it can be expressed as follows, where $\bar{E}_{N'N}$ is the potential difference between the load's neutral point (**N'**) and the supply's star point (**N**) – as measured by the voltmeter shown in Figure 6.16:

$$\bar{E}_{N'N} = \frac{\bar{E}_{AN}Y_{AN'} + \bar{E}_{BN}Y_{BN'} = \bar{E}_{CN}Y_{CN'}}{Y_{AN'} + Y_{BN'} + Y_{CN'}}$$

In this equation, $\bar{E}_{N'N}$ represents the potential of the star-connected **load's neutral point** (**N'**) with respect to the **supply's star point** (**N**), $\bar{E}_{AN}, \bar{E}_{BN},$ **and** $\bar{E}_{CN}$ represent the symmetrical supply phase voltages, and $Y_{AN'}, Y_{BN'},$ **and** $Y_{CN'}$ represent the load's phase admittances.

From this, we can determine the load's *phase voltages*, using the following equations:

$$\bar{E}_{AN'} = (\bar{E}_{AN} - \bar{E}_{N'N}) \quad \bar{E}_{BN'} = (\bar{E}_{BN} - \bar{E}_{N'N}) \quad \bar{E}_{CN'} = (\bar{E}_{CN} - \bar{E}_{N'N})$$

And, from these, the line phase/line currents can be determined as follows:

$$\bar{I}_{AN'} = \bar{E}_{AN'}Y_{AN'} \quad \bar{I}_{BN'} = \bar{E}_{BN'}Y_{BN'} \quad \bar{I}_{CN'} = \bar{E}_{CN'}Y_{CN'}$$

Worked example 4

An unbalanced three-wire star-connected load has the following phase-impedances: $Z_{AN} = (2+j0)\,\Omega$, $Z_{BN} = (2+j2)\,\Omega$, and $Z_{CN} = (0-j15)\,\Omega$, and is connected to a symmetrical three-phase supply having a phase voltage of 250 V. Calculate, (a) the potential of the load's displaced neutral point with respect to the star point of the supply system, (b) the load's phase voltages, and (c) the load's phase/line currents.

Solution

(a) The first step is to change the load impedances into their equivalent *admittances*. At the same time, because we are going to multiply them by the supply voltages, it would be convenient to express them in polar, rather than in rectangular, form:

$$Y_{AN'} = \frac{1}{Z_{AN'}} = \frac{1}{(2+j0)} = \frac{1}{2\angle 0°} = 0.5\angle 0° \text{ siemens}$$

$$Y_{BN'} = \frac{1}{Z_{BN'}} = \frac{1}{(2+j2)} = \frac{1}{2.83\angle 45°} = 0.353\angle -45° \text{ siemens}$$

$$Y_{CN'} = \frac{1}{Z_{CN'}} = \frac{1}{(0-j5)} = \frac{1}{5\angle -90°} = 0.2\angle 90° \text{ siemens}$$

These can now be inserted into Millman's equation:

$$\bar{E}_{N'N} = \frac{\bar{E}_{AN}Y_{AN'} + \bar{E}_{BN}Y_{BN'} + \bar{E}_{CN}Y_{CN'}}{Y_{AN'} + Y_{BN'} + Y_{CN'}} = \frac{\bar{E}_{AN}Y_{AN'} + a^2\bar{E}_{AN}Y_{BN'} + a\bar{E}_{AN}Y_{CN'}}{Y_{AN'} + Y_{BN'} + Y_{CN'}}$$

$$= \frac{\left([250\angle 0°]\times[0.5\angle 0°]\right) + \left([250\angle 240°]\times[0.353\angle -45°]\right) + \left([250\angle 120°]\times[0.2\angle 90°]\right)}{(0.5\angle 0°) + (0.353\angle -45°) + (0.2\angle 90°)}$$

$$= \frac{(125\angle 0°) + (88.25\angle 195°) + (50\angle 210°)}{(0.5\angle 0°) + (0.353\angle -45°) + (0.2\angle 90°)}$$

Both the nominator line and the denominator line require us to *add*, so we must change these values *from* polar form *to* rectangular form. We won't show the calculations required to do this but, instead, present it as having been done:

$$\bar{E}_{N'N} = \frac{(125+j0) + (-85.24-j22.84) + (-43.30-j25)}{(0.5+j0) + (0.25-j0.25) + (0+j0.2)}$$

$$= \frac{-3.54-j47.84}{0.75-j0.05} \times \left(\frac{0.75+j0.05}{0.75+j0.05}\right) \text{ (rationalising)}$$

$$= \left(\mathbf{0-j64}\right) \textbf{ volts} \quad \text{or} \quad \mathbf{64 \angle -90°} \textbf{ volts (Answer a.)}$$

(b) We can now move on to determine the values of the unbalanced load's *phase voltages*:

$$\bar{E}_{AN'} = (\bar{E}_{AN} - \bar{E}_{N'N}) = (250+j0) - (0-j64)$$

$$= \left(\mathbf{250+j64}\right) \textbf{ volts} \quad \text{or} \quad \mathbf{258 \angle 14.36°} \textbf{ volts (Answer b.)}$$

$$\bar{E}_{BN'} = (\bar{E}_{BN} - \bar{E}_{N'N}) = \left(a^2\bar{E}_{AN} - \bar{E}_{N'N}\right) = 250(-0.5-j0.866) - (0-j64)$$

$$= (-125-j216.5) - (0-j64)$$

$$= \left(\mathbf{-125-j152.5}\right) \textbf{ volts} \quad \text{or} \quad \mathbf{197\angle 230.66°} \textbf{ volts (Answer b.)}$$

$$\bar{E}_{CN'} = (\bar{E}_{CN} - \bar{E}_{N'N}) = \left(a\bar{E}_{AN} - \bar{E}_{N'N}\right) = 250(-0.5+j0.866) - (0-j64)$$

$$= (-125+j216.5) - (0-j64)$$

$$= \left(\mathbf{-125+j280.5}\right) \textbf{ volts} \quad \text{or} \quad \mathbf{307 \angle 114°} \textbf{ volts (Answer b.)}$$

The phasor diagram for this situation is illustrated in Figure 6.17. You will notice how the load's '**displaced neutral**' causes the load's phase voltages to lose their symmetry (as shown in blue). Note that the tips of the supply and load phasors *coincide*, but *the load's neutral is displaced from the supply's star point* – making the magnitudes of the load's phase currents, and angles between them, asymmetrical.

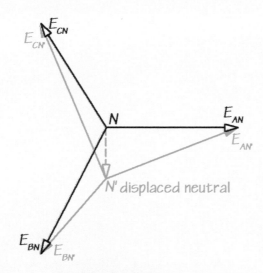

Figure 6.17

(c) Returning to our worked example, we can determine the values of the load's *phase/line currents* (for a star-connected load, the line and phase currents are the same):

$$\bar{I}_{AN'} = \bar{E}_{AN'} Y_{AN'} = (258\angle 14.36°) \times (0.5\angle 0°)$$

$$= \mathbf{129\ \angle 14.36°\ ampere\ (Answer\ c.)}$$

$$\bar{I}_{BN'} = \bar{E}_{BN'} Y_{BN'} = (197\angle 230.66°) \times (0.353\angle -45°)$$

$$= \mathbf{69.4\ \angle 185.28°\ ampere\ (Answer\ c.)}$$

$$\bar{I}_{CN'} = \bar{E}_{CN'} Y_{CN'} = (307\angle 114°) \times (0.2\angle 90°)$$

$$= \mathbf{61.4\ \angle 204.07°\ ampere\ (Answer\ c.)}$$

Conclusion

Now that we've completed this chapter, we need to examine its **objectives** listed at its start. Placing a question mark at the end of each objective turns that objective into a **test item**. If we can answer those test items, then we've met the objectives of this chapter.

CHAPTER 7

Electrical measurements

On completion of this chapter, you should be able to:

1. identify the electrical measuring instruments used to measure current, voltage, and resistance.

2. explain how an ammeter must be inserted into a circuit in order to measure electric current.

3. explain how a voltmeter must be inserted into a circuit in order to measure potential difference.

4. describe the main constructional features of a moving-coil instrument, together with the function of each.

5. describe the principle of operation of a moving-coil instrument, in terms of its deflection torque, control torque, and damping.

6. explain the term 'sensitivity' as it applies to measuring instruments.

7. explain the functions of shunt and multiplier resistors.

8. determine the value of the shunt resistance required to convert a moving-coil instrument to read any given circuit current.

9. determine the value of the multiplier resistance required to convert a moving-coil instrument to read any given potential difference.

10. recognise multiple shunt variable range circuits and multiple multiplier variable range circuits.

11. explain the term, 'loading factor', as applied to voltmeters, together with its significance regarding the choice of voltmeter.

12. explain how moving-coil instruments are adapted to measure a.c., as well as d.c., quantities.

13. describe the principle of operation of a repulsion-type moving-iron instrument.

14. read and interpolate analogue meter scales.

15. explain the main steps required in preparing to measure current, voltage, and resistance, with an analogue multimeter.

16. describe the construction features of an electrodynamic wattmeter,

17. describe the basic principle of operation of an electrodynamic wattmeter.

18. describe the significance of a wattmeter's 'polarity markings', and the correct way of connecting a wattmeter to measure the power of a complete circuit or of an individual component within that circuit.

19. describe how to use an electrodynamic wattmeter using instrument transformers.

20. explain how to protect an electrodynamic wattmeter's current coil, when monitoring the power of an electric motor.

21. explain the basic principle of operation of a digital voltmeter and digital ammeter.

22. describe the main advantages of using a digital multimeter compared with an analogue multimeter.

23. describe the operating principle of a 'clamp-on' meter, and how it is used.

Electrical Science for Technicians. 978-1-138-84926-6 © Adrian Waygood.
Published by Taylor & Francis. All rights reserved.

Introduction

Before writing this chapter, a great deal of thought was given to whether, these days, it is still necessary for technicians to understand the construction, operation, and use of **analogue measuring instruments**. After all, analogue instruments have now largely been superseded by digital instruments, which are very much easier to read, mechanically more robust and, to some extent, less expensive.

This matter was discussed with a number of people within the electrotechnology industry, and the general feeling was that an understanding of analogue instruments provides us with a better appreciation of measurements and instrumentation than would be the case if we *only* dealt with digital instruments.

Furthermore, correspondents reported that, in many cases, their office cabinets are still crammed with numerous analogue instruments, which they are still expected to be able to use – and to use properly.

For this reason, then, it has been decided that this chapter should cover *both* **analogue** *and* **digital instruments**.

Historical background

Whenever we conduct an experiment in a college laboratory to, say, confirm Ohm's Law, we might be forgiven for thinking, *'Why did it take him so long to discover the bleedin' obvious?'*

Well, why did it? Let's consider the following.

In the 1820s, when Ohm was conducting his experiments, *there were no standard measuring instruments. He had to design and build his own!*

Nor were there any standard units of measurement! The 'ampere' and the 'volt', for example, *didn't even exist at that time!* They weren't established for another 60 years. And the 'ohm' wasn't introduced until nearly 30 years after the death of the man it was named in honour of.

That's why it took him so long to discover the *'bleedin' obvious'*. In fact, it's a measure of the stature of the nineteenth-century physicists that they were able to achieve the results that they did!

The history of electrical measuring instruments dates back to 1820, when the deflection of a compass needle by the current passing through a nearby wire was first described by Hans Christian Ørsted (1771–1851), who then went on to study the phenomenon, not only for its own sake, but as *a potential means of measuring electric current.*

By using multiple turns of wire, it was discovered that the strength of the magnetic field could be *increased*, making the compass more sensitive to smaller currents. These 'multipliers', as they were originally called, became known as 'galvanometers', named in honour of the Italian physicist, Luigi Galvani (1737–1798).

The current passing through the galvanometer's coil caused the compass needle to deflect, and the angle through which it deflected increased for greater values of current. These instruments used the earth's natural magnetic field to provide the restoring force on the compass needle and, so, they needed to be accurately aligned with the earth's field before they could be used. Used in this way, these instruments were known as '**tangent galvanometers**'. Later, permanent magnets were used to provide the restoring force, making it unnecessary to align the instruments with the earth's field, and such instruments were known as '**astatic galvanometers**'.

An example of a tangent galvanometer is shown in Figure 7.1 – of course, it would be used horizontally, *not* vertically, as shown.

tangent galvanometer

Figure 7.1

As the design of these galvanometers improved, the compass was eventually replaced by a small magnet, attached to a lightweight mirror, and suspended by a thread. The purpose of the mirror was to deflect a light beam across a scale positioned some distance away; this greatly magnified any tiny movements by the magnet.

These galvanometers all had the major disadvantage that the deflection they produced was *not* proportional to the current being measured, for example doubling the value of current wouldn't double the compass needle's deflection!

This problem, however, was eliminated in 1882, when two Frenchmen, physicist Jacques-Arsène d'Arsonval (1851–1940) and engineer Marcel Deprez (1843–1918), developed what we know, today, as the '**moving-coil**' instrument, in which a coil rotates through a narrow, circular, air gap (which provided a consistent linear magnetic field), against the restorative torque provided by a pair of hair springs.

With the ability to now measure an electric current relatively consistently, the next step was to establish *units* by which it could be measured. This was decided at the *First International Conference of Electricians*, held in Paris in 1881, and the unit chosen was the **ampere**, named in honour of André-Marie Ampère (1775–1836). At the same conference, the **volt** (named after Alessandro Volta) was adopted as the unit for potential difference, and the **ohm** as the unit for resistance.

However, the definitions for these units, adopted by the 1881 Conference, had some serious deficiencies: they were simply *not* reproducible outside specialist laboratories. This problem was addressed by the *Fourth International Conference of Electricians*, held in Chicago in 1893, when an attempt was made to redefine these units in such a way they could be replicated in ordinary laboratories. For example, what was then known as the 'international' **ampere** was redefined as *'that constant current which will deposit a mass of 0.001 118 000 g of silver, per second, from a silver nitrate solution'*.

This definition of the **ampere** remained in force until 1948, when it was abandoned by the *General Conference of Weights and Measures*, and redefined as *'that constant current which, if maintained in two straight parallel conductors of infinite length and negligible cross-sectional area and placed one metre apart in a vacuum, would produce between them a force equal to 2×10^{-7} newtons per unit length'*. Which, of course, is the definition we still use today.

The story of the **ampere**, however, *may* not be quite at an end! As the search for ever increasing accuracy continues, a subcommittee of the *International Committee for Weights and Measures* has proposed revised definitions of the SI base units, for consideration at the 25th *General Conference of Weights and Measures* was held in November 2014. The proposed new definition of the ampere will be based on the numerical value of the elementary charge (i.e. the amount of charge on a single electron), a figure yet to be agreed! So, as far as the ampere is concerned, watch this space!

Measuring current, voltage, and resistance

Although we are able to observe the *effects* of an electric **current** or **voltage**, we cannot, of course, 'see' the actual current or voltage itself. So, in order to *detect* and to *measure* these quantities, together with **resistance**, we need instruments that will do the 'seeing' and 'measuring' for us. The three basic instruments used for this purpose are the ammeter, voltmeter and ohmmeter (Figures 7.2, 7.3 and 7.4):

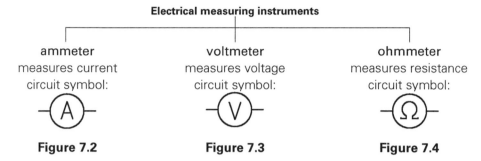

Electrical measuring instruments

ammeter	voltmeter	ohmmeter
measures current	measures voltage	measures resistance
circuit symbol:	circuit symbol:	circuit symbol:

Figure 7.2   **Figure 7.3**   **Figure 7.4**

Most of us are very unlikely to use these *individual* instruments outside a college laboratory. We are *far* more likely to use a **multimeter** (Figure 7.5), either analogue or digital, which *combines the functions of these three separate instruments into one*.

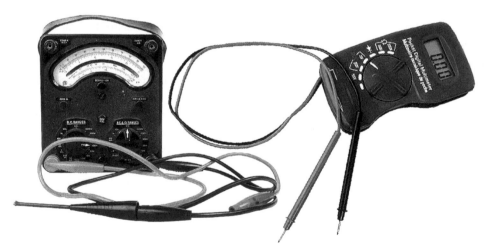

Analogue and digital multimeters
combine the functions of an ammeter, a voltmeter, and an ohmmeter

Figure 7.5

In this chapter, though, we'll first consider the *separate* functions of a multimeter.

We'll start by learning how *analogue* instruments work and how to use them, as they are significantly more complicated to prepare and use. If we can handle an analogue instrument, then we will have no difficulty transferring these skills to a *digital* instrument.

Analogue instruments

Measuring current

Before connecting an **ammeter** into a circuit, we must always start by *de-energizing that circuit* with its load-breaking device, such as a circuit breaker.

To measure **current**, we must disconnect the circuit at the point where we wish to measure that current and, then, insert the ammeter at that point. In other words, *an ammeter must be connected in **series*** with the circuit under test, as illustrated in Figure 7.6 (where *R* represents the load). The circuit can then be re-energised.

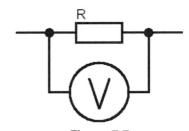

Figure 7.6

As it's very important that inserting an ammeter should have very little effect on the value of the current that *normally* flows in the circuit, ammeters are manufactured with ***very*** *low values of internal resistance*. An 'ideal' ammeter would have no resistance whatsoever, but this is not possible of course.

Because all ammeters have a very low internal resistance, it is vitally important that they are *never*, under any circumstances, connected *in parallel* with the load. Accidentally connecting an ammeter in parallel with the load will result in a short-circuit current that will flow through the instrument which, unless it is internally fused, is very likely to damage the ammeter and may even result in personal injury.

> **Ammeters** must ***always*** be connected in **series** within a circuit. Ammeters have a *very low resistance*.

Measuring voltage

To measure **potential difference**, or voltage, we must connect a **voltmeter** between *two* points which are at different potentials, as shown in Figure 7.7. In other words, *a voltmeter must always be connected in **parallel*** with the part of the circuit under test.

In order to work, a voltmeter must draw *some* current from the circuit under test which can lead to inaccurate results. We call this the '**loading effect**' of a voltmeter.

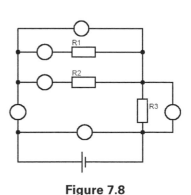

Figure 7.7

To minimise the 'loading effect' and improve its accuracy, a voltmeter must only be allowed to draw a *very* small current from the circuit under test. So voltmeters are manufactured with a ***very*** *high internal resistance*. An 'ideal' voltmeter should have infinite resistance, but this is not possible because it does need *some* current to drive the instrument.

> Voltmeters must *always* be connected in **parallel** in a circuit. Voltmeters have a *very high internal resistance*.

Worked example 1

Examine the circuit shown in Figure 7.8, and identify which circles represent ammeters, and which circles represent voltmeters.

Solution

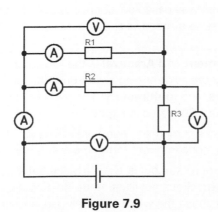

Figure 7.9

121

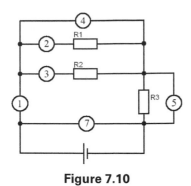

Figure 7.10

Worked example 2

In the circuit shown in Figure 7.10, (a) identify which quantity (current or voltage) is measured by the numbered instruments, and (b) describe what would happen if instrument (4) was an ammeter:

Solution

ammeter 1:   the supply current

ammeter 2:   the current through R_1

ammeter 3:   the current through R_2

voltmeter 4:   the voltage across R_1 and R_2

voltmeter 5:   the voltage across R_3

voltmeter 7:   the supply voltage.

If **instrument (4)** was an *ammeter* then, because of its very low resistance, it would short-circuit the battery.

Types of measuring instrument

Although **digital instruments** are now widely available, there are still a great many **analogue instruments** in use and, so, *we must be able to use both types.*

Analogue instruments are significantly more complicated to use than digital instruments, and the skills required to use these instruments are much more easily transferred to digital instruments than *vice versa.*

In this chapter, then, we will be examining the following types of instrument:

▶ **moving-coil movement:** ammeters, voltmeters, ohmmeters
▶ **moving-iron movement:** ammeters, voltmeters
▶ **electrodynamic movement:** wattmeters
▶ **digital:** ammeters, voltmeters, ohmmeters.

Moving-coil instruments

General construction

The portable moving-coil instrument was designed in the latter half of the nineteenth century by an American engineer by the name of Edward Weston (1850–1936). He called this instrument his 'Portable Instrument' because, up until then, electrical measuring instruments were designed for use in laboratories, and could not be transported. His design was based on a principle already discovered by two Frenchmen, Jacques-Arsène d'Arsonval and Marcel Deprez, which will be explained in this section.

The **moving-coil instrument** or **d'Arsonval** is the most common type of meter movement employed by *precision* analogue instruments such as **ammeters**, **voltmeters**, **ohmmeters**, and **multimeters**. The main components of a typical moving-coil instrument are illustrated in Figure 7.11.

A horseshoe-style permanent magnet, together with a pair of shaped pole pieces and a silicon-steel cylindrical 'concentrator', forms the instrument's **magnetic circuit**.

The concentrator does not rotate, but is *fixed* in place between the pole pieces by means of non-magnetic brackets (not illustrated). Its purpose, together with that of the pole pieces, is to minimise the width of the air gap, and to achieve a uniformly

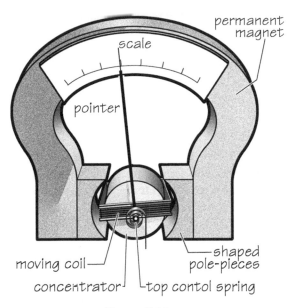

Figure 7.11

distributed magnetic field of maximum flux density within that air gap, as shown in Figure 7.12.

The instrument's **coil** is manufactured from many turns of fine insulated wire, wound around a lightweight rectangular-shaped aluminium 'former', which is pivoted on frictionless (jewelled) bearings, enabling it to rotate through a limited arc (typically around 100°) *within* the magnetic-circuit's air gap. A pointer, attached to the coil assembly, moves across a scale as the coil rotates.

The moving coil is connected to its external circuit via a pair of coiled hairsprings, located at opposite ends of the coil. Their function is to *control* the movement of the coil, and to *restore* the coil and pointer back to their 'rest' position when there is no current flowing around the coil.

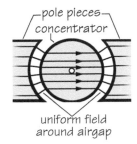

uniform field around airgap

Figure 7.12

Deflecting force and control force on coil

Figure 7.13 represents a front view of a moving coil showing (for the purpose of clarity) just three turns, with a current flowing in the directions shown – i.e. with current flowing *towards us* on the left-hand side of the coil, and flowing *away from us* on the right-hand side of the coil.

Applying **Fleming's Left-Hand Rule** for 'motor action' to opposite sides of the coil, will show that the left-hand side of the coil is subject to an *upward force (F)*, while the right-hand side of the coil is subject to an equal *downward force* of equal. These forces couple together, providing the torque that *drives the coil in a clockwise direction.*

The magnitude of force acting on each conductor is given by:

$$F = BI\,l$$

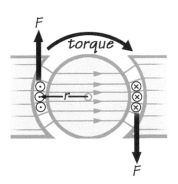

Figure 7.13

where: F = force (N)

B = flux density (T)

I = current (A)

l = length of coil side (m)

As the flux density and the axial length of the coil are constants for the instrument, we can conclude that the force and, therefore, the **torque**, acting on the coil *is proportional to the current flowing around that coil.*

If the moving coil was *only* subjected to this deflecting force, then it would continue to rotate clockwise, just like a motor. However, as the coil rotates clockwise, it *tightens* the two control springs (hairsprings), which then together act to provide a counterclockwise **restraining force** or **torque**.

In fact, it's the control springs that provide the moving-coil instrument with its '**standard**'. *All* measuring instruments must provide some sort of 'standard' by which the quantity being measured (current, in this case) can be compared. For a moving-coil instrument, this 'standard' is the *restraining torque* provided by the control springs, against which the torque provided by the coil must act.

When the restraining torque is *exactly* equal to the deflecting torque, the coil will stop rotating, and **the angle through which it has turned will be proportional to the value of the current flowing through it** – so if we were to, say, *double* the current, we would also *double* the amount of deflection.

The **scale** across which the moving-coil's pointer moves will, therefore, be *linear*. Linear scales are relatively easy to read, and is one of the major advantages of a moving-coil movement compared with other types of analogue meter movement.

All analogue instruments incorporate a mechanical **zero set** adjustment for precisely locating the pointer over the scale's zero position.

With moving-coil instruments, this is normally provided by an eccentric-screw that will slightly adjust the 'relaxed' position of the control springs, enabling the pointer to be brought to hover exactly over its zero position.

Subjecting an analogue instrument to knocks and bumps frequently causes the pointer to move away from its zero position, so *it is very important to always 'zero' the instrument before making any measurements*.

Damping

'**Damping**' describes the elimination of any tendency for an instrument's pointer to oscillate about a mean point, due to a combination of the inertia of its moving parts, the controlling torque of its springs, and any subsequent minor variations in current.

An 'undamped' instrument will be very difficult, if not impossible, to read. So *all* analogue measuring instruments require some means of damping. Moving-coil instruments utilise what is known as **eddy-current damping**.

When the coil rotates within the instrument's permanent magnetic field, an e.m.f. is induced into its rectangular aluminium former, by 'generator action'. This e.m.f. causes a current (eddy current) flow around the former, the direction of which results in a magnetic field which, by Lenz's Law, acts to *oppose the movement of the coil*. This acts to provide the instrument with its damping torque.

Swamp resistor

The instrument's moving coil is manufactured from a length of very fine insulated copper wire that is susceptable to variations in resistance due to any changes in ambient temperature.

To minimise any errors due to such changes in the coil's resistance, the coil is connected in series with a **swamping resistor** whose resistance is large in comparison with that of the coil, and which is manufactured from an alloy with a *very low* **temperature coefficient of resistance** (i.e. a metal whose resistance is hardly effected by quite wide variations in temperature).

For example, if the resistance of a 1 Ω coil increases to, say, 1.1 Ω, when its rated current flows through it, this will introduce a 10% error in the instrument's reading.

But by calibrating the instrument with a series swamp resistor of, say, 49 Ω, the overall resistance at rated current will change from 50 Ω to 50.1 Ω. This will result in an error of just 0.2%.

A **swamp resistor**, then, is an integral part of *all* moving-coil instruments. And whenever we refer to an instrument's 'coil resistance', what we actually *mean* is the *combined* resistance of the coil itself *and* of its swamping resistor, as shown in Figure 7.14.

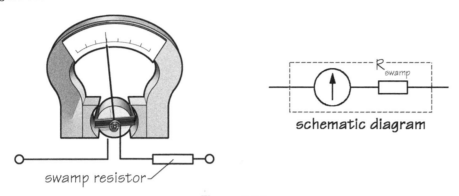

schematic diagram

Figure 7.14

Polarisation

Because the moving coil operates in a permanent magnetic field, its *direction of rotation* depends upon the *direction of the current* flowing through it. Moving-coil instruments, therefore, are described as '**polarised**' instruments. As very few portable instruments have a zero position at the centre of their scale, it is very important that a moving-coil instrument is connected to a circuit so that its pointer moves 'upscale', rather than 'downscale' – i.e. forward, rather than backwards against its stop.

Accordingly, a moving-coil instrument's terminals are always clearly marked either with **colours** (red for its positive terminal; black for its negative terminal) or with **plus** and **minus symbols** engraved adjacent to them. These markings indicate the polarity *to which they must be connected* – i.e. the positive terminal must be connected to a positive potential, and the negative terminal to a negative potential.

For this reason, ***all moving-coil instruments are d.c. instruments***.

Sensitivity

The '**sensitivity**' of a moving-coil instrument *indirectly* indicates *the value of current required for the instrument to achieve full-scale deflection (f.s.d.)*.

An instrument's sensitivity is normally displayed somewhere on the instrument's scale.

However, sensitivity is *not*, as we might expect, expressed in terms of the full-scale deflection current itself but, rather, as '**ohms per volt**' (Ω/V) – in other words, **sensitivity is the *reciprocal* of the value of the instrument's full-scale deflection current**. This might seem odd, but there is a very good reason for doing this, but it is beyond the scope of this book.

To find the value of the full-scale deflection current, given an instrument's sensitivity, we must perform the following calculation:

Given a sensitivity of, for example, 20 000 Ω/V, the full-scale deflection current is worked out as follows:

$$I_{fsd} = \frac{1}{\text{sensitivity}} = \frac{1}{20\ 000} = 50 \times 10^{-6}\,\text{A} \quad \text{or} \quad 50\,\mu\text{A}$$

So, for an instrument having a **sensitivity** of **20 000 Ω/V**, the current that will result in full-scale deflection is **50 μA**.

Moving-coil instrument as a d.c. ammeter

As we have learnt, **ammeters** must be connected in *series* with the circuit whose current is to be measured. But, because the moving-coil winding is very small and light, it is capable of carrying only very small currents (milliamperes or microamperes).

So, if we want it to measure **larger** currents, it's necessary to ensure that *most* of the circuit's current *bypasses the coil*, with only just enough of that current flowing through the movement to result in the necessary deflection. This is achieved by connecting a *very accurate low-value resistor*, called a '**shunt resistor**', in parallel with the coil, as illustrated in Figure 7.15, and calibrating the instrument's scale appropriately.

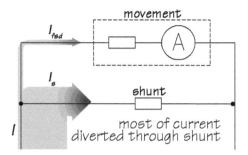

Figure 7.15

Suppose, for example, we want an ammeter having a full-scale deflection current of, say, 1 mA, to measure a circuit current of up to, say, 100 mA.

In this case, when 100 mA is flowing in the main circuit, 1 mA needs to flow through the moving coil to achieve full-scale deflection, while the difference of 99 mA is diverted through the shunt.

The value of the shunt's resistance may be calculated using basic circuit theory, as shown in the following worked example, where:

R_m = resistance of movement (including swamp resistor)
R_s = resistance of shunt
I_{fsd} = full-scale deflection current
I_s = shunt current
I = full-scale current of ammeter and shunt

Worked example 3

A moving-coil instrument has an internal resistance (combined coil and swamp resistance) of 100 Ω and a full-scale deflection current of 1 mA, as shown in Figure 7.16. What resistance must a shunt have in order to convert the instrument to a 1–100-mA d.c. ammeter?

Solution

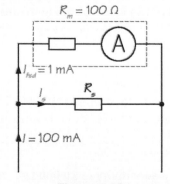

Figure 7.16

The meter movement and shunt resistor are in parallel with each other, so the voltage across each must be the same, so we can start by stating:

$$U_{shunt} = U_{movement}$$
$$(I_s R_s) = (I_{fsd} R_m)$$

where:
$$I_s = (I - I_{fsd}) = (100 - 1) = 99 \, mA$$

Inserting the value of I_s into the second equation, we have:

$$(I_s R_s) = (I_{fsd} R_m)$$
$$(99 \times R_s) = (1 \times 100)$$
$$R_s = \frac{1 \times 100}{99} = \mathbf{1.01 \, \Omega \, (Answer)}$$

Of course, the instrument's scale must then be **calibrated** to *indicate* 100 mA at its full-scale deflection – *regardless* of the fact that only 1 mA is *actually* flowing through it.

As we can see from this example, the resistance-value of a shunt has to be *very* precise, and must be manufactured from an alloy with a very low temperature coefficient of resistance, such as **constantan**, so that it is unaffected by normal variations in temperature.

To increase the versatility of a moving-coil ammeter even further, its current range can be *extended* by using a number of different shunts, of appropriate values of resistance, each of which can be selected using a **range-selector switch**. For example, the circuit illustrated in Figure 7.17 shows *four* shunt resistors, $\boldsymbol{R_{S1}}$, $\boldsymbol{R_{S2}}$, $\boldsymbol{R_{S3}}$, and $\boldsymbol{R_{S4}}$, *any* of which can be connected in parallel with the meter movement according the position of the range switch.

To correspond to each shunt setting, the instrument's scale must have an appropriately calibrated range. Alternatively, it could have just one range to which a 'multiplier' (a multiplication factor) must be applied, according to which shunt is selected.

To prevent a multi-range ammeter's movement from becoming damaged by being accidentally connected directly to the circuit *without* its shunt, whenever the range switch is operated the switch must be of the *'make-before-break'* rotary-type shown in Figure 7.17. For example, in the above example, R_{S2} **must** be connected *before* R_{S1} is disconnected, and so on.

A method of *completely eliminating* the possibility of accidentally connecting a multi-range ammeter to the circuit without a shunt, is to use what is known as the '**Ayrton Shunt**' or '**Universal Shunt**'. This type of shunt is commonly used in multimeters, and is illustrated in Figure 7.18.

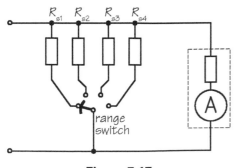

Figure 7.17

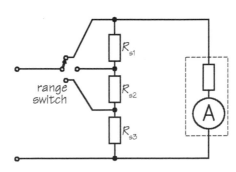

Figure 7.18

Moving-coil instrument as a d.c. voltmeter

Moving-coil instruments are, essentially, *ammeters*. But they can be converted to measure voltage through the use of a 'multiplier resistor'.

Voltmeters measure the *difference in potential* between two points in a circuit and, therefore, must be connected in *parallel* with the circuit components between those two points. To prevent the instrument's movement from drawing too much current from the circuit, it's necessary to connect a resistor, called a **multiplier**, in *series* with that movement.

A multiplier acts to *limit* the current through the movement, ensuring that it does not exceed the coil's full-scale deflection current. The value of the multiplier's resistance may be calculated using basic circuit theory, as shown in the following worked example, where:

R_m	= resistance of movement (including swamp resistor)
$R_{multiplier}$	= resistance of multiplier
I_{fsd}	= full-scale deflection current
R	= total resistance of shunt plus movement

Worked example 4

A moving-coil instrument has an movement resistance (combined coil and swamp resistance) of 100 Ω and a full-scale deflection current of 1 mA. What resistance must a multiplier have in order to convert that instrument to a 1–500 V d.c. voltmeter?

Solution

As usual, we start by sketching the circuit, as illustrated in Figure 7.19.

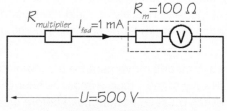

Figure 7.19

We want the voltmeter to give full-scale deflection when connected to 500 V. So, at this voltage, we want the full-scale deflection current (1 mA) to pass through the instrument. Therefore, the necessary combined resistance *(R)* of the movement *and* the multiplier must be:

$$R = \frac{E}{I_{fsd}} = \frac{500}{1 \times 10^{-3}} = 500 \times 10^3 \ \Omega \text{ or } 500 \ k\Omega$$

So, the resistance of the multiplier is the difference between this *total* resistance *(R)* and the resistance of the movement *(R_m)*, calculated as follows:

$$R_{multiplier} = R - R_m = 500\,000 - 100 = 499\,900 \ \Omega \text{ (Answer)}$$

Once again, as we can see from this worked example, the resistance-value of a multiplier has to be *very* precise and, just like a shunt resistor, must be manufactured from an alloy with a very low temperature coefficient of resistance.

To increase the versatility of a d.c. voltmeter, its voltage range can be also *extended* by using a number of multipliers, of appropriate values of resistance, *any* of which can be selected using a **range-selection switch**. For example, Figure 7.20 shows *four* multipliers, R_1, R_2, R_3, and R_4, any of which can be connected in series with the meter movement according to the position of the range switch.

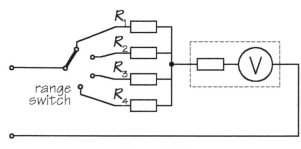

Figure 7.20

An alternative and, in fact, more common arrangement for connecting multipliers is shown in Figure 7.21. In this example, as the range-selector switch is moved, the value of the 'multiplier' is changed by adding or removing the multiplier resistors in series with each other:

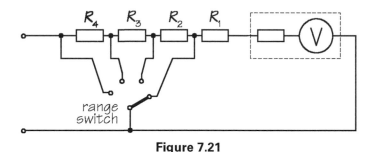

Figure 7.21

For each multiplier position, the voltmeter's scale must have a separate, appropriately calibrated, voltage range. Alternatively, it could have a single range to which a multiplication factor must then be applied, according to which multiplier is selected.

Loading effect of voltmeter

The **sensitivity** (ohms per volt) of a voltmeter is a *very* important factor to consider when selecting the appropriate instrument for measuring voltages in a circuit.

While a relatively low-sensitivity instrument will likely give us accurate results when measuring voltages in low resistance circuits, it is *guaranteed* to give us completely *inaccurate* results if used to measure voltages *in high resistance circuits*.

This is because adding the voltmeter acts to *reduce the equivalent resistance of that part of the circuit to which it is connected*. This is called the **loading effect** of a voltmeter, and is demonstrated in the following examples:

Consider the circuit shown in Figure 7.22, comprising two 100 kΩ resistors, R_1 and R_2, connected in series with each other, across a 20 V supply. This circuit acts as a simple voltage-divider, with 10 V appearing across each resistor:

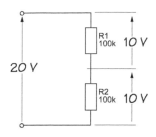

Figure 7.22

129

An **Avometer** model **7X** multimeter, when set to its 10 V range, has an internal resistance of 5 kΩ. Let's see what happens if we connect this instrument across resistor R_1 in Figure 7.23.

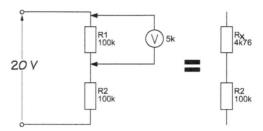

Figure 7.23

$$\frac{1}{R_x} = \frac{1}{R_1} + \frac{1}{R_{meter}} = \frac{1}{100} + \frac{1}{5} = \frac{1+20}{100} = \frac{21}{100}$$

$$R_x = \frac{100}{21} = 4.76\,\text{k}\Omega$$

The circuit has now, essentially, become one with **4.76 kΩ** in series with 100 kΩ, so the overall circuit resistance has now fallen to 104.76 kΩ, and the resulting current will be:

$$I = \frac{E}{R} = \frac{20}{104.76 \times 10^3} = 0.191 \times 10^{-3} = 0.191\,\text{mA}$$

Therefore the voltage indicated by the voltmeter will be:

$$U_x = IR_x = 0.191 \times 10^{-3} \times 4.76 \times 10^3 = \textbf{0.909 V (Answer)}$$

An **avometer** model **9X** multimeter, when set to its 10 V range, has a *much greater internal resistance* of 200 kΩ. So, let's see what happens if we connect this instrument across resistor R_1 in Figure 7.24.

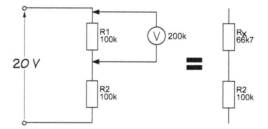

Figure 7.24

The effective resistance of the circuit, using the avometer 9X will become:

$$\frac{1}{R_x} = \frac{1}{R_1} + \frac{1}{R_{meter}} = \frac{1}{100} + \frac{1}{200} = \frac{2+1}{200} = \frac{3}{200}$$

$$R_x = \frac{200}{3} = 66.7\,\text{k}\Omega$$

The circuit has now, essentially, become one with **66.7 kΩ** in series with 100 kΩ, so the overall circuit resistance has now fallen to 166.7 kΩ, and the resulting current will be:

$$I = \frac{E}{R} = \frac{20}{166.7 \times 10^3} = 0.12 \times 10^{-3} = 0.12\,\text{mA}$$

So the voltage indicated by the instrument will be:

$$U_x = IR_x = 0.12 \times 10^{-3} \times 66.7 \times 10^3 = \textbf{8 V (Answer)}$$

In the above examples, the *actual* voltage appearing across resistor R_1, is **10 V**. But an **Avometer 7X** multimeter would indicate a value of **0.909 V**, while an **Avometer 9X** multimeter would indicate a value of **8 V** – which is *far* closer to the *actual* voltage, but not really close enough! So *neither* of these analogue multimeters is really suitable for measuring voltages in circuits with these levels of resistance.

To achieve greater accuracy, we would need to use a voltmeter with *a far higher value of internal resistance.*

The 'ideal' voltmeter should, in fact, have an *infinite* internal resistance that would make it 100% accurate.

But, of course, this is *not* possible with moving-coil instruments, because they need to draw *some* current from the circuit under test in order to drive their movement. However, many *digital* voltmeters *are* able to achieve much more accurate results than are possible with moving-coil type instruments.

Because of its **loading effect**, the 'ideal' internal resistance of a voltmeter should be **infinite**. This, however, is not practical as all moving-coil instruments need to draw current in order to deflect their movement.

In any event, **it's important that a voltmeter's internal resistance is significantly higher than that of the circuit under test**.

To summarise: for accurate results, it's important to select a voltmeter whose internal resistance is *very* much higher than the resistance of the circuit or component across which it will be connected.

Fortunately, while this requirement is *essential* for laboratory work, most commercial voltmeters and multimeters are perfectly adequate for the types of situation that most of us are likely to encounter in the field.

Accuracy and precision

An instrument's '**accuracy'** is expressed as a *percentage of its full-scale deflection reading*, e.g. '1.5% *at full-scale deflection'*.

To keep the mathematics simple, let's see how this relates to an ammeter with a scale that reads up to, say, 100 mA.

At *full-scale deflection*, i.e. when the pointer indicates 100 mA, the accuracy of the reading is **100 mA ±1.5%**. In other words, when the pointer indicates 100 mA, the *actual* current flowing could be up to 1.5 mA either side of 100 mA.

But, unfortunately, this figure of ±1.5 mA then *applies right across the entire scale,* not just to the top end of the scale! So, if the pointer indicates, say, 10 mA, the *actual* value could be anywhere within the range of 8.5–11.5 mA – so, the instrument's accuracy, *at 10 V*, has risen to ±**15%** – *significantly worse than it was at full-scale deflection!*

So, the further 'downscale' (i.e. towards the lower end of the scale) the reading, *the worse the accuracy becomes.*

For this reason, for *any* multi-range instrument, it's important that we *always use whichever scale provides us with the greatest deflection.* For example, if we wanted to accurately measure a current in the region of, say, 23 mA, we will achieve greater accuracy with the ammeter set to its 25 mA scale rather than to, say, its 100 mA scale.

'**Precision**', on the other hand, is *the degree to which repeated measurements under unchanged conditions show the same results.*

Whereas '**precision**' is a measure of an instrument's movement and build-quality, '**accuracy**' depends on us (a) preparing the instrument properly before we use it, and (b) being able to 'read' its scale properly.

For example, let's suppose we use a good-quality moving-coil instrument that we had forgotten to 'zero' before use. And let's assume that the pointer was not hovering exactly over its scale's zero graduation before we started to make our measurements. No matter how *precise* the instrument is at making measurements, *if it hasn't been properly zeroed*, then it will *never* give us a correct reading due to the lack of *accuracy* resulting from our failure to zero the instrument before using it.

In *everyday* language, the terms '**accuracy**' and '**precision**' are frequently used interchangeably whereas, in engineering, *they are quite different.*

The **accuracy** of a measurement system is the degree of closeness of measurements of a quantity to its *actual* (true) value.

The **precision** of a measurement system, on the other hand, is *the degree to which repeated measurements under unchanged conditions show the same results.*

The difference is best explained in terms of the shooting target, illustrated in Figure 7.25.

high accuracy / low precision

low accuracy / high precision

high accuracy / high precision

Figure 7.25

In the top-left example of Figure 7.25, all the shots land within the bull and, therefore, have *high accuracy*. On the other hand, the individual shots are not grouped very close together and, so, they have *low precision*.

In the top-right example, every shot has completely missed the bull and, therefore, we have *low accuracy*. On the other hand, the four shots are grouped very close together and, so, have *high precision*.

In the bottom example, all shots are within the bull and, therefore, have *high accuracy*. At the same time, the individual shots are grouped very close together and, so, they also have *high precision*.

So, how does this relate to using an instrument to take a measurement? Well, an analogue instrument must *always* be zeroed before use. If it *isn't,* then regardless of how *precise* its measurements, it will *never accurately* measure the quantity being measured.

A.C. measurements

We have already established that a moving-coil instrument is a *direct-current instrument*, because the *direction* of the torque applied to its movement depends upon the *direction* of the current flowing through that movement.

If a *very* low (say a fraction of one hertz) frequency a.c. current were to pass through a moving-coil ammeter, then the instrument's pointer (assuming a centre-zero position) would swing lazily back and forth, following the change in current. At mains' frequency (50 Hz), however, the change in current direction is far too rapid for the movement to be able to follow it; instead, the pointer will indicate the **average value** of that current.

The average value of a **sine wave** (the type of waveform supplied by the mains), however, is *zero* – so the instrument would indicate zero! However, the average value of a *full-wave rectified* waveform is **0.636** of its peak value. So, to enable a moving-coil instrument to read a.c. values, its movement can be supplied via a **full-wave bridge rectifier** – as illustrated in Figure 7.26.

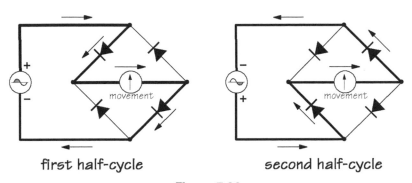

first half-cycle   second half-cycle

Figure 7.26

With this arrangement, the heavy line shows the (conventional) current direction through the diodes as the input voltage changes polarity. Note how the current through the movement acts in the *same* direction *despite* the supply voltage continually reversing polarity, and the waveform through the instrument is as illustrated in Figure 7.27.

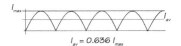

Figure 7.27

Although the instrument *actually* responds to the **average value** of the rectified waveform, the instrument's scale is *calibrated to indicate that waveform's **root-mean-square (r.m.s.)** value*, where:

$$I_{rms} = 0.707\, I_{max}$$

The significance of an r.m.s. value was fully explained in the companion book, **An Introduction to Electrical Science**, so it's *only* necessary to realise that alternating currents and voltages are *always* expressed in r.m.s. values, *never* as peak values! For example, when we talk about a '230 V' mains voltage, that value is an r.m.s. value that has a peak value of 325 V.

We should also be aware that these instruments are calibrated for **sine waves**. For waveforms with any other shape, the instrument will supply completely inaccurate readings.

For the measurement of alternating currents, the instrument's *range* is controlled by adjusting the turns ratio between the primary and secondary windings of a **scaling transformer** – as illustrated in Figure 7.28.

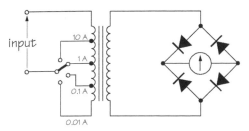

Figure 7.28

Moving-coil instrument as an ohmmeter

An **ohmmeter** is used to measure the *resistance* of a circuit or of a circuit component. It also provides us with a convenient way in which to check for *continuity*. The schematic diagram, illustrated in Figure 7.29, shows the internal circuit for a typical ohmmeter:

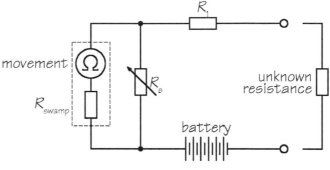

Figure 7.29

The movement (i.e. the moving coil together with its swamp resistor) is connected in *series* with a battery, a fixed-value resistor, R_1, and a pair of terminals to which the unknown resistance will be connected. Connected in *parallel* with the movement is a variable resistor, R_s.

The variable resistor, R_s, shunts the movement, and is used to zero the instrument to compensate for any changes in the battery's voltage as it loses charge over time. 'Zeroing' must *always* be performed *before* carrying out a measurement, and involves temporarily short-circuiting the instrument's terminals, and adjusting R_s, until the pointer hovers directly over the scale's zero mark.

The function of the series resistor, R_1, is to protect the movement from burning out, by preventing the current that flows during the zeroing process from exceeding the movement's full-scale deflection current.

The scale of an ohmmeter differs significantly from that of an ammeter or voltmeter, in *two* important ways. Firstly, its scale is *reversed* – i.e. it reads from right to left – with 'zero ohms' corresponding to full-scale deflection. Secondly, the scale is *non-linear*, as illustrated in Figure 7.30, with its graduations becoming closer and closer together and, therefore, more difficult to read, for higher values of resistance (i.e. towards the left-hand end of the scale).

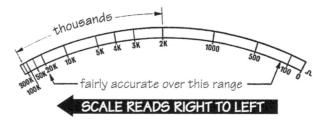

Figure 7.30

Using the built-in variable resistor to obtain a full-scale deflection is called **zero-ohms adjustment**, and *this action **must** be carried out prior to taking any resistance measurement*. This compensates for any variation in the voltage of the instrument's built-in battery; if a zero-ohms adjustment *cannot* be achieved then its voltage is too low and the battery must be replaced.

When using an ohmmeter, we must *always* observe the following rules:

▶ ***Never connect an ohmmeter to a live circuit*** – doing this is not only likely to burn-out the instrument, but it may also be hazardous to us, the user. It's not unusual to come across multimeters with one or other of its ohms' range settings not working because someone has failed to do this!

▶ Beware of measuring resistance of any component that is connected in parallel with other components, or we'll end up measuring the overall resistance of *all* those components! If the component cannot be removed, then it may be necessary to temporarily remove one of the component's connections.

▶ When using a multimeter to measure resistance (or to check continuity), it is important to *switch the instrument off its resistance setting* when the measurement has been completed. Failure to do so might result in the battery becoming discharged should the test-leads accidentally come into contact with each other during storage.

Analogue multimeters

The first **multimeter** is credited to a GPO telecommunications' engineer, named Donald Macadie, who was fed up with having to carry so many different instruments with him when working in the field. As his new instrument could measure amperes, volts, and ohms, he named it the 'AVO'.

An analogue **multimeter** is a moving-coil instrument that can measure direct and alternating currents and voltages, together with resistance. Because of their versatility, multimeters are far more widely used than separate ammeters, voltmeters, and ohmmeters, which are mainly confined to laboratory use.

In North America multimeters are often called '**VOM** meters' – this is the acronym for '**V**olt–**O**hm–**M**illiampere' (few analogue multimeters can read current beyond the milliampere range). Probably the most common analogue VOM meter used in North America was manufactured by the *Conway* company.

One of the most widely used, accurate, and versatile analogue multimeters ever to have been manufactured, is the rugged **Avometer**, shown in Figure 7.31, which was in continuous production, in its various versions, from the early 1920s until it finally went out of production in 2008, when the company was no longer able to source its component parts. Although expensive, these multimeters were the first choice for any electrical engineer, electrical technician, or electrician lucky enough to afford one. The British company the *Megger Group* still manufactures avometers but, these days, they are all digital instruments.

Figure 7.31

Moving-iron instruments

Moving-iron instruments are *analogue* instruments that will measure *both* d.c. *and* a.c. without the need for rectification. However, unlike a moving-coil instrument, the rugged moving-iron instrument is *not* a precision instrument and is often described as being 'cheap and cheerful'! We are unlikely to encounter these instruments in a laboratory, but are very likely to see them as voltmeters or ammeters on **control panels** in substations where they are usually supplied from instrument transformers, and where very accurate readings aren't necessary.

There are, in fact, *two* types of moving-iron instrument: the '**attraction**' **type** and the '**repulsion**' type. In this chapter, we'll only examine the 'repulsion' type.

The **repulsion-type moving-iron instrument** works on a very simple principle. Imagine placing *two* ferrous-metal rods, side by side, inside a de-energized coil

(Figure 7.32a). When the coil is *energized*, the ends of the two rods acquire magnetic polarities that we can determine by applying the 'right-hand grip rule' to the coil. Because adjacent ends of each rod acquire the *same* polarity as each other, they will spring apart (Figure 7.32b) and, the *greater* the current, the *further* apart they will spring.

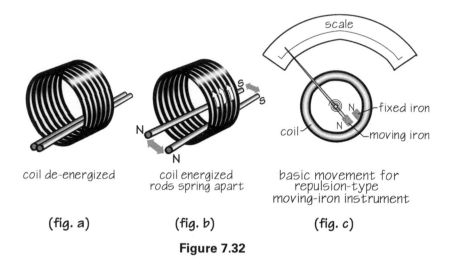

coil de-energized

coil energized
rods spring apart

basic movement for
repulsion-type
moving-iron instrument

(fig. a)   (fig. b)   (fig. c)

Figure 7.32

Imagine, now, that a piece of iron is *fixed* to the inside of the coil, while another piece is *pivoted*. When a current flows through the coil, the pivoted piece of iron will be pushed away from the fixed piece of iron, against the controlling torque of a pair of hairsprings, and an attached pointer moves across a scale that has been suitably calibrated (Figure 7.32c). Since the force between the fixed and moving-iron pieces is not proportional to the current, the distance moved by the pivoted rod is *not* proportional to the current flowing through the coil, the instrument's scale is non-linear, difficult to read at its lower end, and not greatly accurate.

Practical moving-iron instruments use carefully shaped iron 'vanes', rather than rods, as illustrated in Figure 7.33. These shapes are engineered in such a way that the non-linearity of the scale can be somewhat improved, enabling more accurate readings to be made. Not shown, for reasons of clarity, is a **damping** mechanism (usually a light-metal vane or a piston moving through a cylinder), which prevents the pointer from oscillating about its final resting place.

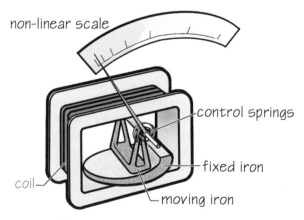

repulsion-type moving-iron instrument

Figure 7.33

The reason that the moving-iron instrument can measure a.c. as well as d.c. is because, whenever the current changes direction through the coil, the polarities of adjacent ends of the iron pieces reverse at the same time – maintaining a force of repulsion between them.

Reading analogue instruments

A good-quality analogue instrument is a *very* accurate instrument – possibly even *more* accurate than an inexpensive, 'consumer-level', digital instrument (but less accurate than a professional digital multimeter)! Analogue instruments, however, are somewhat more difficult to read precisely, and a certain amount of preparation is necessary. *We should, therefore, develop a habit of following these guidelines:*

1   First of all, an analogue instrument's movement is affected by gravitational forces and, therefore, the instrument is designed to give accurate readings *only* when placed in a particular (usually horizontal) position, in which case it should *never* be propped-up at an angle, placed vertically, or held in the hands – where gravitational forces, or movements, will act on the pointer to give an inaccurate reading.

2   Next, the pointer should always be checked to ensure it is hovering *exactly over zero*. If it isn't, then it must be mechanically adjusted, using a small screwdriver, by means of a **zero-adjust screw** located below the instrument's scale —as shown in Figure 7.34. *We mustn't confuse this with the zero-ohms adjustment for an ohmmeter.*

pointer
off zero

zero-adjust
screw

Figure 7.34

3 Multimeters usually have several scales, corresponding to the different functions (voltage, current, resistance) and ranges available, and to whether we are measuring a.c. or d.c. We must *always* ensure that we read from the correct scale – this sounds obvious, but is so often the reason for incorrect readings.

4 Most instruments, and certainly *all* multimeters, offer different **ranges** (e.g. voltage ranges offered might be 0–125 V, 0–250 V, 0–500 V). The *highest* range should *always* be selected first but, then, *whichever scale provides the greatest deflection should* **always** *be used when taking a reading*. We must remember that the **accuracy** of any instrument always applies to its *full-scale deflection*, so the accuracy decreases towards the lower-end of the scale!

5 When taking a reading, the **parallax mirror** (which runs adjacent to the graduated scale, behind the pointer) should *always* be used. Its purpose is *to ensure that our eye is* **exactly** *above the pointer and not to one side or other*. When our eye is *exactly* above the pointer, the reflection of the pointer is not visible – as shown in Figure 7.35.

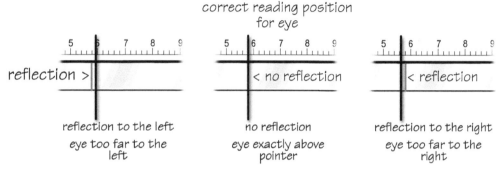

Figure 7.35

6 After completing a resistance measurement, multimeters should *never* be left on the resistance setting – if the test-leads are accidentally shorted, the battery will be discharged.

7 When transporting an analogue multimeter, if there is no 'off' setting, then it should *always be set on one of its* **current** *ranges* – this will enable the instrument to 'self damp' through its shunt resistance, minimising any movement of the pointer during transit. The principle of 'self-damping' is very simple: any movement of the coil induces a voltage into the coil, the resulting current then circulates through the shunt and, by Lenz's Law, acts to *oppose the movement of the coil*.

Exercise

Examine the multi-range voltmeter scales, shown in Figure 7.36. Then, *taking great care to select the correct scale*, write down the readings when the pointer comes to rest at positions *a–f*.

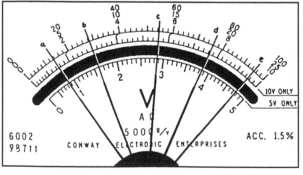

Figure 7.36

Table 7.1

Scale	pointer positions				
	a	b	c	d	e
0 V — 5 V					
0 V — 10 V					
0 V — 25 V					
0 V — 50 V					
0 V — 500 V					

Answers

Did you notice that the scale directly above the parallax mirror *must only be used for the 0–10 V range*? And the scale directly below the parallax mirror *must only be used for the 0–5 V range*? If so, well done! This is *not* a 'trick question' but, rather, typical of the types of scale that we are quite likely to encounter on *real* analogue instruments – **so it's very important to examine the scales and use the correct scale for each range!** Here are the answers:

Table 7.2

Scale	pointer positions				
	a	b	c	d	e
0 V — 5 V	0.98	1.6	2.86	3.85	4.76
0 V — 10 V	1.5	3.0	5.6	7.65	9.55
0 V — 25 V	3.4	7.1	13.6	19	23.75
0 V — 50 V	6.8	14.2	27	38	47.5
0 V — 500 V	68	142	270	380	475

Incidentally, did you notice the **instrument's sensitivity**, of **5000 Ω/V**, shown on the scale, reproduced in Figure 7.36?

Electrodynamic instruments: wattmeters

Principle of operation

In the case of a d.c. circuit, **power**, the *rate of doing work*, may be determined by simply taking the *product* of the simultaneous readings of an **ammeter** connected in series with the load, and a **voltmeter** connected in parallel with its supply voltage:

$$P = EI$$

where:
P = power (W)
E = voltage (V)
I = current (A)

Power, of course, is expressed in **watts** (symbol: **W**).

For steady-state conditions, this ammeter/voltmeter method of determining power is straightforward. However, there are many occasions when the power of a d.c. circuit varies from moment to moment due to changes in load (e.g. d.c. motors). For a.c. loads, the ammeter/voltmeter method will always give us the **apparent power**, expressed in volt amperes; it will *only* furnish us with the **true power** of a load, *providing its power factor is unity*.

So a *single* instrument that measures power by continuously monitoring the instantaneous values of both current and voltage, would be much more useful. This instrument is called a **wattmeter**.

The major advantage of using a wattmeter is that it will *always* indicate the **true power** of a circuit, *regardless of whether the supply is d.c. or a.c.* and, in the case of a.c., *regardless of the load's power factor.*

Construction

An analogue **wattmeter** consists of *two* **field windings**. One of these windings, called the '**current coil**', is fixed and consists of a few turns of relatively thick wire, and is connected *in series with the load*. The other coil, called the '**voltage coil**' (or, in North America, '**potential' coil**) or '**pressure coil**', is pivoted about its axis, and consists of many turns of relatively thin wire, and incorporating a high value of resistance, is connected *in parallel with the load*.

So, in terms of their connections, we can think of the current coil as being 'equivalent' to an ammeter, and the voltage coil as being 'equivalent' to a voltmeter, as illustrated in Figure 7.37.

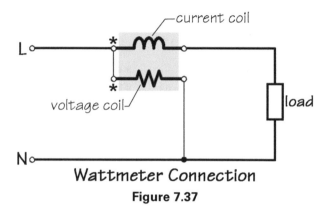

Wattmeter Connection

Figure 7.37

The fixed **current coil**, in fact, is separated into two halves, and its purpose is to provide a uniform magnetic field (Φ_i) across the gap between its two halves, which is *proportional to the load current*.

The **voltage coil**, whose magnetic field (Φ_e) is *proportional to the supply voltage*, is located within the current coil's magnetic field, and is pivoted allowing it to rotate against a counter-torque provided by a pair of hair springs connected to its shaft.

This coil arrangement, which is illustrated in Figure 7.38, is termed an '**electrodynamic**' or a '**dynamometer**' type movement.

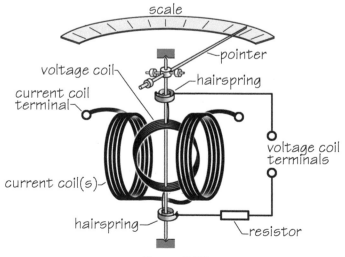

Figure 7.38

When connected to a load, the combined magnetic fields set up by the two coils react with each other, causing the voltage coil to deflect through an angle that is proportional to the *product* of the two fields ($\Phi_i \times \Phi_e$) – i.e. proportional to the *power* of the load.

A pointer attached to the voltage coil then indicates the value of the power as it sweeps across an appropriately calibrated scale. Not shown, for reasons of clarity, is an **air-damping** system, which acts to prevent the instrument's pointer oscillating about its indicated value as it comes to rest. This usually consists of a vane, attached to the moving coil, and it is the movement of this vane through the air which acts to damp the movement.

When correctly connected to the load, the wattmeter's pointer will move in the same direction, *regardless* of whether a direct, or alternating, current is flowing. This is because the currents flowing through the current coil and voltage coil are always in the same direction relative to each other: when the current direction reverses in the current coil, it also reverses at precisely the same instant in the voltage coil, so the direction of the resulting torque is unaffected.

If an excessively high current passes through an ammeter, it's usually quite obvious because its pointer will swing hard across the scale and up against its stop. Unfortunately, this is *not* true for a wattmeter!

So it's *very* important that the values of the currents flowing through the wattmeter's current and voltage coils *are never allowed to exceed either of those coils' rated values* (as listed on the instrument's nameplate) because its reading may not indicate that there's anything amiss!

For example, the combination of an *excessively large* current flowing through the current coil, coupled with a *particularly low* voltage applied to the voltage coil, may result in a *moderate* value of power being indicated by the wattmeter – while, in the meantime, the current coil is actually *in the process of overheating and burning itself out!*

The current coil is far more susceptible to damage from excess current, because the voltage coil circuit has a high resistance, which usually limits the current through it.

Wattmeter terminals

Most wattmeters have *four* terminals: one pair supplying the current coil, and the other pair supplying the voltage coil. The **current coil's** terminals are normally identified with a symbol such as *'I' or 'A'*, while the **voltage coil's** terminals are identified with a symbol such as *'E'* or *'V'* or *'U'* – depending on the manufacturer.

It's *very* important that the current and voltage coils are connected in such a way that the relative directions of the currents through them result in the pointer deflecting upscale and *not* downscale (i.e. backwards).

For this reason, the terminal supplying the 'start-end' of each coil is identified with what is termed a '**polarity marking**', which can be either an *asterisk* (*) or a *plus/minus* (±) sign, or a dot (•) – again, depending on the manufacturer.

Polarity markings *must* be taken into account whenever the wattmeter is installed into a circuit. A wattmeter that is trying to read 'downscale' is usually an indication that the polarity marks have been ignored, and that one or other of the coils has been connected into the circuit in the reverse direction relative to the other! In which case, the leads to one of the windings must be reversed.

The **correct** and **incorrect** ways of connecting a wattmeter, using its polarity markings, are illustrated in Figure 7.39. The rule is that the currents through the current coil and the voltage coil *must flow in the same direction, relative to each other* – as shown in Figure 7.39.

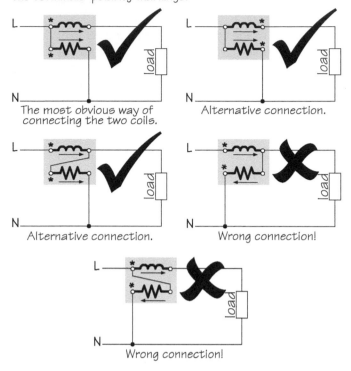

Connection Rule:
Currents must flow in the same directions, relative to the terminals' polarity markings.

The most obvious way of connecting the two coils.

Alternative connection.

Alternative connection.

Wrong connection!

Wrong connection!

Figure 7.39

As we can see, the terminals labelled with a polarity mark must *both* be connected on the *same* side of the supply (*either* to the line side *or* to the neutral side in the case of an a.c. circuit; *either* the positive *or* the negative side in the case of a d.c. circuit). In the examples shown in Figure 7.39, those connections that fail to follow this rule will cause the meter's pointer to be deflected downscale (backwards).

For *extremely* accurate power measurements, whether the voltage coil should be connected on the supply side, or on the load side, of the current coil, is important. However, this only really applies for laboratory work rather than for field work, and is beyond the scope of this chapter.

To avoid confusion, it is highly recommended that, for most purposes, the instrument is best connected as shown in the top-left illustration in Figure 7.39.

To extend their range and versatility, many wattmeters are *multi-range* instruments – e.g. the wattmeter illustrated in Figure 7.40 has *four* **voltage settings** (60 V, 120 V, 240 V, and 360 V), and *two* **range-multiplier** settings (×5 and ×10).

So, *before* using this particular wattmeter, we must *always* set its voltage-selector switch to the appropriate voltage setting (i.e. at, or *above*, the value of the voltage to which the wattmeter is connected) and the reading taken from the corresponding scale. The reading obtained is then multiplied by the range-multiplier setting – according to whichever setting gives the largest deflection. The *greater* the deflection, the *greater* the accuracy.

In common with all other analogue instruments, prior to use, this particular instrument should be placed horizontally (not vertically, as illustrated), and the zero-adjust screw used, if necessary, to accurately set to the pointer to its zero setting.

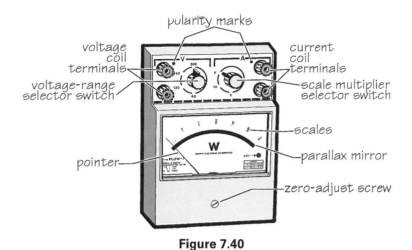

Figure 7.40

The range multiplier on this particular instrument also has a **'k'** setting, which enables the current coil to be temporarily bypassed if, for example, it is to be used to measure the power of a motor whose large starting current might otherwise damage the current coil when the motor is first switched on. Once the starting current has fallen to its normal load current, the range-multiplier switch can be moved away from its 'k' setting.

Many wattmeters *don't* have this useful feature, in which case it is important to incorporate an alternative means of bypassing the wattmeter's current coil during a motor's start-up. This can be achieved, for example, by using a switch connected in parallel with the current coil, as illustrated in Figure 7.41. The switch is *closed* when the motor is started up then, when the starting current then falls to its normal operating current, the bypass switch can then be opened, allowing current to pass through the wattmeter's current coil:

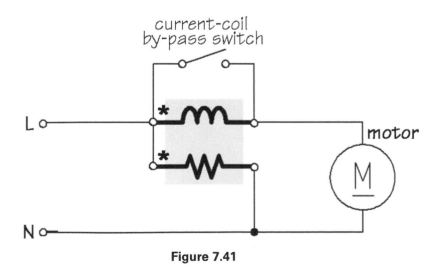

Figure 7.41

How to Connect a Four-Terminal Wattmeter to a Load

- The **current coil** must be in **series** with the load.
- The **voltage coil** must be in **parallel** with the load.
- The correct connection of **polarity markings** must be observed.

(Continued)

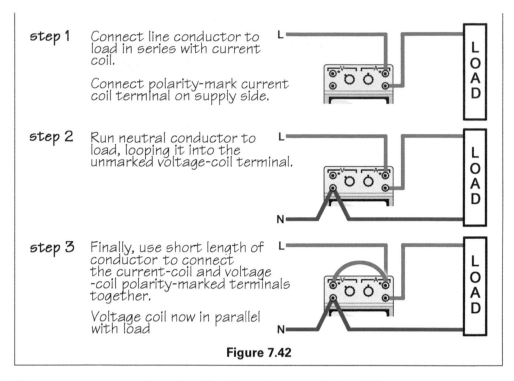

step 1   Connect line conductor to load in series with current coil.

Connect polarity-mark current coil terminal on supply side.

step 2   Run neutral conductor to load, looping it into the unmarked voltage-coil terminal.

step 3   Finally, use short length of conductor to connect the current-coil and voltage-coil polarity-marked terminals together.

Voltage coil now in parallel with load

Figure 7.42

The wattmeter connections described thus far are used when it is necessary to measure the power of the complete load. But we can also connect a wattmeter to measure the power of *individual* components within any circuit.

There are four 'rules' to follow when using a wattmeter to measure the power of separate circuit components:

a   Connect the current coil as we would an ammeter – i.e. in *series* with that component.

b   Connect the voltage coil as we would a voltmeter – i.e. in *parallel* with that component.

c   Ensure that the current ratings of the two coils are *not* exceeded.

d   Observe the requirements for the polarity markings.

For example, note how the wattmeter is connected in order to measure the individual powers of resistors R_1 and R_2, in the circuits shown in Figure 7.43.

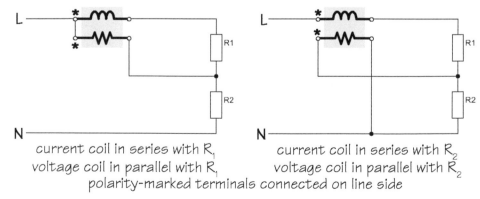

current coil in series with R_1
voltage coil in parallel with R_1

current coil in series with R_2
voltage coil in parallel with R_2

polarity-marked terminals connected on line side

Figure 7.43

Using a wattmeter with instrument transformers

To safely measure the power of a **high-voltage** circuit, it's important to *electrically isolate* the wattmeter from that circuit. This is achieved using instrument **transformers**.

As illustrated in Figure 7.44, the wattmeter's current and voltage coils are supplied from a **current transformer** (**CT**) and a **voltage transformer** (**VT**) respectively. The CT and VT not only act to *isolate* the wattmeter from the hazardous high-voltage system, but also to *reduce* the values of the current and voltage to levels that match the current/voltage ratings of the instrument.

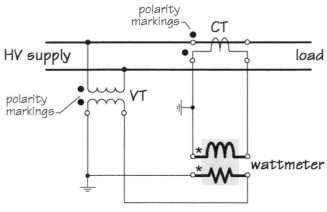

Figure 7.44

Note that it is extremely hazardous to allow an energized current transformer's secondary winding to be open-circuited because a very large voltage can appear across its secondary terminals. When not in use, the CT's secondary terminals *must* always be short-circuited and, usually, a link is provided for this purpose. It's very important to emphasise that this link must only be removed *after* connecting the wattmeter's current coil to the CT's terminals, and it must be reinstalled *before* the current coil is disconnected.

> To avoid a high-voltage shock hazard, a **current transformer's** terminals must ***never***, *under any circumstances*, be allowed to become open-circuited! They must be shorted together *before* disconnecting an ammeter, and the short must always be removed *after* the ammeter has been reconnected.

To ensure that the wattmeter reads 'upscale' and *not* 'downscale', it's very important to observe not only the *wattmeter's* polarity markings, but also those of the *instrument transformers*. The rule is that the wattmeter's polarity-marked terminals should be connected to the instrument transformers' corresponding polarity-marked terminals.

The earth connections play no part in the operation of this circuit, but are required for safety reasons.

Digital measuring instruments

As we learnt at the beginning of this chapter, **digital measuring instruments** have now largely replaced analogue instruments. In fact, with very few exceptions, analogue instruments are no longer being manufactured – for example, the legendary **Avometer** went out of production in 2008, apparently because of its manufacturer's (**Megger Group**) inability to source its component parts. The very first digital multimeter, manufactured by *Non-Linear Systems*, went into production in 1955.

In fact, it's probably safe to say that *analogue instruments have long reached the limits of their development* – their movements and an ease of reading have become about as good as they can possibly become. Digital instruments have now overtaken analogue instruments in terms of their lower cost, robustness, potential accuracy, and the elimination of reading errors.

Figure 7.45

Let's briefly compare digital multimeters with analogue multimeters:

▶ Analogue multimeters have several scales and ranges for measuring current, voltage, and resistance and, often, different scales for measuring a.c. and d.c. They are, therefore, relatively difficult to read, require practice, and prone to mistakes (e.g. reading off the wrong scale!).

▶ Digital multimeters have a simple, digital, readout (in some cases, with automatic range selection) and are very easy to read with less chance of making an error.

▶ Analogue multimeters are relatively delicate and must be treated with care to prevent them from becoming mis-calibrated or even damaged. For example, it's not unusual to come across an instrument whose moving coil has become dismounted from its bearings.

▶ Digital multimeters are tough and robust, and are far less likely to become damaged. In fact, physical damage is unlikely to affect their accuracy.

▶ Analogue multimeters require a careful set-up routine prior to taking a measurement. Gravity affects their accuracy, and they must normally be used in a horizontal position.

▶ Digital meters have to be set to the appropriate function setting (current, voltage, or resistance), but many (but not all) types will automatically select the most appropriate *range* within that setting, will compensate for an aging battery, and all will self-zero prior to resistance measurements. Gravity has absolutely no effect on their accuracy, so these instruments may be used in *any* position.

How digital instruments work

Understanding the operation of **digital measuring instruments** requires a knowledge of advanced electronics and, so, a detailed explanation is well beyond the scope of this book.

Instead, a *greatly simplified* explanation will suffice.

The 'heart' of a digital voltmeter is an electronic component called a '**gate**' that controls the flow of pulses between a **pulse-generator circuit** ('**clock**') and a **counter/readout circuit**.

A gate is a solid-state **switch**, but it should be noted that the terminology we use with a gate is *opposite* to that which we use for a switch! An '*open* gate' is equivalent to a '*closed* switch', and a '*closed* gate' is equivalent to an '*open* switch'! So, as we can see, an 'open gate' will allow pulses through, whereas a 'closed gate' will block the passage of those pulses!

We need to bear this in mind, during the following explanation.

Every measuring instrument requires some sort of '**standard**', or 'yardstick', against which it can compare the quantity (voltage, current, etc.) being measured. In the case of the moving-coil instruments we learnt about earlier, the restraining torque of the *control springs* that act as the 'standard' against which the deflection torque of the movement (coil) is compared.

But, for most digital instruments, the most convenient standard is **frequency** or **time**. Piezoelectric crystals, for example, can be manufactured to provide a highly accurate output-frequency. Such a crystal provides the heart of what is known as a '**pulse-generator circuit**' or '**clock**', which provides a continuous output of square-wave **pulses** whose frequency is a *sub-multiple* of the original crystal-frequency.

These pulses are fed to a '**counter/readout circuit**', via the instrument's **gate**, as illustrated in Figure 7.46. When the gate is 'open' the pulses can flow through; when the gate is 'closed' they cannot. The function of the counter/readout circuit is to (a) *count the number of pulses* that have passed through the gate while it was open, and (b) *provide a digital-readout* expressed in volts.

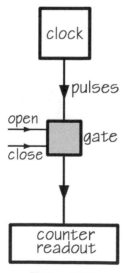

Figure 7.46

The **gate** itself is controlled by a circuit that generates a repetitive, straight-line, negative-slope voltage-waveform, called a '**ramp**' – as illustrated in Figure 7.47. In this simplified example, the ramp potential falls, linearly, from +10 V to −10 V, before continuously repeating itself.

This decreasing **ramp potenial** is fed to a pair of '**AND logic gates**'. An 'AND logic gate' is a solid-state device that provides an output *whenever its two input signals are the same*. We mustn't confuse a *logic* gate with the gate that controls the pulses between the clock and the counter/readout circuit.

The ramp's falling potential is fed to one of the inputs to **AND logic gate 1** and, when its value *exactly* matches the input from the potential being measured (point **A** on the graph), the logic gate emits a pulse that *opens* the gate.

A *reference* potential (0 V), is fed to one of the inputs to **AND logic gate 2** and, when the falling ramp potential *exactly* matches that input (point **B** on the graph), the logic gate emits a pulse that *closes* the gate.

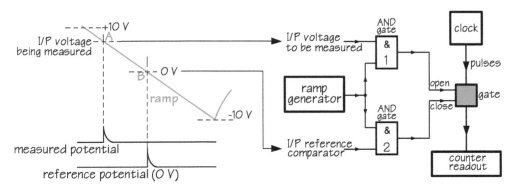

Figure 7.47

So the gate remains *open* for *as long as it takes* for the ramp potential to fall from point **A** to point **B** which, of course, is *proportional to the potential difference between those points* – i.e. the potential difference being measured by the digital voltmeter!

So, the ramp provides a means of converting *potential difference* into *time*: i.e. the *greater* the potential difference, the *longer* the gate remains open.

All that's necessary, now, is to *count* the number of pulses that have passed through the gate while it was open: the number of pulses being (a) *proportional to the length of time the gate is open* and, therefore, (b) *proportional to the potential difference being measured!*

The final step is to calibrate a digital readout to indicate the *voltage* rather than the number of pulses counted by the **counter circuit**.

A digital *ammeter* works in a slightly different way. By passing the test current through a very accurate resistor (built into the meter), the resulting voltage drop *across* that resistor will be proportional to that current, and this voltage drop is then applied to the ramp generator – except that its display output is then calibrated in terms of amperes, milliamperes, or microamperes, rather than volts.

Of course, an *actual* digital instrument is very much more complicated than this explanation – especially a digital multimeter, which must be capable of measuring d.c. and a.c. voltage and current over a wide range of values, as well as measuring resistance and, in some cases, other quantities too.

This, of course, is well beyond the scope of this book.

Features of a digital multimeter

There is an enormous range of digital multimeters available from a great many manufacturers. They *all* measure d.c. and a.c. currents and voltages, as well as resistance, and many will also have other features that enable them to measure capacitance, inductance, etc., as well as to test diodes and transistors.

So let's take a brief look at the basic features that are common to *all* multimeters, by referring to the example shown in Figure 7.48.

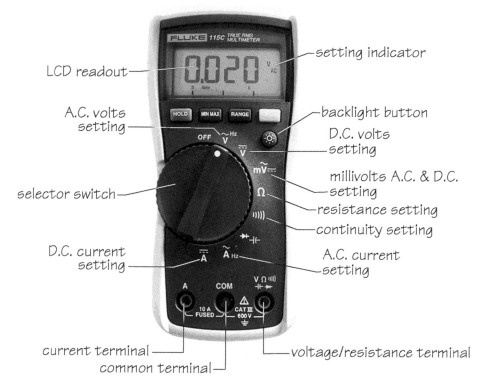

LCD readout

setting indicator

A.C. volts setting

backlight button

D.C. volts setting

millivolts A.C. & D.C. setting

selector switch

resistance setting

continuity setting

D.C. current setting

A.C. current setting

current terminal

common terminal

voltage/resistance terminal

Figure 7.48

Starting with the instrument's terminals: a black test-lead should be inserted into the centre **common (COM) terminal**. To measure *current*, the red test-probe should then be inserted into the left-hand current terminal, marked '**A**'. To measure **voltage** or **resistance**, the red test-lead should, instead, be plugged into the right-hand terminal, marked '**VΩ**'.

We will notice that the **current terminal** is marked to indicate that currents up to **10 A** may be measured, and the voltage-terminal is marked to indicate that voltages up to **600 V** may be measured.

The **selector switch** should *always* be set to '**OFF**' whenever the instrument is not in use, to conserve its battery's power. Turning the switch clockwise will enable each of the following functions:

▶ position 1:   **a.c. voltage** measurements
▶ position 2:   **d.c. voltage** measurements
▶ position 3:   **a.c.** or **d.c. millivolt** measurements
▶ position 4:   **resistance** measurements
▶ position 5:   **continuity** tests (enabling an electronic bleep)
▶ position 6:   **diode** and **capacitance** tests
▶ position 7:   **a.c. current** measurements
▶ position 8:   **d.c. current** measurements.

For each position selected, a confirmatory **setting indicator** will appear on the LCD display ('a.c. voltage', in the above illustration).

This particular instrument has an **automatic range-selection** feature. This means that, with the exception of voltages in the millivolt range, it is *unnecessary* to select an appropriate range before conducting measurement. It's also unnecessary to perform a 'zero-ohms' check, as we would normally have to do *before* measuring resistance with an analogue instrument, because this, too, is fully automatic.

We should, however, be aware that *not all digital instruments have an automatic range-selection feature* (e.g. see the other digital instrument illustrated in Figure 7.45), in which case we should *always* select the highest range initially, then choose the range-setting appropriate to that first reading.

In the field, it's quite common to use a multimeter's 'resistance' setting to check the **continuity** of, say, a fuse – we're not interested in the *resistance* of the fuse, only whether or not it has *melted*. Some multimeters (such as the model, illustrated in Figure 7.48), therefore, have a very useful additional **continuity test** feature: it emits an electronic 'beep' when continuity is confirmed.

'Clamp-on' digital ammeters and multimeters

Measuring **current** normally requires us to *disconnect the circuit* under test, so that we can insert the ammeter *in series with the load*. In the field, it is sometimes difficult to find a suitable location where the circuit can be broken for this purpose. An additional difficulty arises whenever a 'live' circuit cannot be de-energised, preventing us from breaking the circuit to insert the ammeter.

These difficulties can be overcome by using a very useful instrument called a **clamp-on ammeter** or **clamp-on multimeter**, as illustrated in Figure 7.49. These instruments are generally known as '**tong testers**' in North America.

The '**clamp**', referred to in this description, is a laminated **magnetic circuit** (the 'tong'), which may opened-up against a spring, so that the instrument can be clamped *around* a current-carrying conductor. This conductor then acts as the *'bar primary'* for what is essentially a **current transformer**, with the clamp providing the *magnetic circuit* linking to the *secondary winding* located within the instrument. The resulting current flowing in the secondary winding is then measured by the instrument, taking the turns ratio into account.

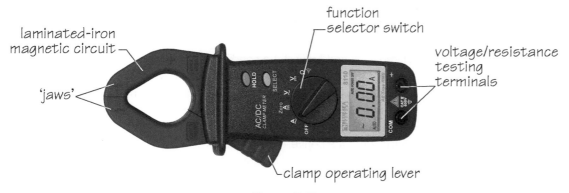

Figure 7.49

The instrument illustrated is an example of a typical modern '**clamp-on**' **multimeter**. With this instrument, the 'clamp' feature is *only* used to measure **current**. To measure voltage and resistance, it provides a pair of terminals for use with a pair of test-leads – just like any other multimeter.

It's important to understand that this instrument can only measure alternating current flowing in an *individual conductor*; it *cannot* measure the current flowing in a *cable* because the magnetic fields set up by the currents flowing in the line *and* neutral conductors, of course, act in *opposite* directions, and effectively neutralising the resultant flux!

Until fairly recently, clamp-on ammeters could *only* be used to measure *alternating currents* because, essentially, they work as **current transformers**, and transformers, of course, *only* work with alternating current. However, new technology has enabled newer models to measure *direct current* as well as alternating current!

This new technology makes use of what's known as the '**Hall Effect**', named in honour of its discoverer, Edwin Hall, as long ago as 1879! These new clamp-on ammeters use a 'Hall-Effect' semi-conductor device to measure the current in the wire.

Hall Effect. If a current passing along a conductor is deflected sideways by a perpendicular magnetic field, then the electrons tend to crowd towards that side of that conductor, making that side more negative than the opposite side. A potential difference, therefore, appears across opposite sides of the conductor, which can be detected by a Hall-Effect detector. The *greater* the magnetic field, the *more* the current will crowd to one side, and the *greater* will be the resulting voltage.

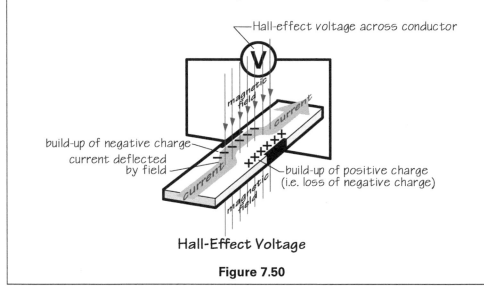

Hall-Effect Voltage

Figure 7.50

So, when the clamp is closed around the conductor, the magnetic field is set up within the clamp's iron core. This magnetic field is then concentrated within an air gap that contains the Hall-Effect detector, and a small current flowing through that detector is deflected by the field (Figure 7.50), creating a voltage at right-angles to the direction of that current. This voltage is then detected and, because it is ultimately proportional to the current in the wire, it's converted into amperes by the ammeter's internal microprocessor, and displayed on the instrument's LCD panel.

Conclusion

Now that we've completed this chapter, we need to examine its **objectives** listed at its start. Placing a question mark at the end of each objective turns that objective into a **test item**. If we can answer those test items, then we've met the objectives of this chapter.

CHAPTER 8

Measuring three-phase power

Objectives

On completion of this chapter you should be able to:

1. explain the significance of a wattmeter's polarity markings.

2. sketch a circuit diagram, showing the 'single wattmeter' method of measuring power of a three-phase a.c. load.

3. explain 'Blondel's Theorem' for using wattmeters to measure the power of a three-phase a.c. load.

4. sketch circuit diagrams, showing the 'three-wattmeter' method for measuring power of a

three-phase, three-wire connected, and four-wire connected, load.

5. sketch a circuit diagram, showing the 'three-wattmeter' method for measuring power of a three-phase a.c. load, and explain how this method measures the total power of that load.

6. given the readings from the 'two-wattmeter' method, determine the power factor of a balanced three-phase a.c. load.

The wattmeter

Introduction

In the chapter on **Electrical Measuring Instruments** (Chapter 7), we learnt the principle of operation and the use of the **wattmeter** to measure the power of a load.

We should recall that, in **a.c. circuits**, we need to concern ourselves with *three* 'types' of power:

▶ **apparent power**, expressed in volt amperes (V·A), which is simply the product of the supply voltage and the load current

▶ **true power**, expressed in watts (W), which is the product of the supply voltage, load current, and the power factor of the load

▶ **reactive power**, expressed in reactive volt amperes (var), which is the vectorial-difference between apparent and true power.

In this chapter, we are going to learn how we can measure the **true power** of a **three-phase load**, using the wattmeter.

Wattmeter connections

Wattmeters have *two* coils: a **voltage coil** and a **current coil**. The voltage coil is always connected in *parallel* with the load, while the current coil is connected in *series* with the load.

Electrical Science for Technicians. 978-1-138-84926-6 © Adrian Waygood.
Published by Taylor & Francis. All rights reserved.

But these connections are complicated by the fact that the way these two coils are connected in relation to each other *may cause the instrument to read 'downscale'* (i.e. backwards)!

To prevent this from happening, and it is particularly important for three-phase circuits, whenever we connect a wattmeter into a circuit, we must take into account its **'polarity markings'**.

Polarity markings were fully discussed in Chapter 7 on *Electrical Measuring Instruments*, but they are so important, a quick review of the topic will not go amiss.

Wattmeter polarity markings

Wattmeters normally have *four* terminals: one pair supplying the current coil, and the other pair supplying the voltage coil. The **current coil's** terminals are normally identified with a symbol such as '*I*' or '*A*', while the **voltage coil's** terminals are identified with a symbol such as '*E*' or '*V*' or '*U*' – depending on the manufacturer.

It's very important that the current and voltage coils are connected in such a way that the relative directions of the currents through the coils result in the pointer deflecting upscale and not downscale.

For this reason, the terminal supplying the 'start-end' of each coil is identified with a **'polarity marking'**, which can be an *asterisk* (*), a *plus/minus* (±) sign, or a dot (•) – depending on the manufacturer.

Polarity markings *must* be taken into account whenever the wattmeter is installed into a circuit. A wattmeter that is trying to read 'downscale' is usually an indication that the polarity marks have been ignored, and that one or other of the coils has been connected into the circuit in the reverse direction relative to the other! In which case, the leads to *one* of the windings must be reversed.

The **correct** and **incorrect** ways of connecting a wattmeter, using its polarity markings, are illustrated in Figure 8.1. The rule is that the currents through the current coil and the voltage coil *must flow in the same direction, relative to each other:*

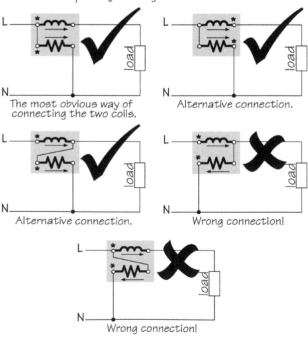

Connection Rule:
Currents must flow in the same directions, relative to the terminals' polarity markings.

The most obvious way of connecting the two coils.

Alternative connection.

Alternative connection.

Wrong connection!

Wrong connection!

Figure 8.1

As we can see, the terminals labelled with a polarity mark must *both* be connected on the *same* side of the supply (*either* to the line side or to the neutral side in the case of an a.c. circuit; *either* the positive *or* the negative side in the case of a d.c. circuit). In the examples shown in Figure 8.1, the connections that fail to follow this rule will result in the wattmeter's pointer being deflected downscale (backwards).

Now that we have reminded ourselves of the importance of polarity markings, we can move on and learn how we use wattmeters to measure the power of a three-phase load.

Measuring the true power of a three-phase circuit

A **wattmeter always** measures the **true power** of a circuit, expressed in **watts**. In the case of single-phase alternating-current loads, this is given by:

$$P = \overline{E}\,\overline{I} \cos$$

where:	$\overline{E}$ = supply voltage (volts)
	$\overline{I}$ = load current (amperes)
	$\cos \phi$ = power factor of load

In this section, we are going to learn how wattmeters may be used to measure the total true power of **three-phase circuits** that supply both **balanced** and **unbalanced** loads.

What a wattmeter 'reads'

To understand what follows throughout the rest of this chapter, it's important to fully understand what a wattmeter actually 'reads' when correctly connected. It reads the

▶ **current** flowing through its current coil, *multiplied by the*
▶ **voltage** appearing across its voltage coil, *multiplied by the*
▶ **cosine** of the phase angle between the current and the voltage.

This is why a wattmeter will always measure the **true power** of a load. It's also important to understand the 'direction' or 'sense' of these currents and voltages as they apply to the wattmeter's coils:

▶ **current flow** – *from* current-coil's polarity-marked terminal *to* its second terminal
▶ **voltage** – the potential of the voltage-coil's polarity-marked terminal, with respect to its second terminal.

'Single-wattmeter' method

For a **balanced, star-connected**, three-phase load, *provided we have access to the neutral point*, we can use a **single wattmeter**, externally, to determine the total power of that load. This is illustrated in Figure 8.2.

For this connection, the wattmeter is measuring the power of the load an *individual* phase (in this case, phase A-N) so, to determine the **total power**, *its reading will need to be multiplied by a factor of **three***.

In the above example, applying 'double-subscript' notation, the wattmeter is measuring:

$$P = \overline{E}_{AN}\overline{I}_A \cos \phi \quad \equiv \quad \overline{E}_{AN}\overline{I}_{AN} \cos \phi$$

Note that although the wattmeter is *actually* monitoring the line current $\left(\overline{I}_A\right)$, for a star-connected system, this current is exactly the same as the phase current $\left(\overline{I}_{AN}\right)$.

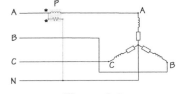

Figure 8.2

In other words, it is measuring the power of phase A-N of the load. So the *total* power of the balanced load will be this value, multiplied by 3. That is:

$$P = 3 \times \text{wattmeter reading}$$

'Three-wattmeter' method

The simplest method of measuring the total power of a **balanced** or an **unbalanced**, *three-phase three- or four-wire*, system is to use the '**Three-Wattmeter Method**'.

With this method, each wattmeter is connected with its **current coil** inserted into a line conductor, and

▶ for a *three-wire* system, one side of each **voltage coil** is connected to the corresponding line conductor, with the opposite end connected to form an 'artificial' star point, as shown in Figure 8.3 (left)

▶ for a *four-wire* system, one side of each **voltage coil** is connected to the corresponding line conductor, with the other end connected to the neutral conductor, as shown in Figure 8.3 (right).

Again, it is essential to carefully observe the wattmeter's polarity markings in order to avoid the instrument from reading downscale.

The power of the load is then *the sum of the three wattmeter readings.*

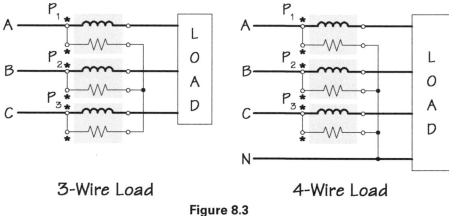

Figure 8.3

$$P = P_1 + P_2 + P_3$$

The fact that we can use *three* wattmeters to measure the total power of a *four-wire* three-phase system complies with what is known as '**Blondel's Theorem**', named in honour of a French engineer, André Blondel, who discovered that to measure the total power of any balanced *or* unbalanced polyphase system, *we can use 'one-less wattmeter than there are wires supplying that load'.*

Blondel's Theorem tells us that, for balanced *or* unbalanced three-phase loads, we can *use one-less wattmeter than there are wires supplying that load.*

Which leads us, rather neatly, on to the '**two-wattmeter**' method.

'Two-wattmeter' method

One of the most commonly used methods of measuring the power of a three-phase load is known as the 'Two-Wattmeter Method'. In accordance with Blondel's Theorem, this method will work for both balanced and unbalanced loads providing there are only three conductors supplying the load. Its connection is illustrated in Figure 8.4.

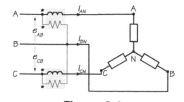

Figure 8.4

The 'Two-Wattmeter Method', then, has the following advantages:

▶ it works for both **star-** *and* **delta**-connected loads
▶ it works for both **balanced** *and* **unbalanced** loads
▶ for balanced loads, the results can be used to determine the load's **power-factor**

First of all, let's see *why* it works for an *unbalanced* system. And, of course, if it works for an *unbalanced* system, then it must also work for a *balanced* system!

The following proof applies to a three-wire star-connected system, but a similar proof can be derived for a delta-connected system.

The three-phase load, shown in Figure 8.4, consists of three different loads connected in star, and supplied from a three-phase supply. Once again, we will be using 'double-subscript notation', and taking the wattmeter's polarity markings into account.

By using *instantaneous* values of voltage and current (represented by lower-case letters, *e* and *i*), rather than r.m.s. values, we can ignore any phase differences because the product of an *instananeous voltage* and an *instantaneous current* will *always* give us the true power of the load.

Firstly, let's ignore the wattmeters, and consider the power of each phase:

▶ the power of the load in **phase A-N** is given by: $p_{AN} = e_{AN} i_{AN}$
▶ the power of the load in **phase B-N** is given by: $p_B = e_{BN} i_{BN}$
▶ the power of the load in **phase C-N** is given by: $p_C = e_{CN} i_{CN}$

So the total power of the three-phase load will be:

$$p = p_{AN} + p_{BN} + p_{CN} = u_{AN} i_{AN} + u_{BN} i_{BN} + u_{CN} i_{CN} \qquad \text{(equation 1)}$$

Now, let's turn our attention to what **wattmeter P_1** is reading:

$$p_1 = e_{AB} i_{AN} = (e_{AN} + e_{NB}) i_{AN}$$

At the same time, **wattmeter P_2** is reading:

$$p_2 = e_{CB} i_{CN} = (e_{CN} + e_{NB}) i_{CN}$$

Adding the two wattmeter readings together, we have:

$$p_1 + p_2 = (e_{AN} + e_{NB}) i_{AN} + (e_{CN} + e_{NB}) i_{CN}$$
$$= e_{AN} i_{AN} + e_{CN} i_{CN} + (e_{NB} i_{AN} + e_{NB} i_{CN})$$
$$= e_{AN} i_{AN} + e_{CN} i_{CN} + e_{NB} (i_{AN} + i_{CN}) \qquad \text{(equation 2)}$$

If we apply **Kirchhoff's Current Law** to the load's star point, we have:

$$(i_{AN} + i_{BN} + i_{CN}) = 0$$
$$(i_{AN} + i_{CN}) = i_{BN}$$

Substituting for $(i_{AN} + i_{CN})$ in equation (2), we have:

$$p_1 + p_2 = e_{AN} i_{AN} + e_{CN} i_{CN} + e_{NB} (i_{AN} + i_{CN})$$
$$= e_{AN} i_{AN} + e_{CN} i_{CN} + e_{NB} (i_{BN})$$
$$= e_{AN} i_{AN} + e_{CN} i_{CN} + (e_{NB} i_{BN})$$
$$= e_{AN} i_{AN} + e_{CN} i_{CN} + e_{NB} i_{BN}$$

This equation corresponds exactly to equation (1), which proves that **the sum of the two wattmeter readings is equal to the total power of the load**. That is:

$$\boxed{P = P_1 + P_2}$$

Although the Two-Wattmeter Method works well for *unbalanced* three-wire loads, it is often used to measure the power of *balanced* loads, such as three-phase induction motors, *where it offers the additional advantage of allowing us to determine the load's* **power factor**. We'll learn more about this a little later.

To show how the Two-Wattmeter Method works with a **balanced load**, Figure 8.5 shows the Two-Wattmeter Method for a balanced star-connected *resistive-inductive* load (typical of a motor), with a lagging power factor.

The current coil of wattmeter, P1, is inserted into line A, and its voltage coil is connected between lines A and B. The current coil of wattmeter, P2, is inserted into a second line – in this case, line C – and its voltage coil is connected between lines C and B.

(It would work equally as well if the wattmeter's current coil were inserted into line B, with its voltage coil connected between lines B and C.)

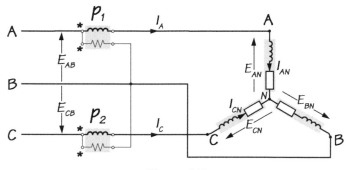

Figure 8.5

The wattmeters' **polarity marks** are *very* important and *must* be observed, and indicate the *sense* (*direction*) of the quantities being is being measured.

To make things absolutely clear, we've constructed the same phasor diagram twice (Figures 8.6 and 8.7) and, in each case, we've highlighted what is read by each wattmeter in bold. For any symmetrical three-phase supply, the line voltages always lead the corresponding phase voltages by 30° and, as the load is resistive-inductive, the phase currents (and, therefore, the line currents) lag the phase voltages by some angle, f ('phi').

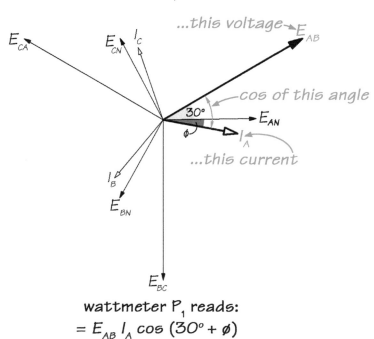

wattmeter P_1 reads:
$$= E_{AB} \, I_A \cos (30° + \phi)$$

Figure 8.6

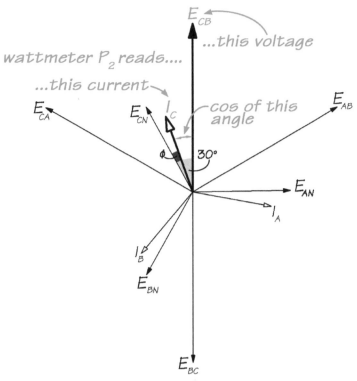

wattmeter P$_2$ reads:
$$= E_{CB}\, I_C\, \cos(30° - \phi)$$

Figure 8.7

From the phasor diagram shown in Figure 8.6, **wattmeter P$_1$**'s current coil reads the **line-current I$_A$**, and its voltage coil measures the **line voltage E$_{AB}$**, and the cosine of **the angle between them**, which is (30° + ϕ). So, wattmeter P1 reads:

$$P_1 = \bar{E}_{AB}\bar{I}_A \cos(30° + \phi)$$

Or, expressed in general terms:

$$\boxed{P_1 = \bar{E}_L\bar{I}_L \cos(30° + \phi)}$$

For **wattmeter P$_2$** (Figure 8.7), the current coil reads the **line-current I$_C$**, and the voltage coil measures the **line voltage E$_{CB}$** (the *reverse* of line voltage EBC – as measured *from* its polarity-marked terminal *to* its other terminal), and the cosine of the **angle between them**, which is (30° – ϕ). So, wattmeter P$_1$ reads:

$$P_2 = \bar{E}_{CB}\bar{I}_C \cos(30° - \phi)$$

Or, expressed in general terms:

$$\boxed{P_2 = \bar{E}_L\bar{I}_L \cos(30° - \phi)}$$

So, if we now *add* the two wattmeter readings together, we have:

$$P_{total} = P_1 + P_2 = \left[\bar{E}_L\bar{I}_L \cos(30° + \phi) + \bar{E}_L\bar{I}_L \cos(30° - \phi)\right]$$
$$= \bar{E}_L\bar{I}_L \left[\cos(30° + \phi) + \cos(30° - \phi)\right]$$

Expanding the content of the square brackets is beyond the level of mathematics required for this book, so we will need to accept it as being:

$$[\cos(30° + \phi) + \cos(30° - \phi)] \equiv 2\cos 30° \cos \phi \equiv \sqrt{3}\cos\phi$$

Giving:

$$P_{total} = \sqrt{3}\,\overline{E}_L\overline{I}_L\cos\phi$$

Which, of course, is the general equation for the *total power* of a balanced three-phase load, thus confirming that *the sum of the two wattmeter readings really is the total power of the load*.

When using the **Two-Wattmeter Method** of measuring power, should one of the wattmeters attempt to read 'downscale' (backwards) there are *two* possible reasons:

1 The *first* reason is that one or other of the wattmeter's current- or voltage-coils may have been incorrectly connected – *so it's essential to strictly observe the instrument's polarity markings* when connecting the wattmeter. If this is the case, then the wattmeter will need to be reconnected correctly, in accordance with its polarity markings.

2 The *second* reason is that the **power factor** of the load may be *less than* 0.5 (i.e. have a phase-angle *greater than* 60°), in which case wattmeter P_1, will give an negative reading (the cosine of an angle greater than 90° is negative). If this is the case, then the connections to one of that wattmeter's coils must be *reversed*, so that the instrument then reads 'upscale', and the resulting value must then be *deducted* from the reading supplied by the other wattmeter.

Worked example 1

A balanced three-phase load, with each phase having a power-factor of 0.8, draws a line current of 30 A from a supply with a line voltage of 400 V. What (a) are the readings of each of the two wattmeters, and (b) the total power of the load?

Solution

a The first step is to calculate the *phase angle* of each load impedance:

$$\angle\phi = \cos^{-1} 0.8 = 36.87°$$

Wattmeter, P_1:

$$\begin{aligned} P_1 &= \overline{E}_L\overline{I}_L\cos(30° + 36.87°) \\ &= 400\times30\times\cos 66.87° \\ &= 12000\times0.393 = 4716 \text{ W } (\textbf{Answer a.}) \end{aligned}$$

Wattmeter, P_2:

$$\begin{aligned} P_2 &= \overline{E}_L\overline{I}_L\cos(30° - 36.87°) \\ &= 400\times30\times\cos(-6.877°) \\ &= 12\,000\times0.993 = 11916 \text{ W } (\textbf{Answer a.}) \end{aligned}$$

b The *total power* of the load is the sum of the two wattmeter readings:

$$P = P_1 + P_2 = 4716 + 11916 + 16632 \text{ W } (\textbf{Answer b}).$$

We can confirm this answer by applying the general equation for the power of a balanced three-phase load:

$$P = \sqrt{3}\,\overline{E}_L\overline{I}_L\cos\phi = \sqrt{3}\times400\times30\times0.8 \approx 16\,628\text{W}$$

(slight difference due to rounding off figures)

Power factor of a balanced load

The **Two-Wattmeter Method** also provides us with a means of determining the power factor of a *balanced* three-phase load. Unfortunately to understand how to *derive* the equation for the power factor, we need a good understanding of trigonometric expansions, which (as stated earlier) is well beyond the level of maths used in this book, so we'll simply reproduce the equation itself.

$$\tan\phi = \sqrt{3}\left(\frac{P_2 - P_1}{P_2 + P_1}\right)$$

In fact, we have to follow *two steps*, to find the power factor. The first step is to **find the phase-angle** (symbol: ϕ), using the above equation, and, then, to *look up* its **cosine**.

When using this equation, it is common practice to assume P_1 as being the smaller of the two wattmeter readings.

Worked example 2

The readings of the two wattmeters used to measure the power of a balanced three-phase load are 2 kW and 1 kW respectively. What is the power factor of the load?

Solution

$$\tan\phi = \sqrt{3}\,\frac{P_2 - P_1}{P_2 + P_1} = \sqrt{3}\,\frac{2-1}{2+1} = \sqrt{3}\,\frac{1}{3} - 0.6$$

$$\phi = \tan^{-1} 0.6 = 30°$$

$$\text{power factor} = \cos\phi = \cos 30° = 0.866 \ (\textbf{Answer})$$

If we were to plot the ratio of the two wattmeters $\left(\dfrac{P_1}{P_2}\right)$, where P_1 is the smaller of the two readings, against power factor, the resulting graph would be as illustrated in Figure 8.8:

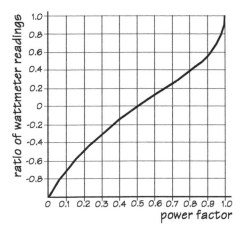

Figure 8.8

From this graph, it becomes clear that, for a power factor of **unity** (purely resistive load), the ratio P_1:P_2 will be **1**.

For a power factor of **0.5**, wattmeter P_1 will be reading **zero**, and P_2 will be **total power** of the load.

For a power factor of **zero** (theoretically purely inductive load), wattmeter P_1 will be reading a **negative value**, and wattmeter P_2, will be reading an **identical, but positive, value** (the sum of the two readings will be zero – as no true power is developed by a purely inductive load).

> **Remember:** A 'negative' reading occurs whenever a wattmeter's pointer tries to deflect 'downscale' (backwards). To find out what it is trying to read, it is necessary to reverse the connections to *either* its current coil *or* its voltage coil (not to both). It will then read 'upscale', but you must remember that this is then a ***negative value***.

Worked example 3

The Two-Wattmeter Method is used to measure the input power of a 400 V three-phase electric motor, and the wattmeter readings are 5 kW and −1.5 kW respectively. Calculate (a) the total input power, and (b) the power factor.

Solution

a Total Power $= P_1 + P_2 = 5 + (-1.5) = 5 - 1.5 = 3.5$ kW (**Answer a.**)

b wattmeter ratio $= \dfrac{P_1}{P_2} = \dfrac{1.5}{5} = -0.3$

Referring to the graph (Figure 8.8), a ratio of −0.3 corresponds to a power factor of approxmately **0.31** (Answer b). As the load is a motor, then it's likely that it's a **lagging** power factor.

Conclusion

Now that we've completed this chapter, we need to examine its **objectives** listed at its start. Placing a question mark at the end of each objective turns that objective into a **test item**. If we can answer those test items, then we've met the objectives of this chapter.

CHAPTER 9

Transformers 1: construction

Objectives

On completion of this chapter you should be able to:

1. recognise and describe the applications of the following types of transformer:
    a. power transformers
    b. distribution transformers
    c. isolation transformers
    d. power-supply transformers
    e. instrument transformers

2. identify, and describe the constructional features and functions of, the major components of a single-phase transformer.

3. describe how transformer terminals are labelled, according to British Standards, North American standards, and using the 'dot convention'.

4. explain the causes and solutions to 'transformer hum'.

5. describe the constructional features of autotransformers and variable output autotransformers.

6. describe how a mutual transformer can be used as an autotransformer.

7. describe typical applications for autotransformers.

Historical background

In the companion book, ***An Introduction to Electrical Science***, we learnt that the physicist Michael Faraday (1791–1867) conducted an experiment in which, by constantly *varying* the current through a coil, he discovered that he was able to induce a voltage into an adjacent coil through the process of **mutual induction**.

Although what Faraday had invented was a primitive **transformer**, he didn't use that term and, as there was no practical use for it at that time, it's unlikely he fully realised the importance of his discovery! For, as we will learn elsewhere in this book, *without transformers, electricity transmission and distribution would quite simply be impossible!*

Working quite independently, the American physicist Joseph Henry (1797–1878) made exactly the same discovery at around the same time, although it was Faraday who published his findings first. Quite properly, both men have been credited with the discovery.

A **transformer** is classified as an 'electrical machine': one that transfers electrical energy from one circuit (called the 'primary' circuit) to another (called the 'secondary' circuit), through inductively coupled conductors – i.e. the transformer's primary and secondary coils or, more properly, 'windings'.

The mechanical analogy of a transformer is a 'balanced beam', that is a horizontal beam supported on a single fulcrum.

Faraday's invention had a long way to go before it became the transformer that we would recognise today. In fact, the first useful 'transformer' was technically an **induction coil**, invented in 1836, by an Irish minister, the Reverend Nicholas Callan (1799–1864). Because a continuously *varying* primary current is necessary for its operation, his induction coil used a vibrating mechanical contact (which worked rather like an electric bell) to continuously interrupt the direct current (batteries being the only practical source of electrical energy in those days!) in the primary winding in order to continuously induce a voltage into the secondary winding.

Callan was also one of the first researchers to realise that the magnitude of the voltage induced into the secondary winding depended on the **turns ratio** of the two windings. For example, by winding more turns on the secondary winding, he found that the voltage induced into that winding could be many times that applied to the primary winding.

> The **induction coil** lives on, and is still used in motor vehicles to raise the 12-V d.c. provided by a lead-acid battery to the thousands of volts necessary to operate a petrol-engine's spark plugs. The mechanical interruption to the primary circuit used to be provided by means of a cam-driven pair of 'contacts' located in the vehicle's distributor head. These days, however, the contacts have been replaced by transistor switching, which eliminates the arcing problems associated with the older, mechanical, system.

Later, in Italy, a French inventor, Lucien Gaulard (1850–1888), working together with his British business partner, engineer John Dixon Gibbs (1834–1912), used induction coils to transmit electrical energy at a voltage of 2 kV from the town of Lanzo, before reducing that voltage to safe levels for use in homes and businesses in Turin about 40 km away. Used in this way, Gaulard's called his induction coils '**converters**', and this feat was only achievable because of their ability to produce high voltages.

After seeing examples of the Gaulard-Gibbs 'convertors' at the 1885 London Inventions Exhibition, the American industrialist, George Westinghouse (1846–1914), against the advice of his engineers, purchased an option on the patent and placed an order for several of them, together with a Siemens alternating-current generator from Germany.

With the introduction of a.c. generators, the need for a vibrating contact to continuously interrupt the primary circuit of an induction coil was no longer necessary and, in 1886, an American engineer, William Stanley (1858–1916), working for the Westinghouse Electric Corporation, redesigned the Gaulard-Gibbs 'transformer' and, in 1886, went on to the produce the first a.c. power system using the Siemens a.c. generator and Stanley's transformers in the town of Great Barrington, Massachusetts.

Meanwhile, in 1885, working quite independently of the Westinghouse Electric Corporation, a distribution transformer was patented by the Ganz Engineering Works in Hungary, having been designed by engineers Károly Zipernowski, Ottó Bláthy and Miksa Déri and, in fact, it was Ottó Bláthy who is credited with introducing the word '**transformer**' into the electrical vocabulary.

The single most important reason for adopting an alternating current (a.c.), rather than a direct current (d.c.) electricity transmission and distribution system, is that, thanks to transformers, *voltage levels in an a.c. system may be changed very easily and, more importantly, very efficiently*. And, as we shall learn, high voltages are absolutely *essential* for the transmission of electrical energy.

Changing voltage levels in d.c. systems is of course possible, but is technically far more complicated and far less efficient, compared with a.c. However, high-voltage d.c. (HVDC) transmission lines have less losses than the equivalent a.c. transmission lines, so high-voltage d.c. transmission lines are used for exceptionally long-distance transmission in countries such as Canada and Russia, or as d.c. 'links' that interconnect otherwise independent grid systems, such as those of the UK and France and the UK and Ireland.

A general 'rule of thumb' used in the electricity supply industry specifies a transmission distance of *one kilometre per kilovolt*. The UK's present maximum standard transmission voltage of 400 kV, first used in Sweden in 1952, therefore, enables electrical energy to be transmitted over a distance of around 400 km.

A **transformer**, then, is classified as an **electrical machine** that will transfer energy between two separate circuits, through mutual induction, and will either *increase* ('**step up**') or *decrease* ('**step down**') an a.c. voltage very easily and extremely efficiently – that is, with very little energy loss.

Schematically, a single-phase transformer is represented as shown in Figures 9.1 and 9.2.

Types of transformer

Transformers of the type we will be examining in this chapter are often referred to as '**static transformers**' in order to distinguish them from another type of transformer called a 'rotary transformer' employed for various specialist applications, such as the spinning recording-head drums used in video cassette recorders. Another widely used 'rotary transformer' is the induction motor, although it is not normally known by this term despite it being an accurate description.

In this book we will be examining static '**power-frequency**' transformers – i.e. those designed to operate within the 40–100 Hz frequency range which, of course, includes 'mains' supply frequency: i.e. 50 Hz in the UK and Europe and 60 Hz in North America.

Transformers used for high-frequency applications in radio and other fields are outside the scope of this book.

There are numerous ways by which transformers may be classified or categorized. **Power-frequency transformers** are, for example, often classified according to their *application*, as being:

▶ power transformers
▶ distribution transformers
▶ isolation transformers
▶ power-supply transformers
▶ instrument transformers.

The transformers described above are correctly known as '**mutual transformers**' (although this term is rarely used in the field) because, of course, they work on the principle of **mutual induction** between their magnetically coupled windings (coils).

Mutual transformers (except for current transformers, described later) are also sometimes called '**constant-voltage transformers**', which simply means that there is no significant difference between their 'no-load' voltage and their 'full-load' voltage.

A completely different type of transformer is the single-winding '**autotransformer**', which works on a somewhat different principle, and which we will also examine separately in a later chapter.

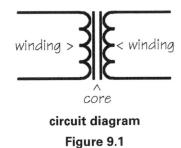

winding >   < winding

^
core

circuit diagram
Figure 9.1

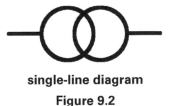

single-line diagram
Figure 9.2

9

Power transformers and distribution transformers

'**Power transformers**' are used in the electricity transmission and primary distribution systems (in the UK: voltage levels between 400 kV and 33 kV), and '**distribution transformers**' are used in the electricity secondary distribution systems (in the UK: 11 kV and below). They vary tremendously in physical size according to their operating voltages and apparent power ratings, as illustrated in Figures 9.3 to 9.6. Practically all power and distribution transformers are enclosed in welded-steel tanks and immersed in oil, which acts to increase their insulation levels and to provide cooling.

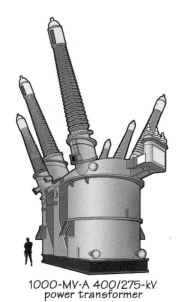

1000-MV·A 400/275-kV
power transformer

Figure 9.3

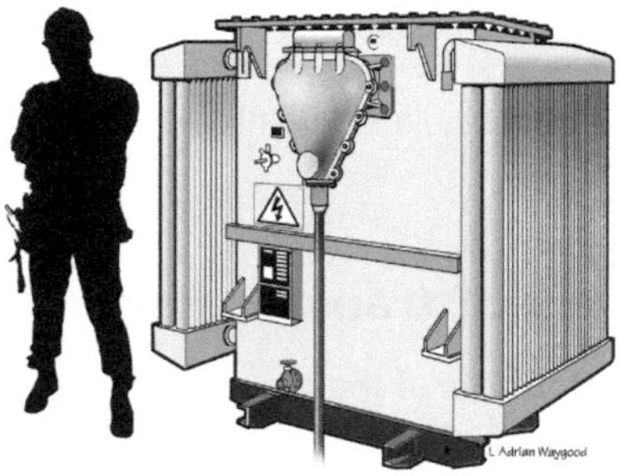

500-kV·A 11-kV/400-V
distribution transformer

Figure 9.4

200-kV·A 11 kV/400-230 V
pad-mounted distribution transformer

Figure 9.5

50-kV·A single-phase 11 kV/230 V
pole-mounted distribution transformer

Figure 9.6

The rated '**capacity**' of a transformer is determined by its rated secondary voltage and its rated secondary current, and determines the maximum load that can be supplied on a continuous basis. By 'capacity' we mean the *product* of voltage and current, which, in a.c. systems, is termed '**apparent power**' expressed in **volt amperes**. Transformer capacities are *not* expressed in watts because the manufacturer has no means of knowing the power factor of the load they will be supplying.

In the UK, the capacities of power transformers used in the **transmission system** are typically within the following ranges:

▶ 400:275 kV ratio: between 1000 MV·A and 500 MV·A
▶ 400:132 kV ratio: up to 240 MV·A
▶ 275:132 kV ratio: between 240 MV·A and 120 MV·A.

For power transformers in the **primary distribution system**:

▶ 132:33 kV ratio: up to 120 MV·A.

For distribution transformers (11 kV/400–230 V) used in the **secondary distribution system**:

▶ ground-mounted transformers: between 1000 kV·A and 200 kV·A
▶ pad-mounted transformers: up to 200 kV·A
▶ pole-mounted transformers: up to 200 kV·A.

Countries such as Canada, the United States, and those located in Central Asia which, due to their vastness, have particularly long transmission lines, use even higher transmission voltages and correspondingly higher transformer capacities.

Currently (2015), the world's highest a.c. line voltage is 1150 kV, used by a 430 km transmission line running between the cities of Ekibastuz and Kokshetau, in the Republic of Kazakhstan in Central Asia. And, currently, the largest three-phase transformer has a capacity of 1630 MV·A.

Isolation transformers

The primary and secondary windings of *all* mutual transformers are electrically isolated from each other, being linked only by a common magnetic field via their silicon-steel cores. '**Isolation transformers**' (Figure 9.7), however, are designed and used *specifically* for isolation purposes, rather than to change voltage levels. For this reason, most isolation transformers have 1:1 transformation ratios – that is, their secondary voltage is *exactly* the same as their primary voltage.

Small, portable, isolation transformers are widely used, for example, on construction sites (where they are known as 'earth-free' transformers) to isolate electrically operated hand tools and, therefore, protect their users from the earthed supply system (they usually also offer a lower and, therefore, safer secondary voltage). In the home, small isolation transformers are used in shaver sockets, to isolate users in the damp environment of a bathroom, from the earthed electrical supply throughout the rest of the house.

on-site power-tool
transformer

shaver socket
isolation transformer

Figure 9.7

Power-supply transformers

These are small, air-cooled ('dry'), general-purpose, transformers (Figure 9.8) used to reduce the residential mains' supply down to whatever voltages are necessary to operate various residential applications such as door bells, low-voltage lighting, etc. These transformers often have multiple tapped secondary windings, in order to provide a useful range of different secondary voltages.

power-supply transformer

Figure 9.8

165

Instrument transformers

'**Instrument transformers**' (Figure 9.9) are so called because their secondary outputs are used to operate **measuring instruments** (e.g. ammeters, voltmeters, wattmeters) and **protection relays**.

They enable voltage or current measurements to be made safely by electrically isolating those instruments from high-voltage or heavy-current systems while, at the same time, reducing the values of those voltages or currents to safe levels to avoid shock hazard to the users.

They are also used to provide the inputs to protection relays used in high-voltage substations. Protection relays are used to monitor high-voltage transmission/distribution lines and their associated equipment, for faults. In the event of a fault, these relays then trip the relevant circuit breakers in order to disconnect that fault from the system.

Instrument transformers are either '**voltage transformers**' (**VT**s) or '**current transformers**' (**CT**s):

▶ **Voltage transformers** (usually called '**potential transformers**' in North America) are used to electrically isolate measuring instruments (voltmeters, the voltage-coil of wattmeters, etc.) and protective relays from the high-voltage system which they monitor, and to reduce those voltages to safe levels.

▶ **Current transformers** are used to isolate measuring instruments (ammeters, the current-coil of wattmeters, etc.) and protective relays from high-voltage systems or from heavy-current systems, and to reduce those currents to safe levels.

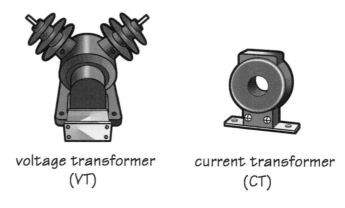

voltage transformer
(VT)

current transformer
(CT)

Figure 9.9

Construction of mutual transformers

A basic single-phase **mutual transformer,** comprises *three* main components: a *magnetic circuit*, more commonly called a **core**, around which are wound two *coils*, more properly called **windings**. One winding is connected to the *supply*, and is termed the **primary winding**, while the other winding is connected to the *load*, and is termed the **secondary winding**.

It's important to understand that the terms 'primary' and 'secondary' do *not* relate to whether the windings are 'high-voltage' or 'low-voltage' windings, but to *how they are connected* – i.e. the primary winding is, by definition, always connected to the *supply*, and the secondary winding is always connected to the *load*.

In the case of power and distribution transformers, these assembled components are usually immersed in oil, within a welded-steel tank. The purpose of the oil is twofold:

(1) it limits the operating temperature of the transformer by providing cooling, and (2) it significantly improves the insulation of the transformer's internal components, thereby reducing its overall physical size and giving it a smaller 'footprint', enabling it to occupy less space in a substation.

The core

The function of a transformer's **core** is to provide a *low-reluctance/high permeability* **magnetic circuit** by which magnetic flux, created by the magnetomotive force in the primary winding, can efficiently couple with the secondary winding, and thereby induce a voltage into that winding through *mutual induction.*

So what are the essential requirements of the material from which a transformer's core is manufactured? At this point, you may wish to review the chapter on ***Magnetic Circuits*** in the companion to this book: ***An Introduction to Electrical Science***. They are to provide a:

▶ *low-reluctance/high permeability* magnetic circuit, in order to maximise the flux density within the core
▶ *low-remanence/low-coercivity/low-area* hysteresis loop: to minimise the energy loss per magnetisation/demagnetisation cycle; this energy loss being termed its *'hysteresis loss'*
▶ *high resistivity* conducting path: to minimise any 'eddy currents' (circulating currents) resulting from undesirable voltages induced into the core; eddy currents are further minimised by manufacturing the core from laminations.

Transformer cores are manufactured from a **silicon-iron** alloy (iron with around 3% silicon content), which is more generally, but inaccurately*, termed '**silicon-steel**', '**electrical steel**', '**transformer steel**', or by some trade name such as '**Stalloy**'. This ferromagnetic alloy is specifically designed to have all of the characteristics described above.

*'Steel' is an alloy of iron and carbon, whereas transformers are actually manufactured from an alloy of iron and silicon. So the widely used term 'silicon-steel' is technically incorrect.

The silicon content of the iron acts to both reduce the core's hysteresis losses and to increase its resistivity by a factor of around 4.5, which acts to reduces the magnitude of any circulating currents ('eddy currents') that result from voltages induced into the core.

There are two general categories of silicon-steel, termed *'grain-orientated'* steel and *'non-orientated'* steel:

▶ **Grain-orientated steel** has a silicon content of around 3%, and is manufactured in such a way that its magnetic properties are optimised *along the direction of its grain,* enabling its flux density to be increased by as much as 20% in that direction.
▶ **Non-orientated steel**, with a silicon content of 2–3.5%, has a randomly orientated grain with similar magnetic properties in *all* directions, but the resulting flux density can be significantly lower than for grain-orientated steel.

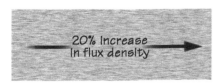

grain-orientated steel   non-orientated steel

Figure 9.10

Laminations

From the point of view of the varying flux, the **core** of a transformer is simply another winding and, therefore, it will induce a voltage into it. If the core was manufactured from a *solid* piece of silicon-steel, then it would behave as a heavy, short-circuited, single winding, and a large current would circulate around it.

This circulating current, known as an '**eddy current**' or '**Foucault current**', whose existence was discovered in the mid-nineteenth century by the French physicist, Léon Foucault (1819–1868), causes the temperature of the core to rise, resulting in energy losses known as '**eddy-current losses**'.

To reduce the eddy-current losses (as well as for ease of assembly), transformer cores are *not* solid but, instead, manufactured from **laminated sheets** – with each lamination lightly insulated from its neighbours. This is achieved by coating one side of each lamination, before assembly, with a thin layer of insulating material, such as varnish (smaller transformers), a dried flour/chalk solution (larger transformers), or by simply allowing a thin layer of corrosion to form on the surface of each lamination prior to assembly.

Laminations have the effect of significantly reducing eddy-current losses.

Let's suppose the core of a transformer were solid, rather than laminated. And let's suppose that the voltage induced into that core is, say, 10 V, and the resistance of the core is, say, 1 Ω. The eddy-current loss, therefore, would be:

$$P = \frac{U^2}{R} = \frac{10^2}{1} = 100 \text{ W}.$$

Now, to keep the maths simple, let's suppose a transformer's core is made up of just *four* laminations. The voltage induced into each of the four laminations would then be *one-quarter* of the voltage induced into a solid core of equivalent size. At the same time, the cross-sectional area of each lamination will be just *one-quarter* that of the solid core, making its resistance *four-times* greater. The eddy-current loss, per lamination, therefore, would be:

$$P = \frac{U^2}{R} = \frac{2.5^2}{4} = 1.56 \text{ W}.$$

So, the total eddy-current loss for four laminations would be:

$$\text{total eddy-current loss} = 4 \times 1.56 = 6.25 \text{ W}.$$

In this example, the total eddy-current loss in a core comprising four laminations, is *one-sixteenth* of the eddy-current loss for a solid core. Or, to put it another way, for a laminated core, **the total eddy-current loss is inversely proportional to the square of the number of laminations.**

So the eddy-current loss per lamination is so small, that the sum of those individual losses is *far* less than the losses that would occur in a solid core of equivalent cross-sectional area!

Theoretically, as shown above, **eddy-current losses** should be *inversely proportional to the square of the number of the laminations*.

Although this figure is based on sound electrical theory, for reasons that are not yet fully understood, measured values of eddy-current losses are actually somewhat higher – by a factor of up to 3.5 in some cases. Despite this, the actual eddy-current losses in a laminated core are still substantially lower than those for an equivalent solid core.

Increasing the number of laminations will enable the eddy-current losses to be reduced to *any* desired level. In practice, however, reducing the thickness of each lamination below around 0.25 mm is undesirable because, taking the laminations' insulation thickness into account, the overall cross-sectional area (i.e. steel plus insulation) of the core will start to become significantly larger than its *effective* cross-sectional area (i.e. the steel alone), resulting in the core taking up too much space without offering any corresponding benefits.

Laminated cores may be manufactured *either* from **stamped laminations** that are simply stacked and clamped together (Figure 9.11, left), *or* from a long **length of silicon-steel** that is first wound into a circular roll, then 'squared off' using an hydraulic press (Figure 9.11, right). These cores are called '**stacked cores**' and '**wound cores**', respectively.

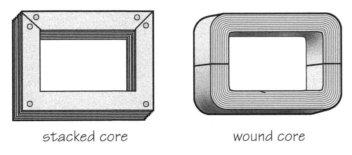

stacked core   wound core

Figure 9.11

Stacked cores must have a removable top 'yoke' (horizontal member) to enable the windings to be installed. **Wound cores** are split horizontally to enable the windings to be installed, after which they are clamped back together.

Stacked cores are relatively easy to manufacture in *any* physical size, and are used for applications ranging from tiny transformers used on electronic printed-circuit boards, right up to enormous power transformers.

Wound cores, on the other hand, are much more complicated to manufacture and assemble, and are almost exclusively manufactured as secondary-distribution transformers. Unlike stacked cores, wound cores are practically always manufactured from 'grain-orientated' transformer steel, in which the metal's grain is orientated *in the same direction as the lines of magnetic flux*, thus significantly improving the flux density within that core. Furthermore, wound cores have lower magnetic leakage, because the flux doesn't have to negotiate its way around any 'sharp' corners.

Figure 9.12 illustrates the manufacturing process for a wound transformer core.

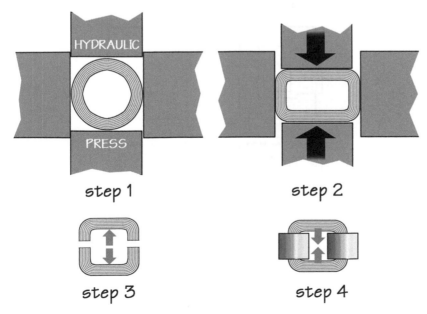

step 1   step 2

step 3   step 4

Figure 9.12

Core designs

The *vertical* sections of a transformer core are called '**limbs'**, and the *horizontal* sections are called '**yokes'**, and the openings between limbs and yokes are termed '**windows'**, as illustrated in Figure 9.13.

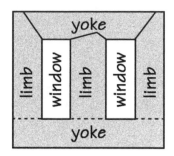

Figure 9.13

Most transformers are manufactured from *stacked* cores while some (usually small distribution transformers) are manufactured from *wound* cores. Figure 9.14 compares stacked and wound core three-phase transformer construction.

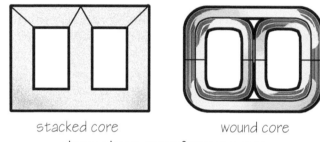

three-phase transformer cores

Figure 9.14

There are *two* common designs for transformer cores, termed a '**core-type'** core and a '**shell-type'** core, as shown in Figure 9.15:

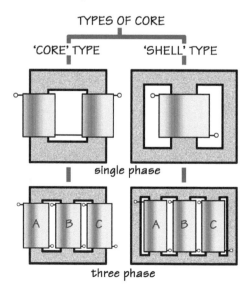

Figure 9.15

As can be seen from Figure 9.15, if the core's outer limbs are *occupied* by windings, then it is a 'core-type' core; if the outer limbs are *unoccupied*, then it is a 'shell-type' core.

The centre limb of a shell-type core typically has twice the cross-sectional area of either of its outer limbs, because *all* of the flux passes through the centre core whereas just *half* of the flux passes through each of the outer limbs. So, for shell-type cores, this has the effect of reducing the core losses compared with the core-type core.

With a single-phase '**core-type**' transformer core, the primary and secondary winding assemblies are distributed equally between the outer limbs. With the single-phase '**shell-type**' core, the primary and secondary winding assemblies are both placed around the centre limb, which results in *lower* flux leakage (i.e. 'wasted' flux that doesn't link both primary and secondary windings).

As can be seen in Figure 9.15, with a three-phase '**core-type**' transformer core, the three phase-winding assemblies are placed around each of the three limbs. With a three-phase '**shell-type**' core, the three phase-winding assemblies occupy the three centre limbs, and the outer limbs are unoccupied.

In the UK, **core-type** transformers are used for *all* power-system applications, while the shell-type core is only used for special applications. However, the shell-type core is generally considered to be the more efficient design, and is commonly used in North America and some parts of Europe for very large power transformers because the height of their limbs can be lower than a corresponding 'core-type' design. This is because their yokes can be narrower than their limbs.

In power and distribution transformers, the cross-section of a stacked core's limbs is made as near to circular as possible in order to fill the space occupied by the windings in order to minimise magnetic leakage. This is achieved by building up stacks of laminations of different widths, as shown in Figures 9.16 and 9.17.

Students are frequently confused about the flux distribution within a **three-phase transformer core**. This is best explained by considering the *four-limbed* core illustrated in Figure 9.18, below. It is reasonable to suppose that the flux generated by each winding must return through the centre core, but this is not necessarily the case!

As each winding current is displaced by 120 electrical degrees then so, too, is the flux generated by each winding. As, theoretically, each flux must 'return' through the centre limb of the core, the flux in that limb must be the phasor-sum of the three fluxes. Well, as we already know, the phasor-sum of three *identical* currents and, therefore, their fluxes, when displaced by 120 electrical degrees; must be *zero*! So there is actually no need for the central limb – which is why three-phase transformers require just three limbs. There is an analogy here, with the fact that no neutral return is required for a three-phase balanced star-connected load.

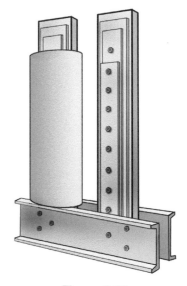

Figure 9.16

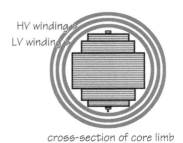

HV winding
LV winding

cross-section of core limb

Figure 9.17

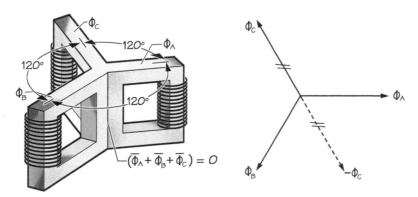

Flux in centre limb is vectorial sum of three fluxes = zero

Figure 9.18

For a transformer supplying an *unbalanced* load, the flux distribution within a three-phase transformer can be rather complicated, and is well beyond the scope of this book. Despite this, three limbs are still perfectly adequate.

The windings

In the electricity industry, coils are more properly termed, '**windings**'. The **primary** and **secondary windings** of most transformers are usually made from insulated *copper* conductors, of sufficient circular or rectangular cross-sectional area to continuously carry the transformer's rated primary and secondary currents without overheating. *Aluminium* is sometimes used instead of copper but, because it requires a larger cross-sectional area to provide the same current rating as copper, its use is restricted to physically large transformers.

The windings' **insulation** must be able to withstand the transformer's primary and secondary rated-voltages at the design operating temperature. On small, dry, transformers, both windings are usually insulated with varnish; on large, oil-filled, power transformers, oil-impregnated paper is widely used to insulate the high-voltage windings.

Windings are not wound directly onto the core, but are prefabricated, having first been machine-wound around tube-shaped fibre 'formers', and the completed assemblies are then placed around individual limbs before the top yoke is installed.

For the purpose of clarity, circuit diagrams usually show the primary and secondary windings installed *separately*, on *different* limbs but, in practice, they are always wound *around each other, concentrically, with their turns distributed equally between different limbs*. This is done to ensure maximum coupling by minimising the amount of flux **leakage**.

In the case of 'core-type' transformers, the lower-voltage windings are usually placed innermost with the higher-voltage windings outermost, with half of each winding-assembly distributed around each limb. This arrangement is termed a '**concentric**' or '**tubular**' type, and is illustrated in Figure 9.19 (left).

In the case of 'shell-type' cores, the high-voltage and low-voltage winding assemblies are often arranged, alternately, as 'ring bundles' around the centre limb. This arrangement is termed a '**sandwich**' or '**pancake**' type, and is illustrated in Figure 9.19 (right).

Both types of arrangement are intended to guarantee maximum coupling/minimum leakage of the flux between the high- and low-voltage windings.

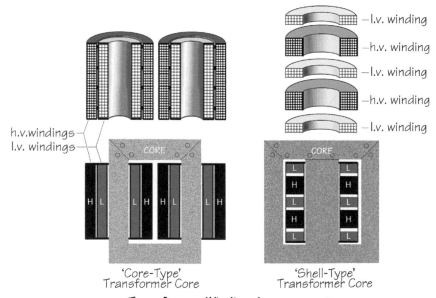

Transformer-Winding Arrangements

Figure 9.19

Some distribution transformers (both single- and three-phase) use windings comprising broad rolled *sheets* of aluminium, that have been interleaved with oil-impregnated paper insulation ('swiss-roll' style – rather like a capacitor), instead of coils of *wire*. Each layer of this aluminium roll is equivalent to one turn of a 'wire' conductor, with the innermost and outermost layers representing the equivalent of the 'start' and 'finish' of the 'winding', and which are connected to their respective terminals.

Tappings and tap-changing mechanisms

Power transformers often have to compensate for minor variations in their supply (primary) voltage in order to maintain their secondary voltage output within the statutory limits. This is achieved by enabling minor changes to be made to the turns-ratio of the transformer and, to allow this, the high-voltage winding is provided with a number of '**tappings**' or '**taps**' (connections) towards one end of that winding, which are connected to a **tap-changing mechanism**.

Tappings are always provided on the *high-voltage winding* so that the tap-changing mechanism's contacts need only handle smaller currents than would be the case if the tappings were installed on the low-voltage winding.

Secondary distribution transformers generally use '**off-load**' **tap-changing mechanisms**, which, as the name suggests, can only be operated with the transformer disconnected from its load. This is because the mechanism (essentially a rotary switch) breaks the high-voltage circuit as it moves from one tap setting to the next. The mechanism is operated using a rotating handle accessible from outside the transformer (see Figure 9.6) which, for the safety of the operator, can only be operated with the transformer *electrically isolated from the system*.

Transmission and primary distribution transformers use '**on-load**' **tap-changing mechanisms** which, as its name suggests, can operate with the transformer connected to the system and supplying its load. On-load tap-changers use a 'make-before-break' switching arrangement, which ensures that there is no break in the high-voltage circuit whenever the tap-changer operates. The switching mechanism is motor-driven, and the complete mechanism is housed in a separate oil-filled chamber (see Figure 9.22), which isolates any contaminated oil (caused by arcing at the contacts) from the oil in the main tank and allows maintenance to take place without having to drain the main tank. On-load tap changers are controlled automatically by **automatic voltage regulator** relays. These relays, located inside the substation building, monitor the secondary voltage and maintain it within the statutory limits by adjusting the tap settings.

Identifying high-voltage and low-voltage windings

In the case of power and distribution transformers, it's easy to recognise the high-voltage terminals and the low-voltage terminals, due to the physical size of their bushings. 'Bushings' are hollow insulators, through which a transformer's high- and low-voltage terminals are connected to the external circuits. High-voltage bushings are significantly longer than low-voltage bushings because they have to withstand greater voltages.

But with smaller power-supply transformers, of the type illustrated in Figure 9.8, its visually less obvious which terminals supply the higher-voltage windings, and which supply the lower-voltage windings.

As we shall learn in the next chapter, the lower-voltage windings carry *a larger current than the higher-voltage windings* as well as having *fewer turns*. Therefore, the lower-voltage winding has a *shorter length* of conductor with a *larger cross-sectional area* than the corresponding higher-voltage winding and, so, has a *lower resistance*.

It is, therefore, possible to identify a **de-energised** and **disconnected** transformer's lower-voltage and higher-voltage terminals by using a multimeter to measure the resistance of the windings. For the reasons explained in the previous paragraph, *the lower-voltage winding's terminals will offer the lower resistance* – so the 'golden rule' is:

lower resistance = *lower*-voltage windings
higher resistance = *higher*-voltage windings

If we are able to access the transformer internally, then the lower- and higher-voltage windings can be easily recognised, visually, because:

▶ Lower-voltage windings carry higher currents, so have the *larger cross-sectional area conductor*, but are *lightly insulated*.
▶ Higher-voltage windings carry lower currents, so have the *smaller cross-sectional area conductor*, but are *heavily insulated*.

Terminal markings

In the UK, the relevant British Standard for terminal markings involves a combination of *letters* together with *numbered subscripts*, as follows:

▶ **High-voltage terminals** are labelled with *upper-case* letters (e.g. **A**, **B**, *or* **C**) and **low-voltage terminals** are labelled with corresponding *lower-case* letters (**a**, **b**, or **c**), where each letter identifies a **line** terminal. A neutral high-voltage terminal is labelled with an upper-case **N**, and a neutral low-voltage terminal is labelled with a lower-case, **n**.
▶ One end (the 'start' terminal) of each winding is identified with *odd-numbered* subscripts (**1**, **3**, **5**, **7**, etc.), while the opposite (the 'finish' terminal) end of each winding is identified with *even-numbered* subscripts (**2**, **4**, **6**, **8**, etc.).

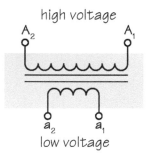

high voltage

A_2 A_1

a_2 a_1

low voltage

Figure 9.20

British and European standards specify that, *when viewed from the high-voltage side* of the transformer, the 'A_1' terminal should be to the *left* and, *when viewed from the low-voltage side* of the transformer, the 'a_1' terminal should be to the *right*.

For example, a two-winding, single-phase distribution transformer's high-voltage winding terminals may be marked as A_1–A_2, and its low-voltage winding terminals may be marked a_1–a_2, as illustrated in Figure 9.20.

At any given instant, the polarity (positive or negative) of low-voltage terminal, 'a_2', is *exactly* the same as the polarity of the high-voltage terminal, 'A_2'.

For a three-phase transformer, the other windings will then be identified as B_1–B_2, C_1–C_2, etc.

For the United States and Canada, the relevant standards for terminal markings also involve a combination of *letters* combined with *numbered subscripts*, but as follows:

▶ **High-voltage** terminals are identified with an *upper-case* letter (**H**) and **low-voltage** terminals are identified with an *upper-case* letter (**X**).
▶ One end (the 'start' terminal) of each winding is identified with *odd-numbered* subscripts (**1**, **3**, **5**, **7**, etc.), while the opposite (the 'finish' terminal) end of each winding is identified with *even-numbered* subscripts (**2**, **4**, **6**, **8**, etc.).

The North American standards also specify that, *when viewed from the **low-voltage** side* of the transformer (i.e. with the high-voltage terminals on the *opposite* or *far* side of the transformer), the high-voltage terminal 'H_1' must be to the left. The corresponding low-voltage terminal, when viewed from the low-voltage side of the transformer, is labelled 'X_1' but, unlike British practice, this can be to the right *or* to the left according to the capacity of the transformer. We'll learn more about the significance of this, later, in the chapter on transformer connections.

At any given instant, the polarity (positive or negative) of low-voltage terminal, 'X_1', is *exactly* the same as the polarity of the high-voltage terminal, 'H_1'.

A typical North American single-phase secondary distribution transformer has a single high-voltage winding, and a *pair* of identical low-voltage windings. So the transformer's high-voltage winding terminals are marked H_1–H_2, and its low-voltage windings' terminals are marked X_1–X_2 and X_3–X_4 respectively, with terminals X_3 and X_4 linked together as shown in Figure 9.21.

The reason for showing a *pair* of secondary windings in Figure 9.21 is because residential supplies in North America are dual-voltage (240/120 V) systems, with 240 V available between the outer terminals X_1 and X_4, and 120 V available between terminals X_1 and X_2 and between X_3 and X_4.

This type of connection is termed a 'split phase' system, and we shall return to this topic in more detail in a later chapter on transformer connections.

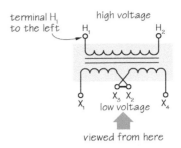

Figure 9.21

Dot convention

An alternative method of identifying a transformer's terminals is by using what is generally termed as the '**dot convention**' or '**dot notation**'. This method is widely used, particularly with instrument transformers and on circuit diagrams, to indicate a transformer terminal's instantaneous **polarity**.

'Polarity' will be dealt with in detail in a later chapter but, essentially, it becomes very important whenever transformers have to be connected in parallel with each other, or when instrument transformers are used to operate instruments, such as wattmeters, where the 'directions' of the voltage and current applied to their terminals is important in order to produce a correct reading.

The 'dot convention' uses a **dot** to indicate the high-voltage terminal otherwise identified as 'A_2' using the British system, or 'H_1' using the North American system.

A *second* dot is then applied to whichever of the low-voltage terminals assumes the same polarity at the same instant. In other words, it identifies the terminal otherwise identified as 'a_2' using the British system, or 'X_1' using the North American system.

At any given instant, the polarities (positive or negative) of all terminals identified with a dot, are identical.

We will learn more about the dot convention in the chapter dealing with transformer connections.

Operating transformers at different frequencies

While on the topic of UK versus North American transformers, it's important to know that a transformer designed to operate at a frequency of 60 Hz *cannot* be operated at 50 Hz at its full rated voltage, because it will overheat and cause damage to its insulation. On the other hand, a transformer designed to operate at a frequency of 50 Hz can operate quite normally at 60 Hz at voltages of up to 120% of its nameplate value. We will examine this in more detail in a later chapter.

Insulation and cooling of power/distribution transformers

Power and distribution transformers are sealed within welded sheet-metal tanks and immersed in oil, which acts both as an *insulator* and as a *coolant*.

These oils are mainly either **mineral-based oils**, or **synthetic liquids** which are typically silicon based, and are normally flame retardant. Unfortunately, mineral oils produce poisonous substances as they oxidize, and their disposal and cleanup after any leakage or spills is difficult and harmful to the environment. As a result, we are now starting to see the introduction of food-grade **vegetable oils**, such as olive oil, soya-bean oil, sunflower oil, and coconut oil. Like mineral oils, vegetable oils have high dielectric strengths, high flash points but, unlike mineral oils, they are *biodegradable*.

The world's first transformer to use **vegetable oil** as an insulator/coolant was commissioned by Siemens in Baden-Würtenberg, in 2014. Vegetable oil has the advantage of being biodegradable and, therefore, environmentally friendly.

The dielectric strength of these oils is significantly higher than that for air, which means that internal components at different potentials can be located very much closer together and, therefore, reduce the area, or 'footprint', occupied by the transformer.

Until well into the 1970s, transformer coolants (together with those used in other electrical equipment, such as power-factor correction capacitors) were often manufactured from oils containing **polychlorinated biphenyls** (**PCB**s), which have since been determined to be extremely hazardous to health: in particular being responsible for certain types of cancer. This initiated a massive programme, worldwide, to replace all PCB coolants with the safer coolants in use today.

All transformers, regardless of their size, become hot when loaded. Transformer oils have a much higher **specific heat capacity** (a measure of the quantity of energy the oil can absorb) than air. This, together with their low **viscosities** (i.e. their ability to flow), make them a far more efficient cooling medium.

For 11 kV/400–230 V distribution transformers, the liquid itself is cooled by natural convection between the windings and the side of the tank, or via fins or tubes welded to the sides of the tank.

In the case of power transformers, the natural convection of the coolant is usually reinforced (once its temperature reaches a pre-determined level) by pumping the coolant through large external heat-exchangers which, themselves, are then cooled by natural or forced air circulation, or even by water circulation.

Figure 9.22, below, shows a typical oil-filled three-phase primary distribution (33/11 kV) transformer, fitted with tank-mounted heat exchangers through which the oil circulates by natural convection. The transformer tank is completely filled with oil, with

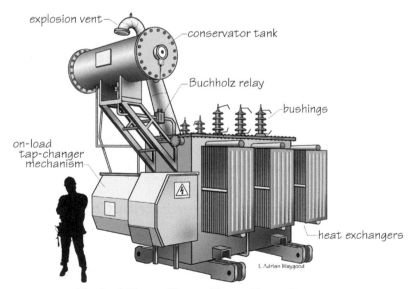

Typical Three-Phase Power Transformer

Figure 9.22

its level being maintained by a cylindrical-shaped **conservator tank** mounted above the transformer.

The conservator tank is typically around half-full of oil, allowing the oil to expand or contract with any temperature change. Air drawn into the conservator tank when the oil contracts is dried by fitting the external air vents with silica-gel 'breathers'. Silica gel is a granular chemical that absorbs water vapour; when dry it assumes a *blue* colour and, when saturated, it assumes a *pink* colour – thus indicating when it should be replaced.

Minor faults within the transformer (e.g. faults between adjacent turns of the same winding) breaks down the oil in the vicinity of the fault and generates gas, which collects in a **gas-detection ('Buchholz') relay** as it bubbles upward through the pipe leading towards the conservator tank. As this gas displaces the oil inside this relay, a float containing a mercury-switch pivots downwards with the falling oil level, triggering an alarm inside the substation. Any excessive loss of oil (e.g. due to a leak) will cause a second float to initiate a disconnection of the transformer, as will a more serious internal fault, which usually causes a surge of oil through the relay, which disturbs the floats.

A catastrophic internal fault will not only cause the Buchholz relay to disconnect the transformer, it may also result in a frangible ('burstable') disc sealing an explosion vent to break into fragments, releasing the internal pressure and protecting the transformer tank from serious damage.

Transformer noise

Ask any old-and-bold electrical engineer why a transformers hums, he'll probably tell us that it's because 'it doesn't know the words'!

The *actual* reason, though, is that transformer noise is mainly due to a phenomenon called '**magnetorestriction**', whereby the cyclic variations in the magnetic flux within that core cause tiny changes in the linear dimensions of the core's lamination.

Magnetorestriction causes the transformer's core to expand/contract at *twice* the supply frequency together with *multiples* ('harmonics') of that frequency. We can think of a transformer core, therefore, as behaving like a giant loudspeaker producing a continuous humming or buzzing noise, which is well within a human's audible frequency range, and extremely irritating.

In residential areas an excessively loud and continuous hum from distribution transformers has resulted in numerous complaints from consumers, in some cases with legal action being taken against the electricity network company. Accordingly, if it is thought that the noise from a newly installed transformer is likely to be excessive, then action *must* be taken to remove the problem. This includes:

▶ siting the transformer away from residential buildings
▶ concealing the transformer behind a wall or a row of trees
▶ enclosing the transformer in a sound-proof housing
▶ specifying a transformer with a lower-than-normal noise standard.

All manufacturers provide information on the noise level of their transformers, expressed in **decibels** (**dB**), and that noise level should always be *lower* than the *ambient noise level* in areas where humans live or work. For example the ambient noise level of a residential property at night is around 25–45 dB, whereas that of a factory can be as high as 80–90 dB.

Although we are *not* going to study the decibel in this book, it's important to understand that this unit represents a *logarithmic* scale, which means that *a change in level of just 3 dB* results in a *doubling or halving of the noise level!* So if, for example, we can reduce the noise level of a particular transformer from, say, 40 dB to 37 dB, then we will reduce its noise level by half!

177

Autotransformers

The term, 'auto', in '**autotransformer**', does *not*, as is often assumed, mean *'automatic'*. Rather, in this context, the word is derived from the Greek prefix, *'autos'*, meaning 'self'. So we use the term 'autotransformer' to describe a transformer having a *single* coil that *'induces a voltage into itself'* in order to produce a secondary voltage.

A 'true' **autotransformer** is a type of transformer that uses a *single tapped winding* rather than the two *separate* and *electrically isolated* windings used by mutual transformers.

However, it is also perfectly possible to configure a **mutual transformer** so that it can also operate as an autotransformer, by simply connecting its *two separate windings* in series with each other. While this configuration allows the mutual transformer to *operate* as an autotransformer, it is not a 'true' autotransformer in the sense of it having a single tapped winding. Furthermore, while the current rating of an autotransformer's winding is the same throughout its entire length, a mutual transformer is likely to have quite different current ratings for each of its separate windings.

Because autotransformers don't have separate windings, unlike mutual transformers *there is no electrical isolation between the primary and secondary circuits*.

Where electrical isolation between the primary and secondary windings is *unimportant*, the use of an autotransformer has a number of advantages over a mutual transformer, which we'll learn about later. On the other hand, where electrical isolation is *important* for safety reasons, then an autotransformer simply *cannot* be used.

For example, in the UK, some parts of **BS 7671** *'Requirements for Electrical Installations'* (i.e. the *'IEE Wiring Regulations'*) apply to step-down autotransformers, specifying that they must *not* be used to supply:

▶ socket outlets
▶ any portable appliance
▶ any extra-low voltages outside the scope of the Regulations – e.g. an electric-bell circuit operating at voltages not above 15 V.

An autotransformer can be used as a **step-down** or as a **step-up** transformer, as illustrated in Figure 9.23.

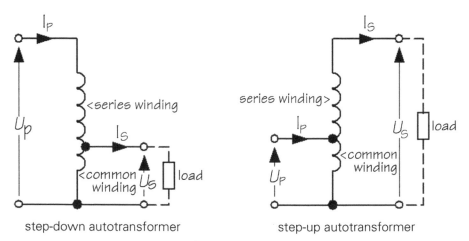

step-down autotransformer   step-up autotransformer

Figure 9.23

As you can see from the circuit diagrams, the bottom part of the winding is common to both the primary circuit *and* the secondary circuit and, for that reason, is termed the '**common winding**'. The upper part of the winding is either in series with the supply (in the case of a 'step-down' autotransformer) or in series with the load ('step-up' autotransformer) and, so, is called the '**series winding**'.

As already mentioned, it's perfectly possible to configure a mutual transformer as an autotransformer, by *connecting its two windings in series*, as shown in Figure 9.24 for a step-up transformer.

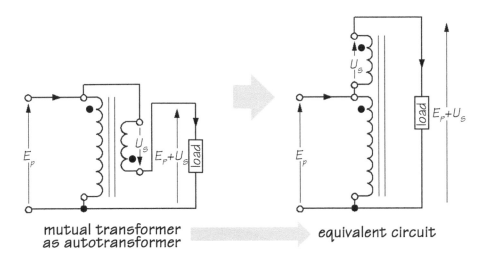

Figure 9.24

Depending on the directions in which the mutual transformer's two windings are physically wound relative to each other, the output voltage from configuration may either be the *vectorial-sum* or the *vectorial-difference* between the voltages induced into each of the windings. In the example shown in Figure 9.24, in accordance with the dot convention, we have assumed that the output voltage is the *vectorial-sum* of the two voltages.

Unlike a true autotransformer, the common and series windings of a mutual transformer configured as an autotransformer will each have *different current ratings*. That's because the higher-voltage winding has a *lower* current rating than the lower-voltage winding.

This means that the lower current rating of the high-voltage winding *may* limit the useful kV·A rating of a mutual transformer used as an autotransformer.

As already mentioned, the major disadvantage of an autotransformer is that the primary and secondary circuits *are not electrically isolated from each other.* And a potential hazard with a step-down autotransformer is that *an accidental break in the common winding will subject the secondary load to the full input voltage.* This should be made clear in Figure 9.25.

Having identified its *disadvantages*, what are the *advantages* offered by an autotransformer? Well, there are *two* advantages:

▶ With only *one* winding, the volume of copper required in its manufacture can be *lower* than for a mutual transformer.

▶ There are no secondary copper-losses, so autotransformers are more efficient than a corresponding mutual transformer.

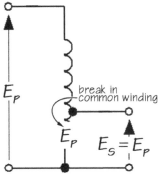

Figure 9.25

Applications for autotransformers

Because autotransformers don't isolate the secondary circuit from the primary circuit, autotransformers should *never* be used for interconnecting high-voltage distribution systems with low-voltage distribution systems (e.g., in the UK, 11 kV to 400 V). However, they are *often* used for connecting high-voltage transmission systems that are operating at different voltage levels – e.g. in the UK, 400 kV to 275 kV.

For some particularly long transmission/distribution lines, **voltage regulators** (basically, autotransformers fitted with automatic tap-changers) are used to compensate for any excessive voltage drop along those lines.

Variable output autotransformers ('variacs')

A **variable output autotransformer**, widely used in laboratories, consists of a stubby cylindrical winding with a strip of insulation removed along its length. A sliding spring-loaded contactor, called a *wiper*, rather like that fitted to a variable resistor, which can be moved along the uninsulated strip then allows a continuously variable secondary output voltage to be obtained by varying the ratio between the series and common parts of the winding. Typically, this output voltage varies from zero to several volts above the value of the primary voltage. A schematic diagram for a variable output transformer is shown in Figure 9.26.

Variable output autotransformers, as illustrated in Figure 9.27, are generally called '**variacs**', although this, in fact, is a brand name dating back to 1934.

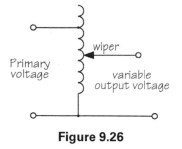

Figure 9.26

variable-output autotransformer: 'Variac'

Figure 9.27

Conclusion

Now that we've completed this chapter, we need to examine its **objectives** listed at its start. Placing a question mark at the end of each objective turns that objective into a **test item**. If we can answer those test items, then we've met the objectives of this chapter.

CHAPTER 10

Transformers 2: theory

Electrical Science for Technicians. 978-1-138-84926-6 © Adrian Waygood.
Published by Taylor & Francis. All rights reserved.

Objectives

On completion of this chapter you should be able to:

1. recognise the *first, second,* and *third approximation* equivalent circuits for a single-phase transformer.

2. explain what is meant by an 'ideal transformer'.

3. apply the general e.m.f. equation to an ideal transformer.

4. describe the relationships between an ideal transformer's voltage ratio, turns ratio (transformation ratio), and current ratio.

5. solve simple problems relating to an ideal transformer's voltage ratio, turns ratio, and current ratio.

6. use the *second approximation equivalent circuit* of a transformer to explain the behaviour of the no-load primary current of an off-load transformer.

7. sketch the phasor diagram representing the no-load condition of a second-approximation equivalent circuit of a transformer.

8. sketch the phasor diagram representing the condition of a second approximation equivalent circuit of a transformer when it supplies (a) a purely resistive load, (b) a purely inductive load, (c) a purely capacitive load, and (d) a resistive-inductive load.

9. use *the third approximation equivalent circuit* of a transformer to explain the terms 'primary and secondary resistance' and 'primary and secondary leakage reactance'.

10. sketch a phasor diagram representing a third approximation equivalent circuit of a transformer supplying a load with a lagging power factor.

11. explain the terms 'inherent voltage regulation' and 'percentage voltage regulation'.

12. explain how to 'refer' a transformer's secondary impedance to its primary circuit, and *vice versa*.

13. calculate a transformer's percentage voltage regulation.

14. summarise and explain the energy losses that occur in real transformers.

15. define the efficiency of a transformer, and specify the condition that results in its maximum efficiency.

16. describe the open- and short-circuit tests performed on a transformer, and summarise what results are obtained from these tests.

17. solve simple problems on transformer efficiency.

18. describe the effect of the supply frequency on the operation of transformers.

19. compare the construction and operation of an autotransformer with a mutual transformer.

20. identify the series and common windings of an autotransformer.

21. explain how a mutual transformer can be reconfigured to operate as an autotransformer.

22. solve simple problems on the operation of an autotransformer.

Note on symbols used in this chapter

Before we embark on this chapter, we need to understand the significance of the **voltage symbols** we will be using.

This can be needlessly confusing because, in *some* situations we use the symbol, *E*, to represent a potential difference or voltage while, in *other* situations, we use the symbol *U*. So we need to establish some ground rules about *how* we will be using these symbols in this chapter and, indeed, throughout this entire book.

▶ The *first* rule is that for a *primary supply-line voltage*, and for a *secondary no-load terminal voltages (e.m.f.)*, we will be using the symbol, *E*.

▶ The *second* rule is that for all *self-induced voltages*, *voltage drops*, and for the *terminal voltage appearing across a load*, we will be using the symbol, *U*.

The subscripts, *p* and *s*, of course, refer to the **primary** and **secondary** circuits, respectively.

Unfortunately, the above rules are by no means universal, as there is *no* agreed standard that specifies *how* engineers and technicians should use these symbols. So we should be aware that this is the way in which *we* will be using these symbols *within this particular book*; other books may choose to use these symbols (together the symbol '*V*', which is often used in place of either '*E*' or '*U*'!) quite differently. Furthermore, some books use the subscripts, **1** and **2**, to represent the primary and secondary circuits.

So Figure 10.1, below, illustrates how *we* will apply these rules to transformer schematic diagrams and phasor diagrams *in this chapter:*

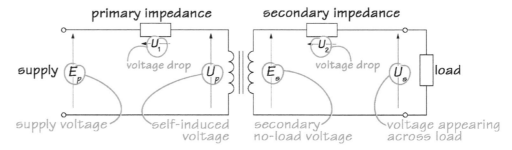

Figure 10.1

Operation of a transformer

Equivalent circuits of a transformer

In this chapter, we are going to learn about the *behaviour* of a static **transformer** under various operating conditions, such as when it is 'off load' and when it supplies various *types* of load such as *purely resistive*, *purely inductive*, *resistive-inductive*, etc.

But, before we begin, we need to understand how we can represent the behaviour of a transformer, by using a '**model**'.

Scientists and engineers often use '**models**' to simulate the complex behaviour of *real* devices, such as electrical machines – including transformers. In most cases, these models take the form of '**equivalent circuits**'. These are *circuit diagrams* consisting, in the case of transformers, of simple resistors and inductors connected in such a way that the circuit will help us to (a) *understand the transformer's behaviour*, and to (b) *predict how the transformer is likely to behave under various operating conditions*.

One of the advantages of an equivalent circuit is that it allows computer modelling of the behaviour of the transformer, enabling software to be developed that will then allow engineers and technicians to apply various *'What if?'* scenarios to the operation of a transformer without subjecting a real transformer to any risk.

For a *basic* understanding of a transformer, its equivalent circuit is very simple, and completely ignores any *energy losses*, *winding resistance*, *leakage reactance* (more on this, later), etc. However, we can also construct equivalent circuits that will take *all* of these factors into account, in which case these circuits need to be somewhat (but not a great deal) more complicated.

An **equivalent circuit** that represents *only* the most basic or fundamental behaviour of a machine is often described as representing a **'first approximation'** of that machine's behaviour. More complicated equivalent circuits, which are necessary to more accurately represent the machine's *real* behaviour, are then described as being a **'second approximation'**, **'third approximation'**, and so on. The *higher* the level of 'approximation', the *more accurately* the equivalent circuit simulates the behaviour of a machine.

Figure 10.2 illustrates the *three* levels of approximation of an equivalent circuit for a transformer that we will be examining in this chapter:

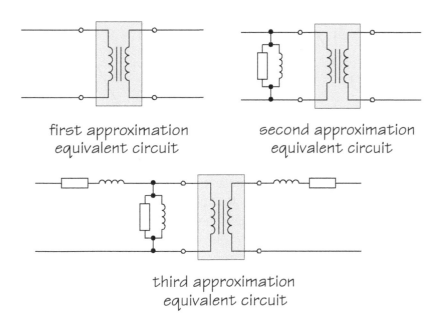

first approximation
equivalent circuit

second approximation
equivalent circuit

third approximation
equivalent circuit

Figure 10.2

As you can see, even the 'third approximation' equivalent circuit is *not* particularly complicated, even though it most closely represents the behaviour of a real transformer.

So, let's start by examining the 'first approximation' of a real transformer, which is better known as an **'ideal transformer'**.

Behaviour of an 'ideal transformer'

An **'ideal transformer'** ('first approximation' equivalent circuit) is one whose windings have *negligible resistance* and, therefore, exhibit *no copper losses*. Furthermore, this 'ideal transformer' hasn't any *iron losses* or any *magnetic leakage* occurring within its core.

10

183

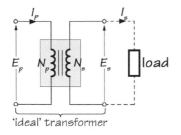

Figure 10.3

'Magnetic leakage' describes any flux generated by the primary and secondary windings that isn't confined within the core of the transformer and which, therefore, fails to link the two windings (as well as their own windings).

Having no **copper** or **iron losses**, an ideal transformer can, therefore, be considered to *be 100% efficient*. And, with no **magnetic leakage**, *its turns ratio and voltage ratio will be identical.*

In fact, the *only* behaviour an 'ideal transformer' exhibits is to either *step up* or *step down*, the **primary voltage**, and to either *step down* or *step up* its **secondary current**. Figure 10.3, shows an 'ideal' transformer whose secondary winding is supplying a load.

Because 'real' transformers are so highly efficient (up to around 98% in the case of large power transformers), in many respects the behaviour of an 'ideal transformer' is not that far removed from the behaviour of a 'real' transformer. In fact, its behaviour is close enough to that of a 'real' transformer that it may be used to explain the basic behaviour of *any* transformer to allow us to develop what's known as the '**general e.m.f. equation**' for a transformer.

General E.M.F. Equation of a Transformer

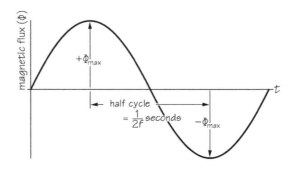

Figure 10.4

Figure 10.4 shows a waveform that represents the variation in magnetic flux (Φ) within the core of a transformer, generated by the primary current over one complete cycle.

If the peak value of this flux is Φ_{max} webers then, over one-half cycle, the flux will vary in value between $+\Phi_{max}$ and $-\Phi_{max}$. That is, it will change by $2\Phi_{max}$ webers over half a cycle.

The time taken, in seconds, for this flux to complete half of one complete cycle corresponds to one-half of that waveform's **periodic time** which, as we have learnt, is defined as *the reciprocal of its frequency*. That is:

$$\text{time taken to complete one half-cycle} = \frac{1}{2} \times \left(\frac{1}{f}\right) = \frac{1}{2f}$$

So the *average* rate of change of flux over this same period of time must, therefore, be:

$$= \left(2\Phi_{max} \div \frac{1}{2f}\right) = \left(2\Phi_{max} \times 2f\right) = 4f\Phi_{max} \text{ (webers per second)}$$

Since, according to **Faraday's Law**, the **instantaneous voltage** induced into an individual winding, is *proportional to the rate of change of flux*, then the **average e.m.f.** induced, *per turn*, must be:

$$E_{average} = 4f\Phi_{max} \text{ volts per turn}$$

And since, over *half* a cycle, the **root-mean-square** value of a sine wave is 1.11 times its average value, then the corresponding *root-mean-square value* of that voltage *per turn* must be:

$$E_{rms} = 1.11 \times E_{average}$$
$$= 1.11 \times (4f\Phi_{max})$$
$$= 4.44f\Phi_{max} \text{ volts per turn}$$

So the **r.m.s. voltage** *(E_p)* induced into a **primary winding**, having *Np* turns, must be:

$$\boxed{E_p = 4.44N_p f\Phi_{max}}$$

We call this very important equation the '**General E.M.F. Equation for a Transformer**', and it applies to both the primary and secondary windings.

And, so, the **r.m.s. voltage** *(E_s)* induced into the **secondary winding**, having *N_s* turns, must be:

$$\boxed{E_s = 4.44N_s f\Phi_{max}}$$

Worked example 1

A transformer's primary winding is supplied from a 230 V, 50 Hz, supply, and results in a sinusoidal magnetic flux having a peak value of 20 mWb being set up within the core. If the secondary winding is required to produce a voltage of 30 V, calculate (a) the necessary number of turns per winding, rounded off to the nearest practical value, and (b) the theoretical number of volts per turn for each winding.

Solution

For the **primary winding**:

$$E_p = 4.44N_p f\Phi_{max}$$
$$N_p = \frac{E_p}{4.44f\Phi_{max}}$$
$$= \frac{230}{4.44 \times 50 \times (20 \times 10^{-3})}$$
$$= 51.8 \text{ turns}$$

The nearest practical number = **52 turns (Answer a.)**

For the **secondary winding**:

$$E_s = 4.44N_s f\Phi_{max}$$
$$N_s = \frac{E_s}{4.44f\Phi_{max}}$$
$$= \frac{30}{4.44 \times 50 \times (20 \times 10^{-3})}$$
$$= 6.76 \text{ turns}$$

The nearest practical number = **7 turns (Answer a.)**

10

Therefore, the **volts per turn** for each winding (based on the *theoretical* number of turns per winding):

$$\text{(volts per turn)}_{primary} = \frac{E_p}{N_p} = \frac{230}{51.8} = \textbf{4.44 V (Answer b.)}$$

$$\text{(volts per turn)}_{secondary} = \frac{E_s}{N_s} = \frac{30}{6.76} = \textbf{4.44 V (Answer b.)}$$

> An important conclusion we can make from this worked example is that *the number of volts per turn on the secondary winding is exactly the same as the number of volts per turn on the primary winding.*

Worked example 2

A 250/3300 V transformer core has its low-voltage winding wound with 70 turns. Calculate the necessary effective cross-sectional area of the core in order to achieve a maximum flux density of 1.29 T, if the transformer is operating at (a) 50 Hz, and (b) 60 Hz.

Solution

$$\bar{E}_{rms} = -4.44 f \Phi_{max} N_p$$

$$\left(\text{since } B_{max} = \frac{\Phi_{max}}{A} \quad \text{then} \quad \Phi_{max} = B_{max} A \right)$$

$$\text{So } \bar{E}_{rms} = -4.44 f B_{max} A N_p$$

(a) At a frequency of 50 Hz:

$$A = \frac{\bar{E}_{rms}}{4.44 f B_{max} N_p} = \frac{250}{4.44 \times 50 \times 1.29 \times 70}$$

$$= \textbf{0.0125 m}^2 \textbf{ (Answer a.)}$$

(b) At a frequency of 60 Hz:

$$A = \frac{\bar{E}_{rms}}{4.44 f B_{max} N_p} = \frac{250}{4.44 \times 60 \times 1.29 \times 70}$$

$$= \textbf{0.0104 m}^2 \textbf{ (Answer b.)}$$

> An important conclusion we can make from this worked example is that a transformer *designed* to operate at 60 Hz requires a core with a smaller volume than one *designed* to operate at 50 Hz.

Effect of supply frequency on transformer design

If we were to repeat worked example 1, above, but specifying a supply frequency of 60 Hz instead of 50 Hz, we might be surprised to discover that the resulting number of primary and secondary turns would turn out to be *lower* than in that worked example. Try it for yourself; you should obtain (theoretical) figures of 43.6 primary turns and 5.63 secondary turns respectively.

Using the *theoretical* number of turns, we see that the turns ratios are *identical* for both frequencies:

At 50 Hz, turns ratio $= \dfrac{51.8}{6.76} \simeq$ **7.66**   At 60 Hz, turns ratio $= \dfrac{43.16}{5.63} \simeq$ **7.66**

So there are practical design differences between transformers that are intended to operate at different frequencies:

▶ Worked example 1, above, shows us that a transformer that is *designed* to operate at 60 Hz will require *less* primary and secondary turns than one *designed* to operate at 50 Hz – even though their rated voltages and turns ratios are identical.

▶ Worked example 2, above, shows us that a transformer designed to operate at a frequency of a 60 Hz transformer can be manufactured with a core having a *lower* (a little less than 20% lower) effective cross-sectional area and, therefore *volume* than one designed to operate at 50 Hz – making it somewhat *smaller and lighter* than a 50 Hz design having otherwise identical ratings.

By 'effective' cross-sectional area, we mean the cross-sectional area of the iron laminations, ignoring the added area due to their insulation.

So there are important design differences between transformers intended to operate at different frequencies.

A transformer that is *designed* to operate at a particular voltage at 60 Hz has primary and secondary windings with *less* turns than one *designed* to operate at 50 Hz. If we were then to operate that 60 Hz transformer at its rated voltage, but at a frequency of 50 Hz, then *the resulting volts per turn would be higher than would be the case if it were to be operated at its rated 60 Hz* – i.e. the *same* rated voltage with *less* turns means *more* volts per turn.

At 50 Hz, volts per turn $= \dfrac{230}{51.8} = 4.44$ volts per turn;

At 60 Hz, volts per turn $= \dfrac{230}{43.16} = 5.33$ volts per turn.

So operating a 60 Hz transformer at 50 Hz, would – in effect – subject that transformer's windings to higher (about 20% higher) primary voltage than it has been *designed* to handle at 60 Hz, and subject its insulation to greater stress.

This *overvoltage* also leads to yet another problem. From the general equation of a transformer, it can be seen that the maximum flux (Φ_{max}) in the core is proportional to the primary voltage. So if operating a 60 Hz transformer at 50 Hz effectively subjects the primary winding to a *higher* voltage, then the core's maximum flux (and, therefore, its flux density) must also *increase* in proportion. Because of the shape of the magnetising curve (see Figure 10.8), a *small* increase in flux density will result in a very much *bigger* magnetic field strength, and that can only be brought about by the transformer drawing a significantly larger magnetising current – and this larger magnetising current will result in the primary winding *overheating*!

So, operating a 60 Hz transformer at 50 Hz would subject its insulation to an over-voltage of 20% together with a larger magnetizing current, the combined effects of which are likely to result in its insulation breaking down.

Accordingly, a transformer designed to operate at 60 Hz should *never* be used with a 50 Hz supply *unless* the 50 Hz supply voltage is about 20% *lower* than that transformer's *rated* primary voltage to compensate for the resulting higher volts per turn and corresponding increase in magnetising current.

10

What we are saying, then, is that a 60 Hz transformer whose primary winding is *designed* to operate at, say, 230 V, can only operate safely at 50 Hz – *provided* the voltage of that 50 Hz supply is no greater than about 185 V.

On the other hand, a transformer designed to operate at 50 Hz can be operated quite safely at 60 Hz (or even higher) because the resulting volts per turn and magnetising current will be *reduced* from their design values at this higher frequency.

So a 50 Hz transformer can be used practically anywhere in the world, whereas a 60 Hz transformer can only be used in those countries that have a 60 Hz supply frequency – *unless* we are prepared to lower its rated primary voltage by about 20%.

International mains supplies operate at either 50 Hz or 60 Hz. Most of the world uses 50 Hz, while 60 Hz is mainly confined to countries within the US sphere of influence. Japan is unique in that western Japan uses 60 Hz (US influence), while eastern Japan uses 50 Hz (German influence) – with both systems interconnected via d.c. links.

Some European electric-railway systems operate at frequencies of 16.7 Hz and 25 Hz, while aircraft systems typically operate at 400 Hz.

A transformer designed and rated for a supply frequency of **50 Hz** can be used *anywhere* in the world, as it can operate without overheating at supply frequencies of 60 Hz.

However, a transformer designed and rated for a supply frequency of **60 Hz** will *overheat* if operated at its rated voltage from a 50 Hz supply and, so, its use is restricted to those countries utilising a 60 Hz supply.

We'll briefly return to this topic, later in this chapter.

Voltage ratio and turns ratio of an 'ideal' transformer

Returning to the **general voltage equation** for a transformer, if we *divide* that for the primary winding by that for the secondary winding:

$$\frac{E_p}{E_s} = \frac{4.44 N_p f \Phi_{max}}{4.44 N_s f \Phi_{max}}$$

$$\frac{E_p}{E_s} = \frac{N_p}{N_s}$$

. . . then we can see that *the primary to secondary **voltage ratio** is equal to the primary to secondary **turns ratio** (also known as the '**transformation ratio**'):*

$$\boxed{\frac{E_p}{E_s} = \frac{N_p}{N_s}}$$

Of course, we had *already* reached this conclusion from the solution to worked example 1, where we found that *the volts per turn must be exactly the same for each winding*, i.e.:

$$(\text{volts per turn})_{primary} = (\text{volts per turn})_{secondary}$$

$$E_p/N_p = E_s/N_s$$

$$\frac{E_p}{E_s} = \frac{N_p}{N_s}$$

From this, then, we can further conclude that, *for an 'ideal transformer'*:

voltage ratio = turns ratio

This means that if the transformer's *secondary* winding has, say, *twice* as many turns as the primary winding, then the induced *secondary* voltage will be *twice* the value of the primary voltage, and the transformer is described as being used as a '**step-up transformer**'.

In the same way, if the *secondary* winding has *half* as many turns as the primary winding then the induced *secondary* voltage will be *half* the value of the primary voltage, and the transformer is described as being used as a '**step-down transformer**'.

It's worth noting, here, that some books define 'turns ratio' as the ratio of *primary turns to secondary turns*, as we have done in this book, while other books define it as the ratio of the *higher to the lower number of turns*, regardless of whether it is used as a 'step down' or 'step-up' transformer.

In practice, *either* definition is fine, just as long as we stick to using one or the other consistently. *However, as the former definition is more intuitive, that's the one we'll be using throughout this chapter.*

Worked example 3

A transformer has a primary winding of 250 turns and a secondary winding of 25 turns. If the primary voltage is 120 V, calculate the secondary voltage.

Solution

$$\frac{E_p}{E_s} = \frac{N_p}{N_s}$$

$$E_s = E_p \frac{N_s}{N_p}$$

$$E_s = 120 \times \frac{25}{250} = 120 \times 0.1 = \textbf{12 V (Answer)}$$

Current ratio of an 'ideal' transformer

Because a transformer doesn't contain any moving parts, we don't have to worry about any energy losses due to friction or 'windage' (air resistance). So, a 'real' transformer is exceptionally efficient compared to other electrical machines: typically around 95–98% at full load, depending on its size! We'll discuss what causes the 2–5% losses later in this chapter.

But, for now, we are still considering the 'ideal transformer' which, as we have learnt, has *no* losses whatsoever and, therefore, has a theoretical efficiency of 100%.

Before proceeding further, we should understand that a transformer's 'power rating' (often termed its 'capacity') describes *the maximum load it can supply without overheating*. This is *not*, as we might expect, expressed in watts but, rather, in **volt amperes** (symbol: **V·A**) – which is usually considered to be the *product of its rated secondary voltage and its rated secondary current*. For an 'ideal transformer', of course, this figure applies to the primary winding as well as to the secondary winding.

As we learned from our study of a.c. theory, the 'true power' of a load is *not* simply the product of its supply voltage and load current. Instead, it is the product of its supply voltage, load current, and *'power factor'*. But, of course, transformer manufacturers have absolutely no means of knowing the power factor of whatever load will eventually be connected to their transformers and, so, *provide their ratings in volt amperes rather than in watts*.

So, a transformer is *always* rated according to the amount of '**apparent power**' that it can supply to its load without exceeding its *rated* secondary and primary voltages and currents.

So, for an 'ideal transformer', which is 100% efficient:

output apparent power = input apparent power

. . . which means:

secondary volt amperes = primary volt amperes

which we can express as follows:

$$E_s I_s \equiv E_p I_p$$

In this respect, a transformer is equivalent to a mechanical **balanced beam**, as illustrated in Figure 10.5.

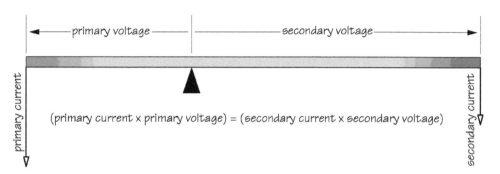

Figure 10.5

For a **mechanical beam** to remain balanced, or in 'equilibrium', the *clockwise* 'moment' (the product of a *downward force* and its *distance* from the fulcrum) acting on that beam must exactly equal its *counterclockwise* 'moment'.

For our '**electrical beam**' (transformer) to be in 'equilibrium', the product of its primary voltage and current must *exactly* match the product of its secondary voltage and current. So, if the secondary current was, say, to *increase* to satisfy the demand of an increasing load, then the primary current would also have to *increase*, in proportion, in order to maintain that equilibrium.

Rearranging, the previous equation, we have:

$$\boxed{\frac{E_s}{E_p} = \frac{I_p}{I_s}}$$

where:

E_p = secondary voltage
E_s = primary voltage
I_p = primary current
I_s = secondary current

So, as you can see, an ideal transformer's current ratio is exactly equal to the *reciprocal* of the voltage ratio.

Worked example 4

A transformer has a primary voltage of 20 V, and a secondary voltage of 50 V. If the load current (i.e. the secondary current) is 30 mA, what is the value of the supply current (i.e. the primary current)?

Solution

$$\frac{E_s}{E_p} = \frac{I_p}{I_s}$$

$$\text{so, } I_p = I_s \frac{E_s}{E_p}$$

$$I_p = (30 \times 10^{-3}) \times \frac{50}{20}$$

$$= 75 \times 10^{-3} \text{ A} \quad \text{or} \quad \textbf{75 mA (Answer)}$$

Of course, if the **current ratio** is the *inverse* of the **voltage ratio**, then it must also the *inverse* of the **turns ratio**.

i.e.

$$\boxed{\frac{I_p}{I_s} = \left(\frac{E_s}{E_p}\right) = \frac{N_s}{N_p}}$$

If we cross-multiply the current and turns ratios, then we'll end up with the following equation:

$$I_s N_s = I_p N_p$$

In the companion book, **An Introduction to Electrical Science**, we learnt that the product of the *current* flowing through a winding, and its number of *turns*, is called the '**magnetomotive force**' of that winding. So we can rewrite the above expression as:

$$\boxed{(\textbf{magnetomotive force})_{\textbf{secondary}} = (\textbf{magnetomotive force})_{\textbf{primary}}}$$

This concept of the primary and secondary magnetomotive forces *being in equilibrium* is very important to grasp, as it helps us understand the behaviour of a transformer on load.

It tells us that any *increase in secondary load current* will cause an increase in the secondary m.m.f. which, in turn, *must* be balanced by a proportional increase in the primary m.m.f. which (since the number of turns in each winding is fixed) *can only be brought about by an increase in the primary current*.

In general terms, then, *any* variation in *secondary* current will cause a proportional variation in the *primary* current. So if, say, the secondary current were to double in value, then so too will the primary current.

Students often believe that a transformer 'steps up' or 'steps down' the primary current. This is a misconception! The secondary current is, in fact, determined by the secondary voltage and the impedance of the load – *not* by the primary current. *So it's the primary current that is then determined by the secondary current, according to the turns ratio*, not the other way around.

Summary of equations for an 'ideal' transformer

$$\frac{I_s}{I_p} = \left(\frac{E_p}{E_s}\right) = \frac{N_p}{N_s}$$

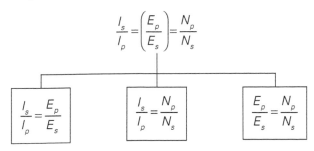

$$\frac{I_s}{I_p} = \frac{E_p}{E_s} \qquad\qquad \frac{I_s}{I_p} = \frac{N_p}{N_s} \qquad\qquad \frac{E_p}{E_s} = \frac{N_p}{N_s}$$

10

Transformer phasor diagrams

So far, we have only examined the behaviour of an '**ideal transformer**' or 'first approximation' equivalent circuit of a transformer. And, as already explained, the behaviour of an 'ideal' transformer is not that far removed from that of 'real' transformers.

But, now, we are going to 'ratchet up' our understanding of transformers by examining their behaviour in rather more detail. And, unfortunately, the 'ideal transformer' ('first approximation') equivalent circuit isn't going to be of much help to us.

Accordingly, we are going to have to move on to the '**second**' and '**third approximation**' equivalent circuits which, progressively, and more accurately, represent the behaviour of 'real' transformers.

Also, in this section, we are going to examine how to construct **phasor diagrams** for transformers, both off- and on-load, and with different types of load.

The purpose of constructing a transformer's phasor diagram is to help us consolidate our understanding of the voltage-, current-, and flux-relationships in transformers.

But, before we start, it's well worth reviewing what we learnt about the basic mechanisms of **self**- and **mutual-induction** in the companion book, *An Introduction to Electrical Science*.

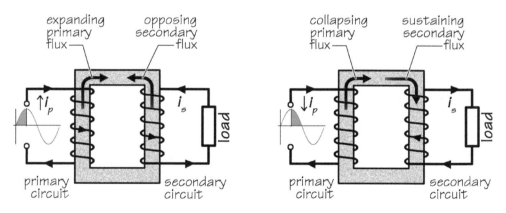

Figure 10.6

We can begin by thinking of a transformer as comprising *two*, completely separate, **self-inductive circuits**: each sharing a common silicon-steel core, with the self-inductive circuit connected to the a.c. mains' supply called the '**primary**' circuit, and the self-inductive circuit connected to a load called the '**secondary**' circuit.

During the *first* quarter-cycle of the applied primary voltage, an *increasing* current (i_p) flows through the primary winding in the direction shown in Figure 10.6a.

As this primary current increases towards its peak value, it creates an *expanding magnetic flux* that acts in a *clockwise* direction around the core. We can confirm this direction is correct by applying the '**right-hand grip rule**' to the primary winding (coil) shown in the diagram.

The effect of this expanding magnetic flux is to induce a varying 'back-e.m.f.' ($-u_p$) into the primary winding through the process of *self-induction*. The maximum value of this self-induced e.m.f. is very nearly equal to that of the supply voltage but, of course, will be 180° out of phase (hence its name, 'back-e.m.f.'), which means that the net primary voltage and, therefore, the resulting primary current must be very small.

In an 'ideal' transformer, *all* of this flux will be contained entirely within the core, and will therefore link with the secondary winding, thereby inducing a secondary potential difference (E_s) into that winding through the process of *mutual induction*.

If the secondary winding is connected to a load, then this mutually induced secondary potential difference will cause a current *(i_s)* to flow through the secondary circuit. In accordance with Lenz's Law, the direction of this secondary current must create a secondary flux that will act *counterclockwise* around the core – in order to *oppose any increase in the primary flux*. Again, we can confirm this by applying the '**right-hand grip rule**' to the secondary winding shown in Figure 10.6b.

During the *second* quarter-cycle of the primary voltage, the primary current starts to *collapse* (remember, the *value* of the primary current is *falling*, it's *not reversing direction*, so its flux, although collapsing, still acts *clockwise* around the core). This *collapse* in the primary flux again induces a potential difference into the secondary winding, by *mutual induction* – but, this time, in the *reverse* direction to the original secondary potential difference. The corresponding reverse in the secondary current now causes the secondary flux to reverse direction, which will now act *clockwise* around the core, trying (again, in accordance with Lenz's Law) *to sustain the primary flux* and, therefore, *the primary current*. Once again, we can confirm this by applying the '**right-hand grip rule**' to the secondary winding.

The action, described above will repeat itself but, of course, in the *reverse direction*, during the *second* (negative) half-cycle of the primary voltage, with the process repeating itself thereafter.

To summarise, as the primary flux tries to increase, the secondary flux tries to oppose it, and as the primary flux tries to collapse, the secondary flux changes direction, and tries to sustain it.

So what should now be obvious is that *the overall magnitude of the flux within the core of a transformer will remain approximately constant despite any variations in the secondary current.*

Now let's move on to examining a transformer's **phasor diagrams**.

We'll start by constructing a phasor diagram from a transformer that is '**off load**' – that is, a transformer *whose secondary winding is open circuited.*

Behaviour of an 'off-load' transformer

In this section, we are going to examine the *'second approximation'* equivalent circuit of a transformer. This equivalent circuit explains the significance of the transformer's **magnetising current**, and also takes into account the '**iron losses**' that occur within the core of the transformer which comprise '**hysteresis losses**' and '**eddy-current losses**':

▶ '**Hysteresis losses**' account for the energy that must be supplied in order to continuously realign the magnetic domains within the iron core, as the flux alternates in direction.

▶ '**Eddy-current' losses** are the I^2R (heating) losses that occur within the core's laminations, due to circulating currents set up by voltages induced into the core itself.

Although *both* these losses are very small, the 'second approximation' equivalent circuit takes these losses into account, making it a somewhat more accurate model of a transformer compared with the 'first approximation' circuit.

This 'second approximation' equivalent circuit consists of a resistance *(R_i)* and reactance *(X_m)* placed *in parallel* with the 'ideal transformer' already described – as illustrated in Figure 10.7.

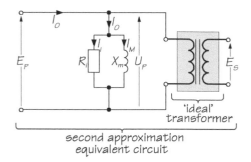

second approximation
equivalent circuit

Figure 10.7

Remember, with no load connected, the 'ideal transformer' part of this circuit *draws no current*; it simply either steps-up or steps-down the voltage applied to its primary winding.

So what's the significance of this resistance and reactance parallel network in our new model?

Well, when we connect the primary winding to a supply voltage, $\overline{E}_P$, this parallel network causes a small **no-load primary current**, $\overline{I}_0$, to be drawn from the supply.

This no-load current does *two* things:

▶ it establishes the magnetic flux within the core, and
▶ it supplies the energy losses that occur within the core (this current is so small that any energy losses in the primary windings are low enough to be ignored).

So this no-load primary current, $\overline{I}_0$, then, can be resolved into *two* components: one, a very small **in-phase component** $(\overline{I}_i)$, is responsible for the *energy losses within the core* and, the other, a much-larger **quadrature component** $(\overline{I}_m)$ responsible for the *formation of the magnetic flux within the core*.

So in this equivalent circuit, the *resistance* (R_i) simulates the path for the *in-phase* component $(\overline{I}_i)$ of the no-load primary current, while the *reactance* (X_m) simulates the path for its *quadrature* (magnetising) component $(\overline{I}_m)$:

no-load primary current $(\overline{I}_0)$

can be resolved into . . .

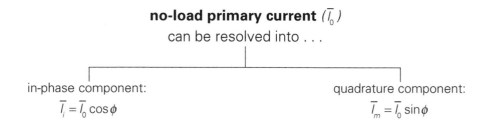

in-phase component:

$$\overline{I}_i = \overline{I}_0 \cos\phi$$

quadrature component:

$$\overline{I}_m = \overline{I}_0 \sin\phi$$

Magnetising current

Let's examine the **magnetising current** $(\overline{I}_0)$ in a little more detail.

For the magnetic flux within the transformer's core to be *sinusoidal* so, too, must the magnetising current that creates it. For this to be so, the core's flux density *must* remain within the *linear* part of the core's **B-H** ('**magnetisation**') **curve** – i.e. *below* the upper 'knee' in the curve, as it approaches saturation.

The shape of a *B-H* (flux density *vs* magnetic field strength) curve (Figure 10.8), of course, entirely depends on the material from which a transformer's core is manufactured and, in order to prevent saturating the core, the flux density *must* remain within the linear part of the curve. In practice, for **transformer steel**, this means that the flux density *mustn't* exceed around 1 T as the curve rapidly loses its linearity beyond this figure.

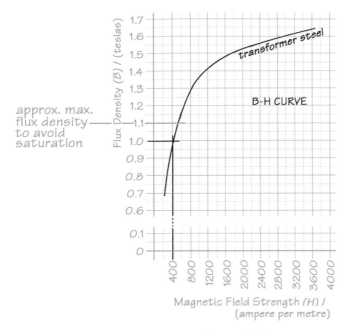

Figure 10.8

To put this particular value of flux density into perspective, the flux density of the earth's magnetic field varies from place to place, but is measured in microteslas (μT). So a flux density of 1 T or more represents a relatively strong magnetic field.

Since **magnetic field strength** *(H)* is defined as the *'magnetomotive force (F) per unit length (l)'*, we can write an expression for the maximum magnetomotive force as follows:

$$\text{since} \quad H_{max} = \frac{F_{max}}{l}$$

$$\text{then} \quad F_{max} = H_{max} l$$

An obsolete, but far more descriptive, term for **magnetic field strength** is '**m.m.f. gradient**'.

But magnetomotive force is *also* the product of the peak value of the magnetising current *(I_{max})* and the number of turns *(N)* in a winding, which is expressed as follows:

$$F_{max} = I_{max} N$$

So, from these two expressions, we can write:

$$I_{max} N = H_{max} l$$

. . . from which we can write an expression for the magnetising current:

$$I_{max} = \frac{H_{max} l}{N}$$

As it's usual to express alternating current as *root-mean-square (r.m.s.) values*, then we need to multiply the above equation by 0.707 or, alternatively, to divide it by $\sqrt{2}$:

$$\overline{I}_0 = \frac{H_{max} l}{\sqrt{2} N}$$

Worked example 5

A small transformer has a transformer-steel core with an effective* cross-section measuring 25 × 25 mm, and a mean path length of 250 mm. If we don't want the peak flux density within the core to exceed 1.0 T, (a) how many turns are required if the primary winding is to be connected across a 230 V, 50 Hz, supply? (b) what is the resulting root-mean-square value of the magnetising current?

Solution

We start by calculating the maximum flux within the core, based on the maximum flux density allowed, and the average cross-sectional area of the core:

$$\Phi_{max} = BA = 1.0 \times ((25 \times 10^{-3}) \times (25 \times 10^{-3})) = 0.625 \times 10^{-3} \text{ webers}$$

From the general e.m.f. equation for a transformer:

$$E_p = 4.44 f N_p \Phi_{max}$$

$$N_p = \frac{E_p}{4.44 f \Phi_{max}} = \frac{230}{4.44 \times 50 \times 0.625 \times 10^{-3}} = \textbf{1658 turns (Answer a.)}$$

Next, from the B-H curve (Figure 10.8) for transformer steel, we find that if the desired flux density is 1.0 T, then the corresponding value of magnetising force is close to 400 A/m. So . . .

$$I_{magnetising} = \frac{H_{max} l}{\sqrt{2} N} = \frac{400 \times (250 \times 10^{-3})}{\sqrt{2} \times 1658} = 0.043 \text{ A or } \textbf{430 mA (Answer b.)}$$

To put things in perspective, *a transformer's magnetising current and, therefore, its* **no-load primary current**, *is typically less than 5% of a transformer's rated primary current* – so it's very small indeed compared with the current that the primary winding is *capable* of handling, i.e. its 'rated' current!

The magnetising current, of course, is the *quadrature* component of the no-load primary current (I_0) and is *very* much bigger than the *in-phase* component (I_i), so the no-load primary current *lags* the supply voltage *by practically 90°*. It's *so* close to 90°, in fact, that, whenever we draw a transformer's phasor diagram, we have to *deliberately* show it somewhat *less* than 90° (Figure 10.9) for the reason explained below.

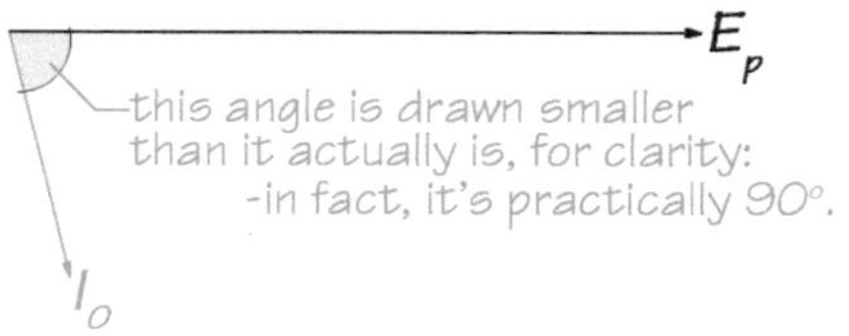

Figure 10.9

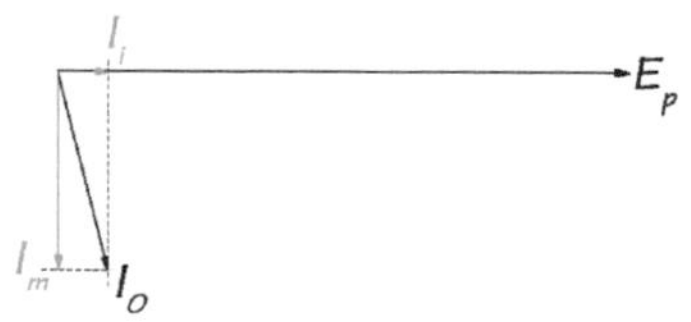

Figure 10.10

Only by *deliberately* drawing the no-load primary current, $\overline{I}_0$, a little *less* than 90°, does it become possible to illustrate how that current can then be resolved into its *two* components: one (I_i), the in-phase component, is *very* small and is exactly *in phase* with the supply voltage, and the other, ($\overline{I}_m$), the quadrature component, is *very* much larger, and *lagging* the supply voltage by *exactly* 90° – as shown in Figure 10.10.

(*By 'effective' we mean the cross-sectional area of the laminated steel, rather than the overall cross-sectional area – i.e. total cross-sectional area *less* that of the insulation between the laminations.)

Once again, it's *important* that we fully understand *three* things here, which we *must* bear in mind whenever we examine any phasor diagram representing a transformer:

▶ Firstly, the **no-load primary current** $(\bar{I}_0)$ is *very* small (less than 5%) compared with the rated primary current the transformer is capable of handling so, whenever we construct a phasor diagram, we have to *deliberately* exaggerate its length or we simply won't be able to see it at all.

▶ Secondly, the **in-phase component** $(\bar{I}_i)$ of the no-load primary current, is *tiny* in comparison with the **quadrature component** $(\bar{I}_m)$ so, again, its length is exaggerated too.

▶ Thirdly, because we have exaggerated the *length* of phasor $\bar{I}_i$ in comparison with $\bar{I}_m$, we have also *reduced* the angle by which the primary no-load current, $\bar{I}_0$, *actually* lags the supply voltage: in reality, *it's only a hair's-breadth off being exactly 90°!*

In other words, if we were to construct the complete phasor diagram accurately, and *to scale*, phasor $\bar{I}_0$ would be only around 5% or so of the rated primary current, and $\bar{I}_i$ *would not be visible*. Yet $\bar{I}_0$ has *two* very important functions: it creates the magnetic flux and supplies the energy losses within the transformer's core, *so it must be accounted for in a phasor diagram*.

It's also important to understand that currents $\bar{I}_m$ and $\bar{I}_i$ don't actually exist as *two separate 'real' currents*, but are simply the *resolved* in-phase and quadrature 'components' of the no-load primary current, $\bar{I}_0$ (see the earlier chapter on resolving phasors into their horizontal and vertical components). In other words, the no-load primary current is simply *behaving* as though a *tiny part* of it is responsible for the iron losses, while *most* of it is responsible for setting up the magnetic flux.

So let's move on. As it is the quadrature **magnetising current** $(\bar{I}_m)$ component that sets up the magnetic flux within the core, this 'current' and that magnetic flux (Φ) must obviously be *in phase* with each other, while *lagging* the supply voltage $(\bar{E}_P)$ by 90°, so we can now draw the phasor representing flux (Φ), as illustrated in Figure 10.11.

Now, as the *magnetic flux* is common to *both* the primary *and* the secondary windings of the transformer, it would make *far* more sense for the **flux**, and *not* the primary voltage, to be the **reference phasor** for the transformer's phasor diagram.

As we have learnt from a.c. theory, for *all* phasor diagrams it is usual to draw the reference phasor along the horizontal, positive axis —so, *let's rotate the complete phasor diagram (Figure 10.12), counterclockwise, through 90°, so that the flux (Φ) now becomes the reference phasor:*

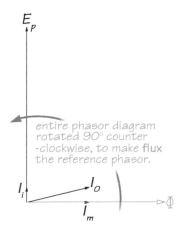

Figure 10.12

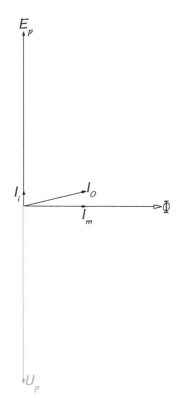

Figure 10.13

In fact, the usual practice is to construct this phasor diagram *having drawn the flux as the reference phasor from the very start*, but we didn't do so simply because we first wanted to make it absolutely clear *how* the magnetizing current $(\bar{I}_m)$ and, therefore, the flux related to the supply voltage.

Now, the effect of the varying magnetic flux is to induce a back-e.m.f. $(-\bar{U}_p)$ into the primary winding through **self-induction**. Because the value of $\bar{I}_0$ is so small, the induced back-e.m.f. will be practically, but not quite, equal to the supply voltage $(\bar{E}_p)$ but 180° out of phase – as shown in Figure 10.13.

This phasor diagram represents the waveforms, illustrated in Figure 10.14.

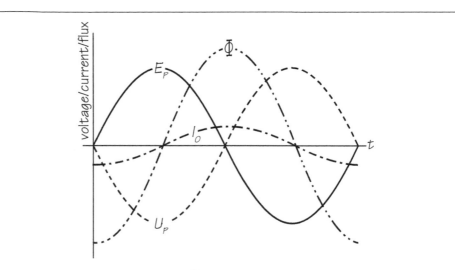

Figure 10.14

The waveform shows the **no-load primary current** *(I0)*, lagging the **supply voltage** *(E_p)* by very nearly 90°, and being mainly responsible for setting up the **magnetic flux** (Φ) in the core which, itself, also lags the supply voltage by practically 90°. The **self-induced voltage** in the primary winding *($-U_p$)* is shown in anti-phase (i.e. 180° out of phase) with the supply voltage.

Continuing, now, with the construction of the phasor diagram. The varying flux will, through **mutual induction**, also induce a voltage $(\bar{E}_s)$ into the *secondary* winding which, by Lenz's Law, must be 180° out of phase with the primary voltage $(\bar{E}_p)$. The value of this voltage, of course, depends on the **turns ratio** of the transformer: for example, if the turns ratio is 1:2 (a 'step up' transformer), then the $\bar{E}_s$ will be *twice* the value of $\bar{E}_p$, and so on. However, to avoid us having to draw a very large phasor diagram, we'll assume that the turns ratio is such that the secondary voltage is just a little larger than the primary voltage. So the **completed phasor diagram** will look as shown in Figure 10.15.

Transformer supplying purely resistive load

Figure 10.16 shows our *'second approximation'* equivalent circuit for a transformer but, this time, with its secondary winding supplying a **purely resistive load**, R_s. As before, the 'ideal transformer' part of this equivalent circuit, enclosed within the grey box, does absolutely nothing *other* than to step-up or step-down the primary voltage and secondary current (remember, the *primary voltage* determines the secondary voltage, but the *secondary current* determines the primary current).

Let's see how this compares with its phasor diagram. We'll start by drawing the **no-load phasor diagram** which, of course, we have already developed (Figure 10.15).

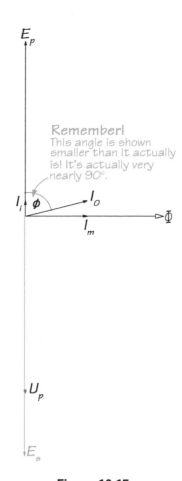

Remember!
This angle is shown
smaller than it actually
is! It's actually very
nearly 90°.

Figure 10.15

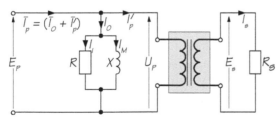

Figure 10.16

Once again, we must constantly remind ourselves that we have *deliberately* exaggerated the length of the no-load primary current $(\bar{I}_0)$ together with its in-phase and quadrature components $(\bar{I}_i$ and $\bar{I}_m)$ as shown in Figure 10.17. In reality, these are *very* small in comparison with the full-load current phasors that we are about to construct.

So, if we now connect a purely resistive load (R_s) across the secondary winding, the secondary voltage will cause a secondary current $(\bar{I}_s)$ to flow around the secondary circuit. The value of this secondary load current is determined from:

$$\bar{I}_s = \frac{\bar{E}_s}{R_s}$$

... and, of course, for a *purely resistive* load, this secondary load current will be *in phase* with the secondary voltage. So we can now draw phasor $\bar{I}_s$ in phase with phasor $\bar{E}_s$ (Figure 10.18).

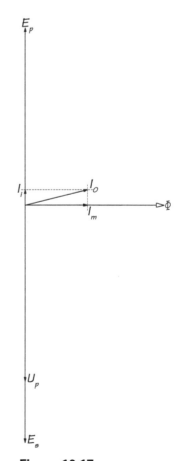

Figure 10.17

10

199

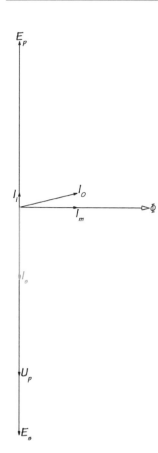

Figure 10.18

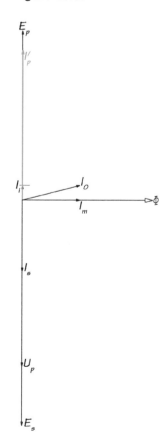

Figure 10.19

By **Lenz's Law**, the direction of $\overline{I}_s$ must act to produce a *demagnetising* flux that tries to *reduce the magnetic flux within the core* which, in turn, will *try* to cause the value of the primary self-induced voltage, $\overline{U}_p$, to reduce as well. As the primary self-induced voltage *opposes* the supply voltage, $\overline{E}_P$, any tendency for $\overline{U}_p$ to *reduce* will cause the primary current to increase and *restore the flux back towards its original value.*

In other words, *any current flowing in the secondary circuit is reflected by a proportional increase in current in the primary circuit* based on the turns ratio of the transformer and, thus, maintaining equilibrium within the windings.

This 'reflected' primary current $(\overline{I}_p')$ is given by:

$$\overline{I}_p' = \left(\frac{N_s}{N_p}\right)\overline{I}_s$$

. . . and is 180° out of phase from the secondary current, thus in-phase with the supply voltage $(\overline{E}_P)$ – as illustrated in Figure 10.19.

The actual, or 'total', **primary current** $(\overline{I}_p)$, therefore, must be the phasor-sum of phasors $\overline{I}_p'$ and $\overline{I}_0$ – as illustrated in Figure 10.20.

As you would expect for a purely resistive load, the secondary current $(\overline{I}_s)$ is *in phase* with its secondary voltage $(\overline{E}_s)$. The primary current $(\overline{I}_p)$, however, actually *lags* the primary voltage $(\overline{E}_P)$ by a small angle, ϕ_p. Because we have exaggerated the length of the magnetising current phasor, in reality *this primary phase angle is very much smaller than it appears to be in the phasor diagram.*

Transformer supplying purely inductive and purely capacitive loads

Now that we've seen how to construct the phasor diagrams for a transformer supplying a *purely resistive load*, we won't need to repeat the entire step-by-step process for a **purely inductive** or for a **purely capacitive** load.

In the case of a **purely inductive load**, the secondary current $(\overline{I}_s)$ will *lag* the secondary voltage by 90°, resulting in a corresponding reflected primary current $(\overline{I}_p')$ which is 180° out of phase with the secondary current, and whose magnitude depends on the turns ratio and which, when vectorially added to the no-load primary current $(\overline{I}_0)$, results in a primary current $(\overline{I}_p)$, which lags the primary voltage by a little less than 90°, as illustrated in Figure 10.21.

In the case of a **purely capacitive load**, the secondary current $(\overline{I}_s)$ will *lead* the secondary voltage by 90°, resulting in a corresponding reflected primary current $(\overline{I}_p')$ which is 180° out of phase with the secondary current and whose magnitude depends on the turns ratio and which, when vectorially added to the no-load primary current $(\overline{I}_0)$, results in a primary current $(\overline{I}_p)$, which *leads* the primary voltage by a little less than 90°, as shown in Figure 10.22.

In *both* cases – purely inductive load and purely capacitive load – the primary phase angle is much closer to a right-angle than it looks in the phasor diagrams because, as always, we have *exaggerated* the length and angle of the no-load primary current $(\overline{I}_0)$.

Transformer supplying a resistive-inductive load

As we have already know, most practical a.c. loads are **resistive-inductive (R-L)**, so we'll finish this section by showing a completed phasor diagram for a transformer supplying an **R-L load**.

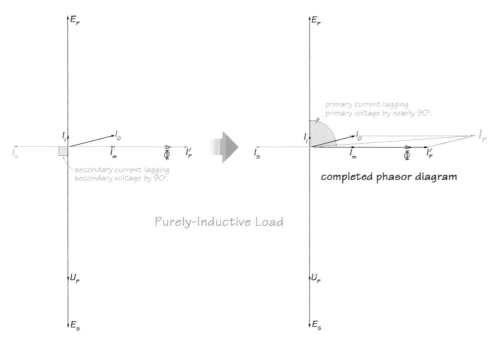

Figure 10.21

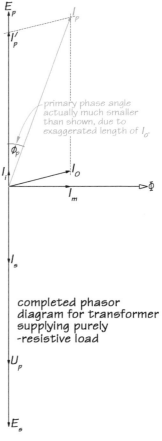

Figure 10.20

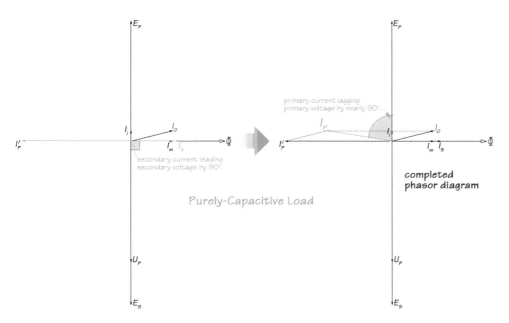

Figure 10.22

Once again, we start by constructing the basic phasor diagram showing the primary and secondary voltages, together with the no-load primary current $(\overline{I}_0)$ and its resolved components $\overline{I}_m$ and $\overline{I}_i$.

Next, we show the secondary current, $\overline{I}_s$, *lagging* the secondary voltage by some angle ϕ_s, followed by the reflected primary current, $\overline{I}_p'$ acting in the opposite direction.

Finally, we can construct the primary current $(\overline{I}_p)$, which is the phasor-sum of $\overline{I}_p'$ and $\overline{I}_0$ (Figure 10.23).

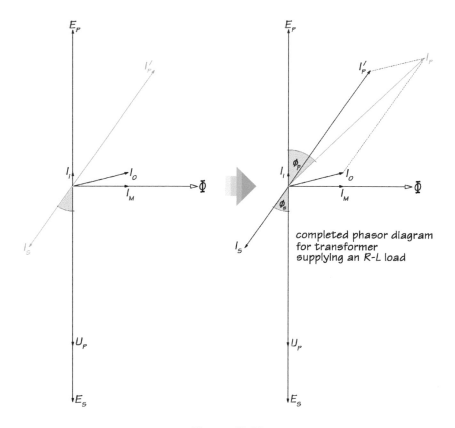

Figure 10.23

The primary current will lag the primary voltage by a somewhat *larger* phase angle, ϕ_s than the secondary phase angle. In reality, the actual primary phase angle is somewhat *smaller* than as shown in the phasor diagram because of the exaggerated length of $\overline{I_0}$.

Third approximation equivalent circuit

In this final section, we are going to examine the *'third approximation'* equivalent circuit of a transformer. This, final, equivalent circuit most accurately represents the behaviour of a 'real' transformer, and accounts for everything that the first and second approximation equivalent circuits have already taken into account, *together with the transformer's copper losses* and *internal voltage drops*.

But, *before* we examine this equivalent circuit, we need to understand another property of a transformer, termed its **'leakage reactance'**.

Leakage reactance

A transformer's **leakage reactance** is a relatively difficult concept to grasp, so let's pause for a moment to learn what it means.

The term, **'leakage'**, in this context, describes any magnetic flux that *fails to link both the primary and secondary windings*. Up to now, we have ignored leakage, and the first and second approximation equivalent circuits do not take the effects of leakage into account.

Leakage actually occurs in *both* the primary *and* in the secondary windings: that is, a small part of the flux produced by the *primary* winding fails to link with the *secondary* winding, and some of the flux produced by the *secondary* winding (when the transformer supplies a load) fails to link with the *primary* winding.

Importantly, not only does the primary leakage fail to link with the secondary winding, but it also fails to link with its *own* winding; similarly, not only does the secondary leakage fail to link with the primary winding but it also fails to link with its *own* winding!

Although, in *both* cases, this leakage is very small compared with the 'useful' or 'effective' flux within the core that links *both* windings, it is *still* enough to affect the behaviour of a 'real' transformer under load.

The primary and secondary leakage flux is often illustrated, schematically, as shown in Figure 10.24. But, in reality, it is much more complicated than this because, as we have learnt, the primary and secondary windings are *not* really wound around different limbs, but are wound, concentrically, *around each other* and *around the same limb!*

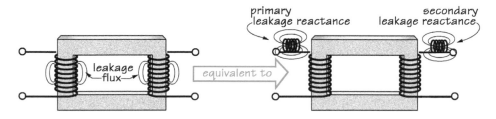

Figure 10.24

Because *some* of the primary flux fails to link with its own winding, the back-e.m.f. induced into the primary winding will be slightly *lower* than it would be had there been no leakage at all. We can account for this 'loss' of induced voltage *as though* it were caused by *a small voltage drop occurring in the primary circuit*, due to *a small inductive reactance connected in series with the primary winding*. We call this the '**primary leakage reactance**' (X_p).

A similar explanation can be applied to the secondary circuit, thus giving the secondary circuit a '**secondary leakage reactance**' (X_s).

As an aside, it's interesting to note that designers of early transformers looked upon leakage, and the resulting leakage reactance, as representing a *failing* in their designs. Since then, however, it has been recognised that a transformer's leakage reactance offers certain *advantages*, because it acts, for example, to *limit* the throughput of any fault current in the event of a short-circuit fault on the secondary side of a transformer – thus limiting any damage that may be otherwise caused by that current before any overcurrent protection device (fuse or circuit breaker) disconnects the fault.

The *'third approximation'* equivalent circuit incorporates these primary and secondary leakage reactances $(X_p$ and $X_s)$, together with the resistances $(R_p$ and $R_s)$ of the primary and secondary windings, as shown in Figure 10.25. The combination of primary resistance and reactance results in the circuit having a **primary impedance** $(R_p + jX_p)$, while the combination of secondary resistance and reactance results in the circuit having a **secondary impedance** $(R_s + jX_s)$.

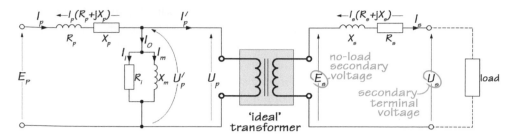

Figure 10.25

The **primary impedance** results in a **voltage drop** within the primary circuit:

$$\text{primary voltage drop} = \overline{I}_p(R_p + jX_p)$$

... which means that the voltage $(\overline{U}_p')$ *actually* applied to the primary windings of the 'ideal transformer', will be equal to the vectorial *difference* between the supply voltage $(\overline{E}_p)$ and the primary voltage drop.

Similarly, the **secondary impedance** will cause a **voltage drop** to appear within the secondary circuit when it is supplying a load:

$$\text{secondary voltage drop} = \overline{I}_s(R_s + jX_s)$$

... which means that the terminal voltage $(\overline{U}_s)$ actually appearing *across the load* will be equal to the vectorial *difference* between the no-load secondary voltage $(\overline{E}_s)$ induced into the secondary winding and the secondary voltage drop.

In the next phasor diagram, we will assume that the transformer is supplying a **resistive-inductive** load. The resulting phasor diagram, therefore, will be based on the phasor diagram that we constructed, earlier, in Figure 10.23.

When constructing this new phasor diagram, however, we must draw the voltage phasor, $\overline{U}_p'$, in place of where the supply voltage $(\overline{E}_p)$ was drawn in all the earlier phasor diagrams – this is because it is $\overline{U}_p'$, and *not* $\overline{E}_p$ that is actually being applied to the primary winding.

To find the *actual* location of the supply voltage, $\overline{E}_p$, we must work *backwards*, and vectorially *add* the primary voltage drops to voltage $\overline{U}_p'$ – that is:

$$\overline{E}_p = \overline{U}_p' + \overline{I}_p(R_p + jX_p)$$

This is most easily done by using the 'tip-to-tail' method of phasor construction shown in Figure 10.26, where the resistive voltage drop is drawn *in phase with the primary current*, and the reactive voltage drop is drawn *leading the primary current by 90°*.

A similar technique is used to locate the position of the secondary terminal voltage $(\overline{U}_s)$ appearing across the load, in the secondary circuit. This time, though, we must vectorially *subtract* the secondary voltage drops from the induced (no-load) secondary voltage $(\overline{E}_s)$ – that is:

$$\overline{U}_s = \overline{E}_s - \left[\overline{I}_s(R_s + jX_s)\right]$$

In this case, the resistive voltage drop is *in phase with the secondary current*, while the reactive voltage drop *leads the secondary current by 90°*.

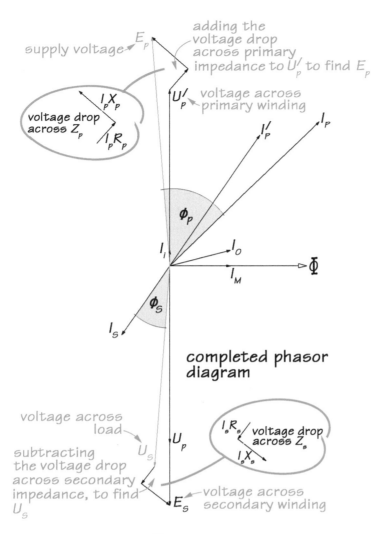

Figure 10.26

The 'tip-to-tail' method of constructing phasor diagrams, shown above, is also known as the '**funicular**' (meaning 'chain-like') method.

While it has been convenient for us to construct the phasor diagram (Figure 10.26) based on the previous phasor diagrams we have already constructed, we must constantly remind ourselves that it is the *supply voltage, $\bar{E}_p$, which is the constant factor for all of these diagrams*. Although *all* the other voltages, voltage drops, etc., might vary in value, *the supply voltage is fixed* and is determined by the electricity supply, and *not* by the behaviour of the transformer.

Determining the values of a transformer's primary and secondary impedances – i.e. their resistances and leakage reactances – is achieved by conducting two, relatively simple, experiments. These experiments, as we shall see, are identical to those used to determine the efficiency of a transformer, and we call them the '**open-circuit test**' and the '**short-circuit test**' respectively, and we will deal with these later in this chapter.

Voltage regulation

We already know that the **internal resistance** of a *battery* causes its terminal voltage to *fall* whenever a load current is drawn from it. And, of course, the *larger* the load current, the *greater* will be the fall in terminal voltage. We see the effect of this, for example, whenever we start our car having inadvertently left the headlights switched on: as the starter motor draws a very large current (hundreds of amperes), the battery's terminal voltage falls, causing the headlights to dim.

The reason for this is that the very large starting current causes an *internal voltage drop* across the battery's internal resistance, and the terminal voltage that then appears across the load is the *difference* between the battery's open-circuit (or 'no-load') terminal voltage and its internal voltage drop.

As we have seen from the phasor diagram in Figure 10.26, *exactly* the same thing happens whenever a load draws current from a transformer. Thanks to the combined effects of the primary and secondary impedances, the load current (which is reflected into the primary) will cause voltage drops to occur on *both* the secondary *and* primary sides of the transformer, resulting in a cumulative reduction in the secondary terminal voltage. And the *greater* the load current, the *greater* will be the resulting fall in that secondary terminal voltage.

But as we shall learn, the **power factor** of the load *also* affects the change in the secondary terminal voltage.

Compared with **unity power factor** (a purely resistive load), a **lagging power factor** (an inductive load), for example, results in a *greater* fall in terminal voltage. Perhaps rather surprisingly, however, a **leading power factor** (a capacitive load) can act to *increase*, rather than to reduce, the terminal voltage!

The effect of the load's power factor on a transformer's terminal voltage, with an increase in load current, is illustrated in Figure 10.27:

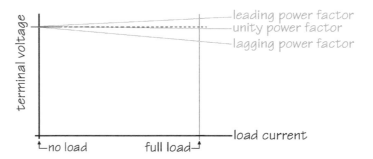

effect of power factor on voltage regulation

Figure 10.27

The reason *why* a load having a *leading* power factor (i.e. a capacitive load) can result in an *increase* in terminal voltage at full load should be made clear by comparing the phasor diagrams shown in Figure 10.28, which show the secondary no-load terminal voltage *(Es)*, the secondary full-load voltage *(Us)*, and the secondary load current *(Iload)*. In each case, we've assumed that the value of the load current and the no-load terminal voltage *(Es)* are constant values, and only the phase angle has changed.

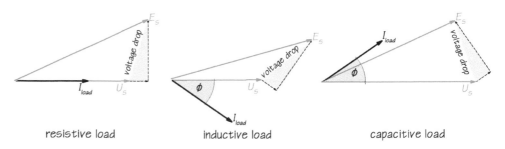

resistive load   inductive load   capacitive load

Figure 10.28

For a load with a *leading* power factor (capacitive load), it is evident from these phasor diagrams that the value of the resulting secondary full-load terminal voltage (U_s) is *greater* than the secondary no-load terminal voltage (E_s).

This change in secondary terminal voltage, from no load to full load, for a specified power factor, is termed the transformer's '**inherent voltage regulation**':

Inherent voltage regulation = (no-load voltage) − (full-load voltage)

However, it's more usual to describe voltage regulation as a '**percentage voltage regulation**', which can be defined in *either* of two ways.

▶ as *'the change in terminal voltage* expressed as *a percentage of the **no-load** terminal voltage'* – in which case it's termed '**percentage voltage regulation down**', or

▶ as *'the change in terminal voltage* expressed as *a percentage of the **full-load** terminal voltage'* – in which case it's termed '**percentage voltage regulation up**'.

British practice favours using 'percentage voltage regulation *down*', whereas North American practice appears to favour using 'percentage voltage regulation *up*'. Intuitively, 'percentage regulation down' appears to be the more intuitive of the two definitions, but arguments can be made in favour of either – but a discussion of this is beyond the scope of this book.

In accordance with British practice, then, we will define '**percentage voltage regulation**', as follows:

> A transformer's **percentage voltage regulation** is defined as *'the change in secondary voltage, from no load to full load, at a specified power factor, expressed as a percentage of its no-load voltage, with the applied primary supply voltage being held constant'.*

In other words,

$$\% \text{ Voltage Regulation} = \frac{(\text{no-load voltage}) - (\text{full-load voltage})}{(\text{no-load voltage})} \times 100\%$$

It's important to note that this definition is based on **r.m.s. scalar values**, *not* phasor, quantities – in other words, *we ignore any phase differences between the two voltages* and take into account only the *modulus* (i.e. the *length* of the phasor) of each voltage expressed in r.m.s. values.

$$(\text{percentage voltage regulation})_{\text{down}} = \frac{(E_s - U_s)}{E_s} \times 100\%$$

where: E_s = modulus of no-load secondary voltage
U_s = modulus of full-load secondary voltage

Of course, **voltage regulation** can also be expressed on a 'per unit' basis where, for example, 3% becomes 0.03, and so on.

As mentioned in an earlier chapter, transformers are generally considered to be '**constant voltage**' devices. This isn't meant *literally*, of course, as the secondary voltage of *all* transformers is bound to change to some extent as their load increases; what it *actually* means is that this change in secondary voltage should be *insignificant* when the transformer is supplying its *rated* load – and the **percentage voltage**

regulation provides us with a means of quantifying or expressing this, as the following two examples will illustrate:

A transformer, with a no-load secondary voltage of 400 V, is found to have a secondary voltage of 395 V at full load. Determine its percentage voltage regulation (down):

$$= \frac{(E_s - U_s)}{E_s} \times 100\%$$

$$= \frac{(400 - 395)}{400} \times 100\%$$

$$= \frac{5}{400} \times 100\%$$

$$= 1.25\%$$

Another transformer, with a no-load secondary voltage of 400 V, is found to have a secondary voltage of 375 V at full load. Determine its percentage voltage regulation (down):

$$= \frac{(E_s - U_s)}{E_s} \times 100\%$$

$$= \frac{(400 - 375)}{400} \times 100\%$$

$$= \frac{25}{400} \times 100\%$$

$$= 6.25\%$$

So, as you can see from these examples, the transformer with the *lower* percentage voltage regulation *is better able to maintain its terminal voltage at full load*.

This, of course, is highly desirable in most cases, which means we usually *want* transformers to have *low* values of percentage voltage regulation. Ideally, a transformer's percentage voltage regulation would be *zero* but, of course, this isn't going to happen with a *real* transformer supplying *real* load!

As a guide, power transformers typically have a percentage voltage regulation of 3% or less when supplying unity power factor loads; this figure increases with inductive loads.

So the **power factor** of the load plays an important part in establishing the actual percentage voltage regulation – but we shall return to this later in this chapter.

However, there are *some* applications in which 'poor' regulation is desirable. An example is a transformer that supplies **discharge lighting** (see chapter on *'Electric Lighting'*). A discharge lamp requires a high voltage, initially, to ignite the lamp, but that voltage should reduce in value once the lamp starts to draw current. A transformer with a deliberately poor voltage regulation will match this requirement. Methods used to deliberately introduce poor regulation are beyond the scope of this chapter.

So *what* causes a transformer's secondary terminal voltage to *fall* when a load current is drawn? The answer is that, as we learnt in the previous section, the load current causes an *internal voltage drops* within the transformer's primary and secondary windings – in just the same way as the load current drawn from a battery causes an internal voltage drop within the battery.

But, with a transformer, of course, it's rather more complicated than is the case with a battery.

This is because variations in the *secondary load current* cause *proportional changes in the primary current*, which then results in an overall voltage drop, which is the *cumulative* effect of both a *secondary* and a *primary voltage drop*.

In order to examine this, we have to re-examine the 'third approximation' equivalent circuit of a transformer, and we should be aware that the 'third approximation' equivalent circuit can actually be *simplified*, in a sense, by 'lumping' together the transformer's primary and secondary impedances. Furthermore, these 'lumped values' can be assumed to exist either entirely on the *primary* side of the transformer or entirely on its *secondary* side – as illustrated in Figures 10.29b and 10.29c (we've temporarily ignored the parallel network on the primary side of this circuit).

In the case of Figure 10.29b, we say that the secondary impedance is being *'referred to the primary'* while, in the case of Figure 10.29c, we say that the primary impedance is being *'referred to the secondary'*.

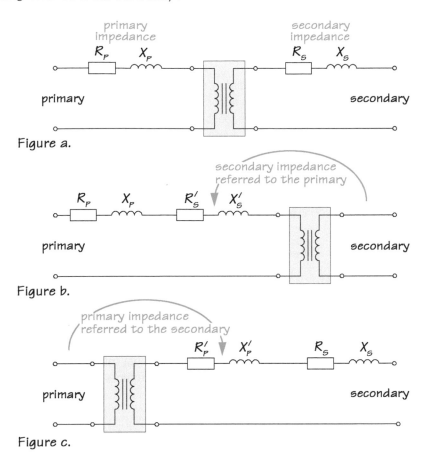

Figure a.

Figure b.

Figure c.

Figure 10.29

Of course, we cannot simply transfer the secondary impedance, as it stands, to the primary (or *vice versa*) because, as we shall now explain, the turns ratio of the transformer will act to *modify its value*.

Referring the secondary impedance to the primary

Let's explain how we do this. To keep life simple, let's start by considering only the secondary winding's *resistance* . . .

We can replace the resistance (R_s) of the *secondary* winding, by inserting additional resistance (R_s') into the *primary* circuit, such that the power developed by R_s', when carrying the *primary* current, is *exactly* the same as the power developed by R_s, due to the *secondary* current. That is:

$$I_p^2 R_s' = I_s^2 R_s$$

$$R_s' = \left(\frac{I_s}{I_p}\right)^2 R_s$$

$$\approx \left(\frac{E_p}{E_s}\right)^2 R_s$$

If we define 'turns ratio' in terms of the *primary to secondary voltage*, using the symbol k, then we can rewrite the above equation as follows:

secondary resistance referred to the primary, $R'_s = k^2 R_s$

We can apply a similar process to the **secondary leakage reactance**, in which case:

secondary leakage reactance referred to the primary, $X'_s = k^2 X_s$

We can *further* simplify the equivalent circuit, as shown in Figure 10.29b, by adding together the primary resistance and the secondary resistance referred to the primary, and doing exactly the same thing for the leakage reactances. We call these the 'equivalent primary resistance' ($R_{p \text{ equivalent}}$) and the 'equivalent primary reactance' ($X_{p \text{ equivalent}}$), respectively:

$$R_{p \text{ equivalent}} = \left(R_p + k^2 R_s\right) \quad \text{and} \quad X_{p \text{ equivalent}} = \left(X_p + k^2 X_s\right)$$

When the secondary impedance is 'referred to the primary', the effective primary impedance becomes:

$$Z_{p \text{ equivalent}} = (R_p + R'_s) + j(X_p + X'_s)$$

Referring the primary impedance to the secondary

In the same way that we can refer the secondary impedance to the primary, *we can also, if we wish, refer the primary impedance to the secondary.*

We needn't pursue this any further, other than to say that when we refer the primary resistance and leakage reactance to the secondary, we must multiply them by the *reciprocal* of the turns ratio squared (or *divide* them by the turns ratio squared, which is the same thing!):

$$R_{s \text{ equivalent}} = \left(R_s + \frac{R_p}{k^2}\right) \quad \text{and} \quad X_{s \text{ equivalent}} = \left(X_s + \frac{X_p}{k^2}\right)$$

$$Z_{s \text{ equivalent}} = (R_s + R'_p) + j(X_s + X'_p)$$

Worked example 6

A 5:1 ratio step-down transformer's primary winding has a resistance of 0.17 Ω and a leakage reactance of 0.25 Ω. Its secondary winding has a resistance of 0.0085 Ω and a leakage reactance of 0.02 Ω. Determine the total equivalent impedance of the transformer, (a) *referred to its primary*, and (b) *referred to its secondary*.

Solution

(a) Total Equivalent Impedance Referred to the Primary:

$$R_{p \text{ equiv}} = \left[k^2 R_s\right] + R_p = \left[\left(\frac{5}{1}\right)^2 \times 0.0085\right] + 0.17$$

$$= 0.213 + 0.17 = 0.383 \ \Omega$$

$$X_{p \text{ equiv}} = \left[k^2 X_s\right] + X_s = \left[\left(\frac{5}{1}\right)^2 \times 0.02\right] + 0.25$$

$$= 0.50 + 0.25 = 0.75 \ \Omega$$

So $Z_{p \text{ equiv}} = (0.383 + j\,0.75) \ \Omega$

Applying Pythagoras's Theorem:

$$Z_{p\,equiv} = \sqrt{(R_{p\,equiv})^2 + (X_{p\,equiv})^2} \quad = \sqrt{0.383^2 + 0.50^2}$$
$$= \sqrt{0.397} = \textbf{0.63 } \Omega \textbf{ (Answer a.)}$$

(b) Total Equivalent Impedance Referred to the Secondary:

$$R_{s\,equiv} = \left[\frac{R_p}{k^2}\right] + R_s = \left[\frac{0.17}{5^2}\right] + 0.0068$$
$$= (0.0068 + 0.0068) = 0.0136\ \Omega$$

$$X_{s\,equiv} = \left[\frac{X_p}{k^2}\right] + X_s = \left[\frac{0.25}{5^2}\right] + 0.01$$
$$= (0.01 + 0.01) = 0.02\ \Omega$$

$$Z_{s\,equiv} = (0.0136 + j0.02)\ \Omega$$

Applying Pythagoras's Theorem:

$$Z_{s\,equiv} = \sqrt{(R_{s\,equiv})^2 + (X_{s\,equiv})^2} \quad = \sqrt{0.136^2 + 0.020^2}$$
$$= \sqrt{0.019} = \textbf{0.138 } \Omega \textbf{ (Answer b.)}$$

Worked example 7

A transformer with a primary winding rated at 5000 V has an equivalent resistance referred to its primary of 6.5 Ω and an equivalent inductance referred to its primary of 0.25 H. Calculate the current drawn by the transformer when connected to (a) a 50 Hz supply and (b) to a 60 Hz supply.

Solution

At 50 Hz:

$$X_{p\,equiv} = 2\pi fL = 2\pi \times 50 \times 0.25$$
$$= 78.5\ \Omega$$

$$Z_{p\,equiv} = \sqrt{R^2 + X_L^2} = \sqrt{6.5^2 + 78.5^2}$$
$$= 78.8\ \Omega$$

$$\overline{I}_p = \frac{\overline{E}_p}{Z_{p\,equiv}} = \frac{5000}{78.8} = \textbf{63.5 A}$$

At 60 Hz:

$$X_{p\,equiv} = 2\pi fL = 2\pi \times 60 \times 0.25$$
$$= 94.2\ \Omega$$

$$Z_{p\,equiv} = \sqrt{R^2 + X_L^2} = \sqrt{6.5^2 + 94.2^2}$$
$$= 94.4\ \Omega$$

$$\overline{I}_p = \frac{\overline{E}_p}{Z_{p\,equiv}} = \frac{5000}{94.4} = \textbf{53.0 A}$$

This worked example further illustrates why a transformer *designed to operate at 60 Hz* should *not* be used on a 50 Hz supply without reducing its rated voltage.

As you can see, this particular transformer draws a primary current of 53 A at 60 Hz, but a current of 63.5 A at 50 Hz. So if the transformer's primary winding was *designed* to handle just 53 A at 60 Hz, then it would overheat if it were to be *operated* at 50 Hz. On the other hand, if it was *designed* to handle 63.5 A at 50 Hz, then it can be operated quite safely at 60 Hz without any fear of overheating.

So a transformer designed to operate at 60 Hz has *fewer turns per winding*, a *smaller core volume* and, as we can now see, it also has *lower rated winding currents* than one with exactly the same rated voltages but designed to operate at 50 Hz.

10

Voltage regulation calculations

Now that we know how to *refer the secondary impedance to the primary* of a transformer, we are in a position to determine its **voltage regulation**.

For what follows, it doesn't really matter whether we refer the secondary impedances to the primary, or the other way around – so, we'll choose the former.

Returning, now, to our 'third approximation' equivalent circuit, using what we have just learnt, we should now be able to see how it can be simplified as illustrated in Figures 10.30a–d:

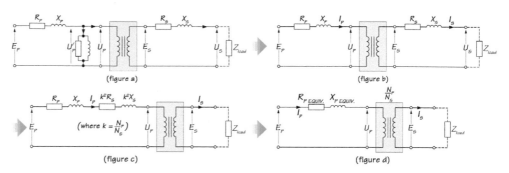

(figure a)   (figure b)   (figure c)   (where $k = \frac{N_p}{N_s}$)   (figure d)

Figure 10.30

In Figure 10.30a, the no-load primary current (i.e. the magnetising and core-loss current) drawn by the parallel *R-L* network is less than 5% of the *rated* primary current and, so, is insignificant enough to be ignored when the transformer is supplying its rated load. So we can safely ignore the parallel network, and redraw the equivalent circuit as shown in Figure 10.30b.

In Figure 10.30c, we have simplified the equivalent circuit even further, by referring the secondary resistance and leakage reactance across to the primary. Of course, this requires us to multiply these secondary values by k^2 – the square of the turns ratio N_p/N_s.

In Figure 10.30d, we have added the primary and referred secondary resistances together, and have done the same thing with the leakage reactances. The 'ideal transformer' then only acts to step up or step down voltage U_p to give us a secondary terminal voltage, E_s.

Now, let's look at the phasor diagram for the primary circuit shown in Figure 10.31.

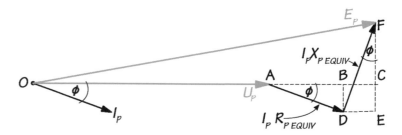

Figure 10.31

For a *lagging* power factor, ϕ, the difference between the *primary supply voltage* $\bar{E}_p$ and the *voltage appearing across the primary winding*, $\bar{U}_p$, can be expressed as follows:

$$\bar{E}_p - \bar{U}_p = I_p R_{p\,equiv} \pm j(I_p X_{p\,equiv})$$

+j indicates applies to a *lagging* power factor, while **−j** applies to a *leading* power factor.

Referring to Figure 10.31, in practice the angle between phasors $\bar{E}_p$ and $\bar{U}_p$ is actually quite small at full load, so there's not a great deal of difference between lengths OC and OF. So we can say that:

$$OF \simeq OC = OA + AB + BC$$

$$OF \simeq OA + AB + BC$$

$$OF \simeq OA + (AD\cos\phi) + (DF\sin\phi)$$

$$E_P \simeq U_p + \left(I_p R_{p\,equiv}\cos\phi\right) + \left(I_p X_{p\,equiv}\sin\phi\right)$$

Since the **inherent regulation** is simply the *difference* between $\boldsymbol{E_p}$ and $\boldsymbol{U_{p'}}$ that is:

$$\text{inherent voltage regulation} = \left(\bar{I}_p R_{p\,equiv}\cos\phi\right) + \left(\bar{I}_p X_{p\,equiv}\sin\phi\right)$$

*Remember, the equations for voltage regulations do **not** take phase differences into account, only the magnitudes of those voltages.*

Worked example 8

A 40 kV·A, 6600:250 V, step-down transformer has a primary resistance of 10 Ω and 0.02 Ω respectively. If the equivalent leakage reactance, referred to the primary, is 35 Ω, neglecting the no-load current, calculate the full-load regulation at a power factor of 0.8 lagging.

Solution

We start by *referring the secondary resistance to the primary*, as follows:

$$R_s' = k^2 R_s$$

$$= \left(\frac{6600}{250}\right)^2 \times 0.02$$

$$= 13.94\ \Omega$$

Now we can find the equivalent resistance, referred to the primary:

$$R_{p\,equiv} = R_p + R_s' = 10 + 13.94 = 23.94\ \Omega$$

We can also determine the full-load primary current, as follows:

$$I_{p\,full\,load} = \frac{S_{rated}}{E_p} = \frac{40\,000}{6600} = 6.06\ \text{A}$$

To determine the inherent voltage regulation with a power factor of 0.8 lagging,

$$E_p - U_p' \simeq \left(I_p R_{p\,equiv}\cos\phi\right) + \left(I_p X_{p\,equiv}\sin\phi\right)$$

$$\simeq (6.06 \times 23.94 \times 0.8) + (6.06 \times 35 \times \sin 36.87°)$$

$$\simeq (6.06 \times 23.94 \times 0.8) + (6.06 \times 35 \times 0.6)$$

$$\simeq 116.06 + 127.26$$

$$\simeq 243\ \text{V}$$

So the voltage appearing across the primary winding, at full load, will be:

$$U_p' = E_p - (\text{voltage drop across primary equivalent impedance})$$

$$= 6600 - 243$$

$$= 6357\ \text{V}$$

10

Which means that the secondary full-load voltage ($E_{s\ full\ load}$) must be:

$$E_{s\ full\ load} = U'_p \times \frac{250}{6600} = 6357 \times \frac{250}{6600} = 240.8\ V$$

Now, finally, we can determine the percentage voltage regulation:

$$(\text{percentage voltage regulation})_{down} = \frac{(E_{s\ no\ load} - U_{s\ full\ load})}{E_{s\ no\ load}} \times 100\%$$

$$= \frac{(250 - 240.8)}{250} \times 100\%$$

$$= \frac{9.2}{250} \times 100\%$$

$$= \textbf{3.7\% (Answer)}$$

Transformer losses

The **energy losses** from an individual transformer are very low – *far* lower than for any other electrical machine mainly, of course, because it has no moving parts. Despite this, if we were to add up *all the losses* for *all the transformers* in the electricity transmission/distribution system, the total can be considerable in terms of wasted energy. Manufacturers, therefore, have put a great deal of effort into minimising their transformers' losses.

The **energy losses** that do occur in a transformer may be summed up as being due to **heat**, **vibration**, and **sound** – with *heat* representing, by far, the greatest loss. In fact, for all practical purposes, we can disregard the losses due to vibration and sound.

Heat losses occur whenever the temperature of the transformer's components rise above the surrounding ambient temperature, with 'heat' being defined as '*energy in transit from a higher temperature to a lower temperature*'.

These losses occur from *both* the windings *and* the core. Those from the windings are traditionally called '**copper losses**' (even when the windings are manufactured from aluminium!) and those from the core are called '**iron losses**', as shown below:

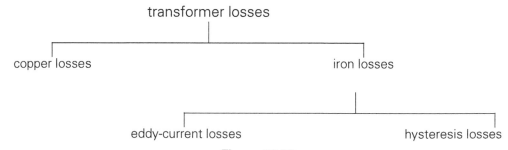

Figure 10.32

Copper losses

It's often been said that '*the consequence of resistance is heat*' so, whenever current flows through a conductor, thanks to its resistance, its temperature rises above its surrounding ambient temperature and energy is lost to the surroundings through heat transfer. These losses occur from both the primary *and* the secondary windings, and are termed '**copper losses**', where:

$$\text{copper losses} = \left(\overline{I}_p^2 R_p\right) + \left(\overline{I}_s^2 R_s\right)$$

For any given value of current, copper losses can *only* be reduced by reducing the resistance of the windings, which is *only* achievable by increasing the cross-sectional area of the windings' conductors – the limiting factor being, of course, the resulting volume, weight, and cost of the transformer.

Iron losses

The energy losses that occur in the transformer's core are collectively called '**iron losses**'. These losses are independent of the load, and are more or less constant. These losses fall into *two* categories: '**eddy-current losses**' and '**hysteresis**' losses.

Eddy-current losses

The core, being a conductor, has voltages induced into it by the alternating magnetic field. In fact, we can think of the core as being a 'third winding' as far as the transformer is concerned! These voltages cause circulating currents in the core, which are called '**eddy currents**'.

The resulting energy losses are proportional, of course, to the *square* of these eddy currents. These eddy currents are caused by voltages induced into the core which, themselves, depend upon the rate of change of flux in the core which, in turn, depend on the maximum flux density and the frequency of the supply. From this, for any given core, it can be shown that eddy-current losses must be proportional to the *square of the supply frequency* and the *square of the flux density*:

$$P_{\text{eddy current}} \propto f^2 B_{\text{max}}^2 \text{ (per unit volume)}$$

Eddy currents act to raise the temperature of the core, causing energy loss to the surroundings through heat transfer. As we have already learnt, in order to minimise these currents, the core is **laminated** – with each lamination insulated from its neighbours. The eddy currents cannot cross between laminations which, therefore, severely restricts the current paths through the core and the thinness of the individual laminations significantly reduce the magnitude of the eddy currents, thus reducing the losses. In addition to this, the silicon-iron alloy from which the laminations are manufactured have a resistivity of around 4.3 times that of iron, which helps to even further reduce the eddy currents.

Hysteresis losses

Hysteresis is the *property* of a ferromagnetic material that determines the energy required to magnetise that material to saturation, demagnetise it, remagnetise it in the opposite direction, and demagnetise it again, over one complete cycle of alternating current.

This amount of energy is represented by the *area* of the **B-H loop** which we discussed in the chapter on magnetic circuits in the companion book, **An Introduction to Electrical Science**. In Figure 10.33, we have illustrated two *B-H* loops —one representing a silicon-iron alloy (a 'soft' ferromagnetic material), and the other representing carbon steel (a 'hard' ferromagnetic material). As we can see, the silicon-steel *B-H* loop has a very much smaller area, showing that it is very suitable for the manufacture of a transformer's core.

In order to reduce hysteresis losses as much as possible, manufacturers have designed their own variations of silicon-iron alloy, and market them under various trade names, such as *Stalloy*, *Permall,* and *Mumetal*.

As already mentioned, it can be shown that hysteresis losses are *proportional to the supply frequency*, and *approximately proportional to* B_{max}^2 (maximum flux-density squared):

$$P_{\text{hysteresis}} \propto f B_{\text{max}}^{\approx 2} \text{ (per unit volume)}$$

10

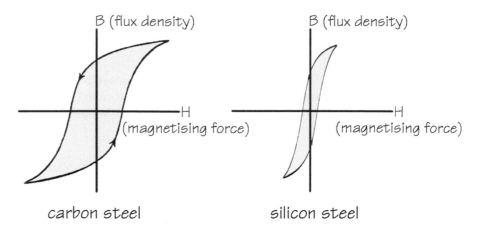

Figure 10.33

Efficiency of a transformer

We define the **efficiency** (symbol: η, pronounced 'eta') of *any* machine as '*the ratio of its output power to its input power*', which can be expressed either as a *per unit* (p.u.) or, more commonly, a *percentage* (%) value. As a machine's output power can *never* match its input power, its efficiency will always be less than 100% (or less than 1.0 p.u.). That is:

$$\eta = \frac{P_{output}}{P_{input}} \text{ per unit} \quad \text{or} \quad \eta = \left(\frac{P_{output}}{P_{input}} \times 100\right) \text{per cent}$$

Since the input power must, of course, be equal to the output power *plus* any losses, we can rewrite the above equation as follows:

$$\eta = \frac{P_{output}}{P_{output} + losses} = \frac{P_{output}}{P_{output} + (\text{copper losses} + \text{iron losses})}$$

If we refer to the (simplified) third approximation equivalent circuit of a transformer, in Figure 10.34. . . .

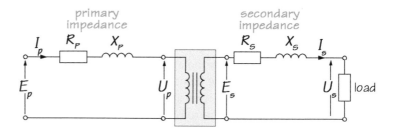

Figure 10.34

. . . then the above equation can be rewritten as follows:

$$\eta = \frac{\left(\bar{U}_s \bar{I}_s \cos\phi_s\right)}{\left(\bar{U}_s \bar{I}_s \cos\phi_s\right) + \bar{I}_p^{\,2} R_p + \bar{I}_s^{\,2} R_s + P_{\text{iron loss}}} \times 100\%$$

Although the derivation of what follows is beyond the scope of this book, it can be shown that *maximum efficiency occurs when the copper losses and iron losses are equal to each other.*

> A transformer's maximum efficiency occurs when the copper losses and iron losses are equal to each other.

Since transformers are normally operated at somewhat *less* than full load, it is usual for them to be designed so that their maximum efficiency also occurs at somewhat less than full load.

So, how can we determine the copper losses and iron losses and, therefore, the efficiency of any particular transformer? Well, we can do this relatively easily by conducting two practical tests, called a '**short-circuit test**' and an '**open-circuit test**'.

Transformer tests

Transformer tests are conducted in order to determine the characteristics of a transformer, such as its **turns ratio**, **percentage impedance**, **winding resistance**, **leakage reactance**, etc., as well its **energy losses** from which we can determine its efficiency.

It must be understood that these tests are not without their dangers, particularly if the transformer's turns ratio is high, in which case it will be necessary to place appropriate barriers and warning placards around the transformer to prevent anyone approaching and coming into contact with its terminals during the test.

Open-circuit test on a transformer

With this test, it's necessary to match the rated voltage of one or other of the windings to that of the supply. If this results in the transformer then acting as a *step up* transformer, then special care *must* be taken to prevent anyone approaching the secondary terminals across which a high voltage may appear.

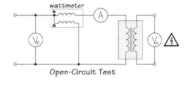

Open-Circuit Test

Figure 10.35

The transformer is connected as illustrated in Figure 10.35, and to a supply voltage that either matches, or that can be adjusted to match, the primary winding's rated voltage and frequency (as specified on the transformer's nameplate). With no-load current flowing, the instruments will then provide us with the following information:

▶ The ratio of the two **voltmeter** readings will provide us with the *turns ratio* of the transformer.

▶ The **ammeter** reading will indicate the value of the *no-load primary current*.

▶ The no-load primary current is so low that the copper losses will be insignificant and, so, the **wattmeter** reading will indicate only the *iron losses*.

Short-circuit test on a transformer

Because, in this test, we will be applying a relatively low voltage to the primary winding, it is best to connect the transformer as a step-down transformer, which reduces the shock hazard appearing at the secondary terminals. Despite this, it's still important to take measures to avoid anyone approaching the transformer's terminals during the test.

With the transformer's secondary winding short-circuited through an ammeter, as illustrated in Figure 10.36, and the primary winding connected to a variable voltage supply (using, for example, a variable autotransformer or 'variac'), the primary voltage is gradually increased until the secondary and primary currents reach their rated full-load values. Typically, this should occur when the supply voltage reaches a value of around 5% or less of the rated primary voltage:

▶ The two **ammeter** readings will tell us when the *rated secondary* and *primary load currents* are reached.

▶ The **voltmeter** will then indicate the value of primary voltage that causes full-load currents to circulate in the secondary and primary windings.

217

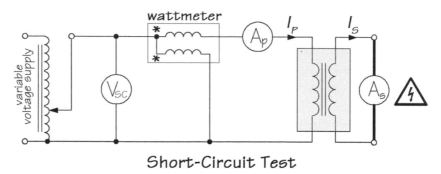

Short-Circuit Test

Figure 10.36

▶ Since the primary voltage and, therefore, the flux, is around just 5% of their rated values, the resulting iron losses will be insignificant and, so, the **wattmeter** reading will indicate the total *copper losses*.

Worked example 9

A 5-kV·A transformer, whose windings are rated at 400 V and 200 V respectively, is subjected to an open- and short-circuit test, each with the 400 V winding as the primary, which reveal the following data:

Table 10.1

open-circuit test:	400 V	1 A	60 W
short-circuit test:	15 V	12.5 A	50 W

Calculate:

(a) the values of R_i and X_m

(b) the equivalent resistance and reactance referred to the primary

(c) the iron and copper losses at full load

(d) the efficiency at a power factor of 0.85.

Solution

(a) From the **open-circuit test**, the no-load primary current, $\overline{I}_0$, as indicated by the ammeter can be resolved into *two* component currents: $\overline{I}_i$ and $\overline{I}_m$. The first (in-phase), responsible for the iron losses, can be determined from the wattmeter and voltmeter readings:

$$\overline{I}_i = \frac{P}{\overline{E}_s} = \frac{60}{400} = 0.15 \text{ A}$$

The second (quadrature) can be determined using Pythagoras's Theorem, as follows:

$$\overline{I}_m = \sqrt{\overline{I}_0^2 - \overline{I}_i^2} = \sqrt{1^2 - 0.15^2} = 0.989 \text{ A}$$

We can now determine the values of R_i and X_m, as follows:

$$R_i = \frac{\overline{E}_s}{\overline{I}_i} = \frac{400}{0.15} = \textbf{2666.7 } \Omega \quad \text{and} \quad X_m = \frac{\overline{E}_s}{\overline{I}_m} = \frac{400}{0.989} = \textbf{404.4 } \Omega$$

(b) From the **short-circuit test**, the equivalent resistance, referred to the primary, can be determined as follows:

$$\text{since } \overline{I}^2 R_{p\,equiv} = P, \quad \text{then } R_{p\,equiv} = \frac{P}{I^2} = \frac{50}{12.5^2} = \textbf{0.32 } \Omega$$

And the equivalent impedance, referred to the primary, can be determined as follows:

$$Z_{p\,equiv} = \frac{\bar{E}_p}{\bar{I}} = \frac{15}{12.5} = 1.2\,\Omega$$

... from which, using Pythagoras's Theorem, we can determine the equivalent leakage referred to the primary:

$$X_{p\,equiv} = \sqrt{Z_{p\,equiv}^2 - R_{p\,equiv}^2} = \sqrt{1.2^2 - 0.32^2} = \mathbf{1.16\,\Omega}$$

(c) As the *iron losses* are unaffected by load, they are given by the **open-circuit test wattmeter reading**: i.e. **60 W**; the *copper losses*, on the other hand, are given at full load by the **short-circuit test** wattmeter reading: i.e. **50 W**.

(d)

$$\eta = \frac{P_{output}}{P_{output} + losses} = \frac{P_{output}}{P_{output} + (\text{copper losses} + \text{iron losses})}$$

$$= \frac{5000 \times 0.85}{(5000 \times 0.85) + 50 + 60} = \frac{4250}{4250 + 50 + 60} = \frac{4250}{4360}$$

$$= \mathbf{0.9748\,p.u.} \quad \text{or} \quad \mathbf{97.48\%}$$

Worked example 10

What value of secondary current must be supplied by the transformer in the previous worked example for maximum efficiency?

Solution

We know that the iron losses remain approximately constant for variations in load, and that the open-circuit test wattmeter reading tells us that the *iron losses* are **60 W**.

For maximum efficiency,

$$\text{copper losses} = \text{iron losses} = 60\ \text{W}$$

$$\bar{I}_p^2 R_{p\,equiv} = \text{iron losses} = 60\ \text{W}$$

$$\bar{I}_p = \sqrt{\frac{60}{0.32}} = \mathbf{13.69\ A}$$

Worked example 11

A transformer having a primary winding made up of 200 turns, when connected to a supply voltage of 230 V at 50 Hz, draws a no-load primary current of 5 A at a power factor of 0.25.
Calculate:

(a) the maximum flux within the core (neglecting leakage)
(b) the total iron losses
(c) the magnetising current.

Solution

(a) Maximum flux within the core:

$$\bar{E}_{rms} = 4.44f\Phi_{max}N$$

$$\Phi_{max} = \frac{E}{4.44fN} = \frac{230}{4.44 \times 50 \times 200} = \mathbf{5.18\ Wb}\ \textbf{(Answer a.)}$$

(b) Total iron losses:

$$\text{iron losses} = \overline{E}_p \overline{I}_0 \cos\phi_0$$

$$= 230 \times 5 \times 0.25 = \textbf{287.5 W (Answer b.)}$$

(c) Magnetising current, which is the quadrature component of the no-load primary current:

$$\overline{I}_m = \overline{I}_0 \sin\phi_0 = 5\sin(\cos^{-1} 0.25)$$

$$= 5 \times 0.9682 = \textbf{4.84 A (Answer c.)}$$

Behaviour of an autotransformer

As explained in the previous chapter, the term, '**autotransformer**', describes a transformer having a single winding which *'acts upon itself'* in order to produce a secondary voltage.

In other words, whenever we apply a voltage to one part of a single coil, another voltage is induced into the *other* part of the coil through **mutual induction** between the two parts of the same coil.

Unlike a mutual transformer, in which *all* the energy from the primary circuit is transferred to the secondary circuit *only* through mutual induction, an autotransformer transfers *part* of this energy by mutual induction with the remainder supplied *directly from the primary circuit*. The energy supplied directly from the primary circuit is sometimes called '**conducted energy**'.

A 'true' **autotransformer** uses a *single tapped winding* rather than the two *separate* and *electrically isolated* windings used by mutual transformers. In this context, the terms '**tap**' or '**tapping**' simply refers to an electrical connection at some point along the length of a winding.

But, as explained in the previous chapter, it is also perfectly possible to configure a **mutual transformer** so that it can operate as an autotransformer, by simply connecting its *two separate windings* in series with each other. While this configuration allows the mutual transformer to *operate* as, and have the characteristics of, an autotransformer, it is not a 'true' autotransformer in the sense of it having a *single* tapped winding with a uniform current rating along its entire length.

Figure 10.37, below, allows us to compare a 'true' step-down autotransformer with a mutual transformer connected to operate as a step-down autotransformer. In the case of the mutual transformer, note how the two windings are connected in series, the primary voltage is applied to both windings, while the secondary voltage is obtained from the low-voltage winding.

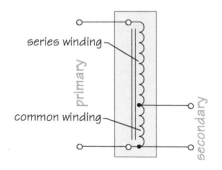

'true' autotransformer

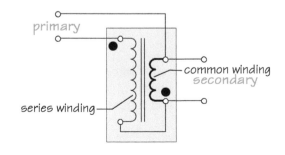

mutual transformer
connected as an
autotransformer

Figure 10.37

Note how, in both cases, the '**series winding**' is so called because (for a step-down autotransformer) it is in *series* with the supply, while the '**common winding**' is so called because it is *common* to both the primary and the secondary.

Note how the autotransformer uses just one uniform coil, which means that the series and common windings have exactly the *same* current ratings. However, in the case of the mutual transformer, the two windings most likely have *different* current ratings. *This is an important difference*, which we must compensate for whenever we apply a load to a mutual transformer connected as an autotransformer.

A true autotransformer, then, consists of a single winding, which is tapped at some point along its length, as illustrated in Figure 10.38. The part of the winding common to both the supply *and* the load, is termed the '**common winding**', with the remainder of the winding termed the '**series winding**'.

If the **series winding** is connected in series with the *supply*, then the autotransformer becomes a '**step-down transformer**'; if the series winding is connected in series with the *load*, then it becomes a '**step-up transformer**'.

Exactly the same terminology applies to a mutual transformer connected as an autotransformer.

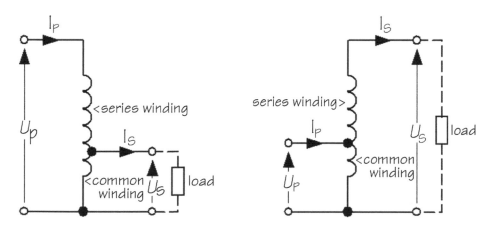

Figure 10.38

The four *'golden rules'* which we must remember for *all* autotransformers (ignoring any losses) are:

▶ The **apparent power** (expressed in volt amperes) supplied *to* the load by the secondary circuit is *always* equal to the **apparent power** demanded *from* the mains supply by the **primary circuit**.

▶ The **apparent power** (expressed in volt amperes) developed by the **series winding** will *always* equal the **apparent power** developed by the **common winding**.

▶ To prevent either the series or common windings from overheating, the *actual* **current** in either winding *must never* be allowed to exceed the *rated* **current** for those windings. This is particularly important for mutual transformers connected as autotransformers.

▶ The current in the common winding is always the difference between the supply current and the load current.

So, for a **step-down autotransformer** (Figure 10.39):

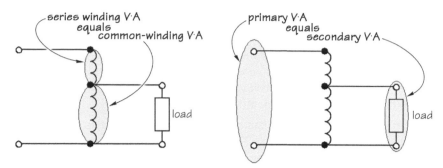

Step-Down Autotransformer

Figure 10.39

And, for a **step-up autotransformer** (Figure 10.40):

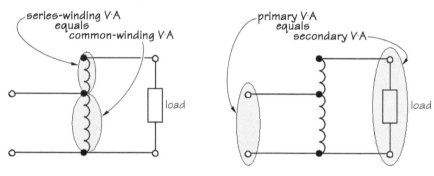

Step-Up Autotransformer

Figure 10.40

As explained, the apparent power developed by the series winding is the same as that developed by the common winding. But care must be taken to prevent the resulting current in either winding from exceeding their rated values.

The *rated* apparent power developed by each winding is the product of the winding's *rated voltage* and *rated current* and, of course, *that rated current must never be exceeded or the winding will overheat*. So if it turns out that the resulting apparent power of each winding is *different*, then *the lower value must be applied to* **both** *windings*.

So if the combination of rated voltage and rated current of the series winding results in a rated apparent power of, say, 100 V·A, and the corresponding rated values for the common winding results in a rated apparent power of, say, 75 V·A, then *neither* winding must be allowed to exceed 75 V·A.

For a **true autotransformer**, the series and common windings will, of course, have the *same* rated currents but, for a **mutual transformer** *configured as an autotransformer*, they will be *different*.

The reason for this is that, in the case of a 'true' autotransformer, the winding is usually wound from a single length of conductor of *constant* cross-sectional area, whereas a 'mutual' transformer has two separate windings, each manufactured from conductors of *different* cross-sectional area. The *lower*-voltage winding always, of course, has the *larger* cross-sectional area.

Worked example 12

Determine the magnitudes and directions of the currents within each part of a 100/50 V step-down autotransformer shown in Figure 10.41, which is supplying a load of 2.5 Ω. Assume that each winding is fully capable of carrying the calculated currents without overheating.

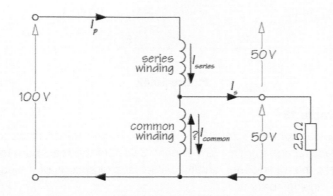

Figure 10.41

Solution

In this example, at full load, the transformer will supply a secondary load current of:

$$I_s = \frac{50}{2.5} = 20 \text{ A}$$

So the load must have an apparent power of:

$$S_s = 50 \times 20 = 1000 \text{ V} \cdot \text{A}$$

Since the apparent power *supplied by* the primary must be the same as the apparent power *supplied to* the load, then the primary apparent power must also be 1000 V·A.

So the primary current can be determined as follows:

$$I_p = \frac{S_p}{E_p} = \frac{1000}{100} = 10 \text{ A}$$

Which means that the apparent power developed by the series winding must be:

$$S_{series} = I_{series} E_{series} = 10 \times (100 - 50) = 10 \times 50 = 500 \text{ V} \cdot \text{A}$$

Since the apparent power developed by the common winding must be the same as that developed by the series winding, then

$$I_{common} = \frac{S_{common}}{E_{common}} = \frac{S_{series}}{E_{common}} = \frac{500}{50} = 10 \text{ A}$$

So we now know the currents in every part of the circuit. If we apply Kirchhoff's Current Law to the junction between the series and common windings, then it should be apparent that the current in the common winding must be acting upwards. We can now label the schematic diagram with the currents and their directions as shown in Figure 10.42.

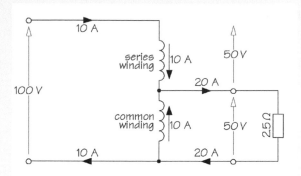

Figure 10.42

Worked example 13

Show how you can connect a 240/120 V, 500 V·A, *mutual* transformer, to provide a 360 V secondary voltage from a 120 V supply. Calculate the maximum load that can be connected to the transformer, and the values and directions of the supply-, load-, and common-winding currents.

Solution

Firstly, we must determine the *rated currents* of the two windings when used as a mutual transformer. We need to do this to ensure than the *actual* current in each winding doesn't exceed their rated value.

For the **240 V winding**:

$$(\text{Apparent Power})_{240\text{-V}} = \bar{E}_{240\text{-V}}\, \bar{I}_{240\text{-V}}$$

$$\bar{I}_{240\text{-V}} = \frac{(\text{Apparent Power})_{240\text{-V}}}{\bar{E}_{240\text{-V}}} = \frac{500}{240} = \textbf{2.08 A}$$

Next, for the **120-V winding**:

$$(\text{Apparent Power})_{120\text{-V}} = \bar{E}_{120\text{-V}}\, \bar{I}_{120\text{-V}}$$

$$\bar{I}_{120\text{-V}} = \frac{(\text{Apparent Power})_{120\text{-V}}}{\bar{E}_{120\text{-V}}} = \frac{500}{120} = \textbf{4.17 A}$$

*We must **not** allow either of these two values to be exceeded, or the transformer windings will overheat and, possibly, damage their insulation.*

Next, we should construct a schematic diagram of the required circuit. As the supply voltage is 120 V, we must configure the mutual transformer connection so that the common winding is the transformer's 120 V winding, and the series winding is the transformer's 240 V winding. This gives us a primary voltage of 120 V and a secondary voltage of 360 V. Remember that the windings must be connected such that the two voltages are *additive*.

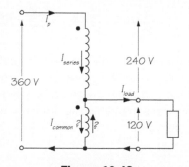

Figure 10.43

Since $\bar{I}_{series}$ which, of course, is the current through the 240 V winding, *mustn't* exceed 2.08 A without overheating, we can determine the maximum allowable apparent power *provided by the supply:*

$$\text{(apparent power)}_{supply} = \bar{E}_{supply}\bar{I}_{series}$$
$$= 360 \times 2.08$$
$$= 749 \; V \cdot A$$

Since the apparent power supplied to the load must be the same as the apparent power provided by the supply, we can now determine the current supplied to the load:

$$\bar{I}_{load} = \frac{749}{120} = \textbf{6.24 A} \;\; (\textbf{Answer})$$

Now let's determine the apparent power developed by the series winding:

$$\text{Apparent Power}_{series} = I_{series}U_{series}$$
$$= I_{load}U_{series}$$
$$= 2.08 \times 240 \approx 500 \; V \cdot A$$

Since the apparent power developed in the series winding must equal the apparent power developed by the common winding, we can now find the value of the current in the common winding:

$$I_{common} = \frac{S_{common}}{U_{common}} = \frac{500}{120} = \textbf{4.12 A (Answer)}$$

This current *doesn't* exceed the rated current of the common winding, so this is fine.

The *direction* of the current in the common winding can now be determined by applying **Kirchhoff's Current Law** to the junction between the common and series winding, from which it is obvious its direction must be ***upwards***.

The interesting thing about the above worked example is the way in which this 500 V·A mutual transformer can, when connected as an autotransformer, supply a load of very nearly 750 V·A, without either of its windings exceeding their rated currents. This demonstrates a method of significantly upgrading the rated apparent power of a mutual transformer without over-rating its windings – providing, of course, that the primary/secondary voltages requirements can be satisfied and the resulting lack of electrical isolation is not an issue or contravenes any regulation.

Conclusion

Now that we've completed this chapter, we need to examine its **objectives** listed at its start. Placing a question mark at the end of each objective turns that objective into a **test item**. If we can answer those test items, then we've met the objectives of this chapter.

Transformers 3: connections

Electrical Science for Technicians. 978-1-138-84926-6 © Adrian Waygood.
Published by Taylor & Francis. All rights reserved.

Objectives

On completion of this chapter you should be able to:

1. describe UK and North American single-phase residential transformer connections.

2. describe UK and North American three-phase residential transformer connections.

3. describe the open-delta three-phase transformer connection, its applications and advantages.

4. list and describe the four requirements for paralleling single-phase transformers.

5. describe the importance of knowing the polarity of a single-phase transformer.

6. describe the simple test for determining the polarity of a single-phase transformer.

7. describe the British Standard requirement for labelling transformer terminals.

8. describe the 'dot convention' method for labelling transformer terminals.

9. list and describe the additional two requirements required for paralleling three-phase transformers.

10. define the term, 'angular displacement', as it applies to three-phase transformer connections.

11. determine the angular displacement of a three-phase transformer bank, for any given connection.

12. determine the connection requirement for a three-phase transformer bank necessary to achieve any given angular displacement.

13. identify a three-phase transformer connection, together with its angular displacement, given its vector group data.

14. construct a phasor diagram for a three-phase transformer under load, given its primary and secondary connections, showing all phase and line voltages and currents.

Introduction to residential supplies

British and North American low-voltage distribution systems are rather different from each other, with North America tending to use a much greater number of distribution transformers, *per capita*, than is the practice in the UK.

Typically, in North America, distribution transformers tend to have a lower capacity (i.e. volt ampere rating) than those used in the Britain and typically single-phase, supplying just 1–10 residences whereas, those used in Britain are more likely to be three-phase, tend to have a much higher capacity and are more likely to supply a whole neighbourhood.

To some extent, this is because North American residential distribution line voltages are limited to 240 V whereas, in the UK, distribution line voltage is 400 V – and higher voltages, of course, allow greater distribution distances for any given conductor size.

The advantage with the North American system is that, in the event of a transformer fault, *fewer* consumers will be affected. This can be important in rural areas served by overhead distribution lines that are subject to damage by electrical storms.

The advantage with the British system is essentially one of economics, in that an increasing number of consumers *doesn't* require a corresponding increase in the capacity of the distribution transformer. This is because of what is called '**after diversity maximum demand**' (**ADMD**). ADMD is defined as *'the maximum demand calculated for a distribution substation, expressed in volt amperes per consumer'*.

The AMDM figure is *significantly lower* than a consumer's maximum load, and reflects the fact that *all* the consumers served by a particular substation will *never demand their maximum load at the same time* – e.g. when consumer *a* is at home, consumer *b* is at work; when consumer *x* cooks for a midday meal, consumer *y* cooks for an evening meal, etc. This is a fact derived from many years of monitoring consumer loads.

The ADMD *reduces* as the number of consumers supplied by a substation *increases*. Accordingly, the capacity of a distribution transformer (a) *never* has to meet the maximum load of *all* the consumers that it supplies and (b) an increase in the number of consumers is reflected by a *far lower* corresponding percentage increase in that capacity.

Residential distribution: single-phase supplies

Single-phase, pole-mounted, distribution transformers are widely used in rural areas to provide energy supplies to remote individual or small groups of residences, such as farmhouses, etc.

UK residential supplies

In the UK, a single-phase distribution transformer's primary winding is connected between any two line conductors of an 11 kV high-voltage three-phase line, while its secondary windings provide its load with a rated nominal voltage of 230 V (+10%/–6%). This transformer connection is shown in Figure 11.1.

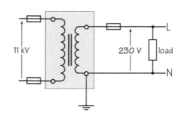

Figure 11.1

The secondary winding's **neutral (N)** is established by connecting one of its secondary terminals to earth. The **line (L)** conductor is connected to the other terminal, and acquires a nominal potential of 230 V with respect to the neutral. To provide overcurrent protection, the line conductor is always fused. The transformer itself is also protected with high-voltage fuses inserted into each of the two high-voltage line conductors.

North American residential supplies

In the United States and Canada, a *completely different* system is used for residential supplies. This is a dual-voltage system, and is termed a '**split phase**' system, although it is sometimes (but quite incorrectly) referred to as a 'two-phase' system. This transformer connection is shown in Figure 11.2.

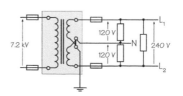

Figure 11.2

A 'two-phase' system is a now-obsolete, polyphase system, provided by an alternator having two armature windings, located 90 electrical degrees relative to each other. This is not the case for North American residential supplies, so using the term, 'two-phase' system is incorrect.

In North American homes, 'light loads' (i.e. most domestic appliances, such as electric kettles, toasters, television sets, etc.) are rated at 120 V, whereas 'heavy loads' (such as water heaters, etc.) are rated at 240 V. Some 'heavy load' appliances, such as washing machines and tumble dryers, are fitted with 120 V motors, but have 240 V heating elements.

As in the UK, the primary winding of a distribution transformer is connected between any two line conductors of a three-phase high-voltage distribution line. In North America, however, high-voltage distribution line voltages are *not* standardised to the same extent as they are in the UK where voltages were established when the electricity supply industry was nationalised; instead various line voltages are available, typically ranging between 7.2 kV – 34.5 kV.

The transformer's secondary consists of *two* windings, each having a nominal voltage of 120 V (±5%). These two windings are connected in series, with the connection between them connected to earth ('ground') and providing the neutral. This system provides *two* voltages: 240 V between the outer terminals (**L$_1$** and **L$_2$**), and 120 V between either of these outer terminals and the earthed (grounded) centre neutral terminal (**N**).

To provide overcurrent protection, both line conductors are fused. The transformer itself is also protected with high-voltage fuses inserted into each of the two high-voltage line conductors.

Residential distribution: three-phase supplies

Three-phase, pole-mounted distribution transformers are widely used in rural areas, whereas pad-mounted, or ground-mounted transformers tend to be used in urban areas where low-voltage distribution is typically by means of underground cables.

UK residential supplies

The practice in the UK is to use *delta-star* connected three-phase distribution transformers, with the secondary *star-connected* windings providing a combination of 400 V line-to-line (i.e. line voltage), and 230 V line-to-neutral (i.e. phase voltage). This arrangement is shown in Figure 11.3.

The main reason for using a delta high-voltage connection is as follows.

> *The high-voltage supply often contain* **harmonics** *(components of the supply current which are multiples of the fundamental supply frequency), which can result in distortion to the supply waveform. In a three-phase system, odd-numbered harmonics are in phase with each other and do not cancel at the neutral point of a star-connection in the same way that balanced 50 Hz phase currents do and, therefore, necessitate the installation of a neutral return conductor if a star connection is used. This can be avoided by always using a* delta *high-voltage connection, within which these harmonic currents are allowed to circulate and, in simple terms, become 'trapped' and not reflected into the secondary. This avoids any distortion to the supply waveform caused by those harmonics while, at the same time, removing the need for a return neutral. Further discussion on the causes and effects of harmonics in electricity supplies are beyond the scope of this book.*

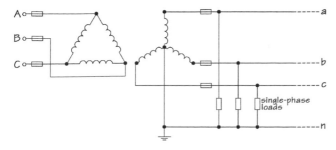

Figure 11.3

In order to *balance* the load, the electricity network companies always try to connect single-phase loads (e.g. individual residences) between alternate line conductors and the neutral conductor (i.e. **a–n**, **b–n**, **c–n**, **a–n**, etc.). Businesses requiring three-phase supplies are, of course, provided with three lines and a neutral. Three-phase supplies to residences are rare in the UK, but quite common in some European countries such as Cyprus.

Three-phase supplies can be provided by using *either* an individual three-phase transformer, *or* (less commonly) by using three individual single-phase transformers (Figure 11.4) connected to form a three-phase, delta-star, transformer '**bank**'.

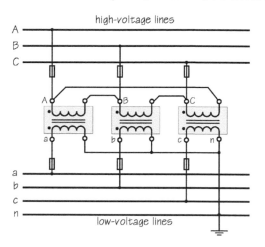

Figure 11.4

North American residential supplies

Because of the 'split-phase', dual-voltage, single-phase system used in North America, it is *not* practical to use a star-connected secondary. Instead, the standard practice in North America is to use *delta/delta*-connected distribution transformers, with the secondary of one of them centre-tapped to provide a split-phase secondary supply – as illustrated in Figure 11.5.

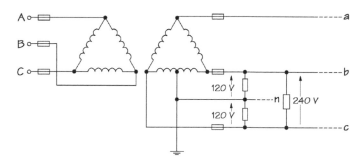

Figure 11.5

In Figure 11.5, phase **b–c** provides a single-phase split-voltage 240–120 V supply, while the line conductors **a–b–c** provide a three-phase supply having a line voltage of 240 V.

In this example, line **a** *plays no part in providing the single-phase supply*. A phasor analysis will reveal that the potential of line **a**, measured with respect to the neutral will be:

$$\bar{E}_{an} = \sqrt{240^2 - 120^2} = \sqrt{43\,200} \approx 208\,\text{V}$$

For this reason, American and Canadian electricians often describe this line as being the system's '**high leg**', and they must take great care to prevent loads being accidentally connected between this line conductor and the neutral when installing three-phase distribution panels.

Three-phase transformer *banks* (as shown in Figure 11.6) are far more widely used in North America than they are in the UK where, as already mentioned, the usual practice is to use three-phase transformers rather than individual single-phase transformers.

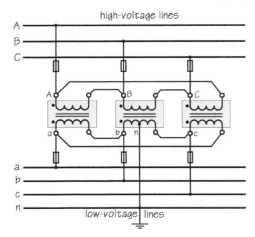

Figure 11.6

Open-delta connection

An advantage with the North American three-phase transformer bank is that, in the event of an electrical fault in which one of the three transformers is damaged, linesmen can quickly restore the three-phase supply (albeit temporarily and at a lower capacity) to a consumer by *removing* the damaged transformer, and *rewiring* the remaining two in what is known as an '**open-delta**' connection – until a replacement transformer becomes available. This arrangement is illustrated in Figure 11.7.

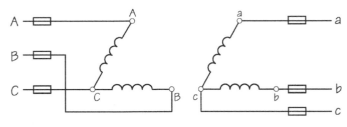

Figure 11.7

Let's examine the phasor diagram for the high-voltage side of this connection, as illustrated in Figure 11.8.

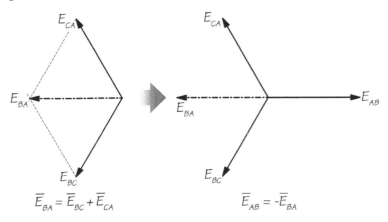

$$\overline{E}_{BA} = \overline{E}_{BC} + \overline{E}_{CA} \qquad \overline{E}_{AB} = -\overline{E}_{BA}$$

Figure 11.8

As you can see, the phasor-sum of $\bar{E}_{BC}$ and $\bar{E}_{CA}$ is $\bar{E}_{BA}$, which is exactly equivalent to a line voltage, $\bar{E}_{AB}$, *acting in the opposite direction!* In other words, the voltage phasor diagram for an open-delta connection is *exactly the same as that for a closed-delta connection,* which means that an open-delta arrangement will continue to provide a three-phase supply to its load.

However the *drawback* with an open-delta connection is that its capacity is *lower* than that for a normal, closed-delta, connection. So the open-delta connection can only be thought of as *a temporary solution* to the loss of one transformer.

Let's examine *why* there is a loss in capacity for this connection.

The 'capacity', or rated 'apparent power', of a closed-delta connection supplying a balanced load is given by:

$$S_{closed\,delta} = \sqrt{3}\,\bar{E}_L\bar{I}_L$$

However, for an 'open-delta' connection, the two line conductors (lines A and C in Figure 11.7) are *in series* with their corresponding windings. Accordingly, the line currents (normally $\sqrt{3}$ times larger than the phase currents) must not exceed the rated phase currents for those windings – in other words, *the line currents must not exceed the rated phase currents.* This has the effect of lowering the capacity of the transformer bank, as follows:

$$S_{open\,delta} = \sqrt{3}\,\bar{E}_L\bar{I}_P = \sqrt{3}\,\bar{E}_L\left(\frac{\bar{I}_L}{\sqrt{3}}\right) = \bar{E}_L\bar{I}_L$$

So, if we examine the ratio between the capacities of the open- and closed-delta connections:

$$\frac{capacity_{open\text{-}delta\,connection}}{capacity_{closed\text{-}delta\,connection}} = \frac{\bar{E}_L\bar{I}_L}{\sqrt{3}\,\bar{E}_L\bar{I}_L} = \frac{1}{\sqrt{3}} \approx 0.58$$

$$capacity_{open\text{-}delta\,connection} \approx 0.58 \times capacity_{closed\text{-}delta\,connection}$$

> So, while an **open-delta** connection allows us to *provide a temporary three-phase supply to a consumer* following the loss of one of the three transformers, the drawback is that *it can only supply 58% of rated capacity of the original closed-delta connection.* In other words, *the consumer will have to temporarily reduce his load by 42%, until the missing transformer is replaced.*

As well as providing a temporary three-phase supply to a consumer following the loss of one transformer, the open-delta connection may also be used to supply an area in which projections indicate a future increase in load. Initially, the capacity of the open-delta transformer bank must be sufficient to supply the existing load but, as that load increases over time, a third transformer can be added to satisfy that increase in demand.

Parallel operation of transformers

Introduction

There are a number of requirements that must be satisfied before two or more **single-phase transformers** can be 'paralleled' – i.e. before they can be connected in parallel with each other, in order to supply the same load. These requirements are:

▶ same **voltage ratio (turns ratio)**
▶ similar **percentage impedance**
▶ similar **kV·A rating**
▶ known **polarity.**

In order to parallel two or more **three-phase transformers**, all the above requirements must be satisfied, together with two *additional* requirements:

▶ same **phase sequence**
▶ same **angular displacement.**

Let's start by looking at the requirements for paralleling single-phase transformers.

Paralleling single-phase transformers

Before we can parallel two or more **single-phase transformers**, the following requirements must be satisfied.

Same voltage ratio

If two single-phase transformers with different **voltage ratios** (or **turns ratios**) are connected in parallel across a common primary voltage, then their secondary voltages will obviously be different. Under 'no-load' conditions (i.e. with no load connected), this will result in a circulating current between the loop formed by the two secondary windings. As the impedance of a transformer's windings is low, this circulating current can be quite high, resulting in unnecessarily high I^2R losses.

When 'loaded', this circulating current will prevent the transformers from sharing that load equally. When fully loaded, this can result in one or other of the transformers becoming overloaded (i.e. exceeding its rated capacity).

Similar percentage impedance

The term '**percentage impedance**' *(%Z)* is a little confusing, as it's defined in terms of voltages – i.e. *'the percentage of a transformer's rated primary voltage that will cause rated current to circulate through the secondary winding when that winding is short-circuited'*.

A transformer's percentage impedance can be determined by short-circuiting the secondary winding with an ammeter, and gradually increasing the primary voltage until rated current flows in the secondary. The percentage impedance is then simply the ratio of that particular primary voltage to the *rated* primary voltage, expressed as a percentage.

So, for example, if a particular transformer has a percentage impedance of, say, **'5%'**, then it would take just 5% of the rated primary voltage to cause 100% of the rated secondary current to flow through the short-circuited secondary winding.

A transformer's percentage impedance acts to limit the value of fault current that could circulate in its secondary winding if that winding were to be inadvertently short-circuited. For example, if a transformer has a percentage impedance of 5%, and its rated secondary current is, say, 100 A, then the resulting secondary short-circuit current would be limited to:

$$\overline{I}_{short\ circuit} = \frac{\overline{I}_{rated}}{\%Z} = \frac{100}{0.05} = 2000\ A$$

A transformer's percentage impedance is shown as part of the specifications listed on its nameplate, and is related to the kV·A-rating of a transformer.

If two transformers, **Tx1** and **Tx2**, as shown in Figure 11.9, each with identical kV·A ratings, but having slightly different percentage impedances (Z_1 and Z_2), are paralleled, then the transformer with the *lower* percentage impedance will contribute the *greater* percentage of the total secondary load current. So the transformer with the *lower*

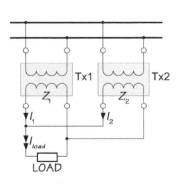

Figure 11.9

233

percentage impedance will tend to reach its rated capacity *before* the transformer with the *higher* percentage impedance.

In other words, the transformer with the *lower* percentage impedance will become overloaded *before* the *combined* kV·A output reaches the sum of the rated kV·A values of the two transformers. So the maximum load the two transformers can supply must be *lower* than their combined rated capacities.

Closely related to percentage impedance is the ratio of **reactance-to-resistance ratio** for individual transformer's windings. Any significant difference in this ratio between two transformers will result in currents with different phase angles flowing in those transformers. So, with the two transformers operating at different power factors, the load's true power (expressed in watts) will not be proportionally shared between them and, again, their combined capacity will be reduced. Fortunately, for matched transformers whose percentage impedances are close, this isn't really an issue.

Identical percentage impedances are difficult to achieve in practice, so the international standard published by the **IEC (International Electrotechnical Commission)** allows a variation or tolerance of ±7.5% for of percentage impedances of 10% or more, and ±10% for percentage impedances values less than 10%.

Similar kV·A rating

Transformers with *different* kV·A ratings will share the load more-or-less in proportion to those ratings (i.e. with each transformer carrying roughly its own share of the load), *providing* their voltage ratios are *identical* and their percentage impedances are *close*. However, it is generally recommended that the kV·A-rating of any two transformers should never differ by more than a ratio of 2:1.

Known polarity

The '**polarity**' of a transformer describes the *instantaneous direction of the potential difference induced across the secondary terminals of that transformer, relative to that across the primary terminals.*

The polarities of transformers connected in parallel must be known in order to avoid connecting the transformers incorrectly, which will result in very large circulating currents flowing between the secondary windings of the two transformers.

Polarity of single-phase transformers

At this point, we need to understand *what* we mean by the '**polarity**' of a single-phase transformer, *why* it is important, and *how* it can be determined.

We already know that, if we want to connect two identical batteries in parallel, we must *identify, and take into account*, the polarities of their terminals. Failure to do so could result in a very large circulating current between the two batteries, which is likely to do significant damage! The left-hand illustration in Figure 11.10 shows the *correct* connection, whereas the right-hand illustration shows the *incorrect* connection which results in a large, circulating, current.

The 'correct' connection is when *like* polarities are connected together (i.e. positive-to-positive, and negative-to-negative).

Well, *exactly* the same principle applies when we connect a pair of transformers in parallel. We *must* take into account the polarities of their secondary terminals in order to avoid large circulating currents flowing around their secondary windings.

But, with alternating current, it's not immediately obvious what we *mean* by 'polarity' because, surely, the polarity of a transformer's secondary terminals is continuously reversing? So how, we might ask, can a transformer possibly have 'polarity'?

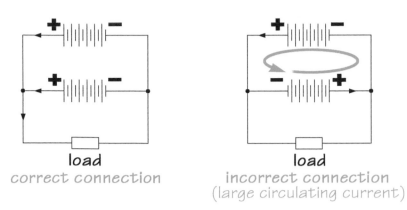

correct connection

incorrect connection
(large circulating current)

Figure 11.10

The answer is that, whenever we talk about the 'polarity' of a transformer, we are describing the *instantaneous* polarities appearing at its secondary terminals. In other words, if we could freeze time, what would the polarities of its secondary terminals be, relative to the polarities of its primary terminals, at that particular instant? Or, to put it another way, in which direction will the induced secondary voltage be acting, relative to the primary voltage?

Well, the answer is that it depends on the *direction* in which the secondary winding has been wound, relative to the primary winding. The left-hand illustration in Figure 11.11 shows the instantaneous polarities if the secondary winding had been wound in one direction, while the right-hand illustration shows the instantaneous polarities if the secondary winding had been wound in the *opposite* direction.

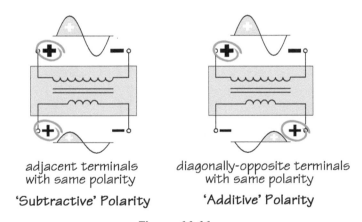

adjacent terminals
with same polarity

'Subtractive' Polarity

diagonally-opposite terminals
with same polarity

'Additive' Polarity

Figure 11.11

If, at any given instant, the polarities of the *adjacent* high- and low-voltage terminals are identical – e.g. both 'positive', as in Figure 11.11 (left) – then we describe the transformer as having '**subtractive polarity**'. If, on the other hand, the polarities of the *diagonally opposite* high- and low-voltage terminals are identical, as in Figure 11.11 (right), then we describe the transformer as having '**additive polarity**'. The significance of the terms, 'subtractive' and 'additive' polarity, will become apparent, shortly.

If we want to connect two transformers in parallel, exactly the same rule applies as for batteries: i.e. the 'correct' connection is when terminals having *like (instantaneous) polarities are connected together*.

So *how* do we determine whether a particular single-phase transformer has '**subtractive**' or '**additive**' polarity, and *what* exactly do these two terms actually mean?

235

If a transformer's terminals are *already* labelled, then whether that transformer is 'subtractive' or 'additive' isn't really an issue (although you may wish to confirm those labels are correct)! It only becomes an issue if the terminals are *unlabelled*, in which case we need to be able to label them ourselves.

In the UK, the practice is to label a single-phase transformer's terminals with a *letter*, followed by a *numerical subscript*, where:

▶ the **high-voltage** terminals are labelled with the *upper-case* letter '**A**', and the **low-voltage** terminals with the corresponding *lower-case* letter '**a**'. When three single-phase transformers are used to form a three-phase transformer bank, the other transformers' high-voltage terminals are labelled '**B**' and '**C**' respectively, and their corresponding low-voltage terminals are labelled '**b**' and '**c**'.

▶ one end of a winding (arbitrarily termed the '**start**' end) is labelled with an *odd-numbered* subscript (e.g. 'A_1'), while the other end (termed the '**finish**' end) is identified with an *even-numbered* subscript (e.g. 'A_2').

So, for a straightforward, single-phase, two-winding transformer, the high-voltage terminals will be labelled 'A_1–A_2', and the low-voltage terminals will be labelled 'a_1–a_2'. If the transformer has any *additional* windings, then their terminals can be labelled 'A_3–A_4', 'a_3–a_4', etc., as necessary.

> In North America, the practice is to label the high-voltage terminals with the upper-case letter, '**X**', and the low-voltage terminals with the upper-case letter, '**X**', each with numerical subscripts that follow the same rule as the UK system.

In accordance with the **BS 171/EN 60076** standard for transformers, *when viewed from the high-voltage side of the transformer*, terminal A_1 is always to the **left**; *when viewed from the low-voltage side of the transformer*, terminal a_1 is always on the **right**. So, as you can see from Figure 11.12, the UK practice is for *all* transmission/distribution transformers to have **subtractive polarity**. Again, the reason why subtractive polarity is preferred will become apparent shortly.

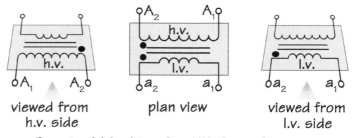

viewed from   plan view   viewed from
h.v. side   l.v. side

Terminal Marking for UK Transformers

Figure 11.12

> In North America, terminal H_1 is always to the *right*, when viewed from the high-voltage side of the transformer – i.e. opposite to UK practice. The standard in North America specifies that, *when viewed from the low-voltage side,* i.e. with the high-voltage windings on the *far* side of the transformer, the H_1 terminal should be to the *left*.

If a single-phase transformer's terminals are unmarked, then it's necessary to *assign* labels to those terminals.

The first step is to *identify the high- and low-voltage terminals*. We can do this by using an ohmmeter to conduct a **continuity test** to confirm which terminals are connected together via a winding and, then, by *measuring the resistance* of each winding, as shown in Figure 11.13.

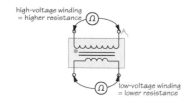

high-voltage winding
= higher resistance

low-voltage winding
= lower resistance

Figure 11.13

Because the high-voltage winding has *a greater number of turns* than the low-voltage winding, and carries a *lower* current, its conductor must obviously be *longer* and have *a lower cross-sectional area*, giving it *a greater* resistance than the low-voltage winding.

A **resistance test**, therefore, allows us to identify and label each high-voltage terminal (higher resistance) with the upper-case letter '**A**', and to identify and label each low-voltage terminal (lower resistance) with the lower-case letter '**a**'. Furthermore, we can label the *left-hand* high-voltage terminal, *as viewed from the high-voltage side of the transformer*, as '**A₁**', and we can label the other high-voltage terminal '**A₂**' – as illustrated in Figure 11.14 below:

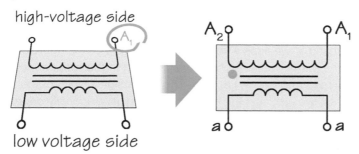

Labelling the terminals following continuity/resistance test

Figure 11.14

Although we can now label the low-voltage terminals using the lower-case letter '**a**', at this point we *can't* apply any subscripts to them yet, because we don't know the polarity of the transformer. This requires us to perform a '**polarity test**'.

To conduct this test, we start by applying a short length of conductor between high-voltage terminal **A₁** and its *adjacent* low-voltage terminal. This, essentially, places the high- and low-voltage windings *in series with each other*. A voltmeter is then connected between the opposite pair of high- and low-voltage terminals, as shown in Figure 11.15.

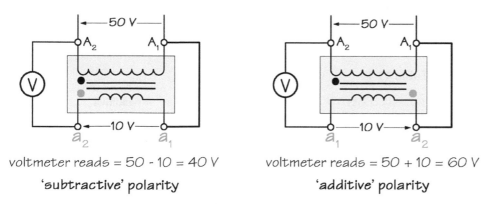

voltmeter reads = 50 - 10 = 40 V   voltmeter reads = 50 + 10 = 60 V

'subtractive' polarity   **'additive' polarity**

Figure 11.15

Next, we apply a small a.c. voltage across the high-voltage winding, which induces a smaller voltage into the low-voltage winding, determined by its turns ratio. In Figure 11.15, for the sake of simplicity, we're assuming that our transformer has a turns ratio of 5:1 and so, if we apply, say, 50 V a.c. to the high-voltage winding, then a voltage of 10 V a.c. will appear across the low-voltage winding.

> In UK practice, when energised, *even*-numbered terminals are considered to be at a *higher* potential than odd-numbered terminals. This is reflected in Figure 11.15, in which the voltage sense arrow representing the 50 V applied to the primary

winding *points from terminal A_1 towards terminal A_2* and is, therefore, 'read' as *'the potential of terminal A_2 measured with respect to terminal A_1'.*

Throughout this chapter, we have also used the '**dot convention**' to identify those terminals that are at the higher potential. So the terminals identified with a dot correspond to the terminals marked 'A_2' (high voltage) and 'a_2' (low voltage), etc. Strictly speaking, it's *not* necessary to use *both* conventions.

If we now apply **Kirchhoff's Voltage Law** to the loop formed by the two windings and the voltmeter, we see that the primary and secondary voltages in the left-hand diagram *oppose each other* and, so, the voltmeter will read their *difference* – i.e. 40 V. This is the significance of the term '**subtractive** polarity' and, so, we can label the low-voltage terminal *adjacent* to terminal 'A_1' as 'a_1'.

On the other hand, if we apply Kirchhoff's Voltage Law to the right-hand diagram, we see that the primary and secondary voltages act in the *same direction*, so the voltmeter will read their *sum* – i.e. 60 V. This is the significance of the term '**additive** polarity' and, so, we can label the low-voltage terminal *diagonally opposite* terminal 'A_1' as 'a_1'.

The term, '**additive**', describes the way in which, as we work our way, clockwise, around the transformer, the windings' voltages *add together*. The term, '**subtractive**', describes the way in which, as we work our way, clockwise, around the transformer, the winding's voltages *subtract from each other*.

Modern practice dictates that power and distribution transformers should be of **subtractive polarity** on the basis that the voltage appearing between any adjacent pair of terminals *should never exceed the voltage applied across the high-voltage terminals*. In the example of the additive transformer shown in Figure 11.15 the voltage appearing between adjacent high- and low-voltage terminals exceeds that applied to the high-voltage terminals, which is considered unacceptable.

While this is the certainly the case in the UK, in North America (for historical reasons) **NEMA** and **EEMAC (National Electrical Manufacturers Association** and **Electrical Equipment Manufacturers Association of Canada)** recommend the following standards:

Table 11.1

200 kV·A or less	h.v. windings at 8660 V or less:	**additive**
200 kV·A or more	h.v. windings at 8660 V or less:	**subtractive**
any value of kV·A	h.v. windings at 8660 V or more:	**subtractive**

Now that we have learnt how to *determine whether a transformer is subtractive or additive*, which enables us to correctly *label its terminals*, we can learn how to connect it in parallel with another transformer.

As already explained, if we want to connect two transformers in parallel, exactly the same rule applies as for batteries: i.e. the 'correct' connection is when terminals with *like (instantaneous) polarities – i.e. with identical labels – are connected together.*

Despite the standards for subtractive- and additive-polarity transformers, it's useful to know how to parallel transformers that have different polarities.

Figure 11.16 (left) shows how we can connect two *subtractive* transformers in parallel, and how (right) we can connect an *additive* transformer in parallel with a *subtractive* transformer. Fuses have been omitted for clarity.

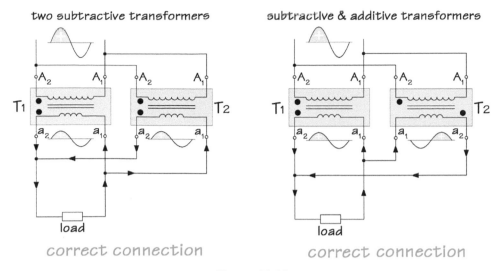

Figure 11.16

Note that, for both connections, similarly labelled terminals are connected together: i.e. **A₁** to **A₁**, and **a₁** to **a₁**.

Figure 11.17 shows what happens if the two transformers' secondary windings are *incorrectly* connected – i.e. with terminals having *opposite* polarity (**a₁** connected to **a₂**) connected together. This particular schematic diagram shows an additive and a subtractive transformer, but the result would be the same if the two transformers were identical. The result is a very large circulating current through the secondary windings, which may cause severe damage to those windings.

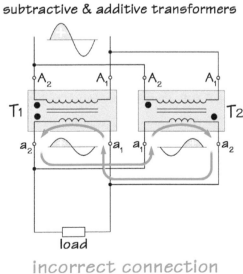

Figure 11.17

Finally, regardless of how *confident* we are that the two transformers have been wired together correctly, no transformers should *ever* be paralleled without *conducting a final test to* **confirm** *that they have been connected correctly.*

This test, which is *essential* to prevent damage to the transformers and possible harm to nearby personnel, is performed as shown in Figure 11.18.

If the transformers have been labelled and installed *correctly*, then the voltmeter will read 'zero' volts, and it is safe to replace the voltmeter with a permanent conductor. If

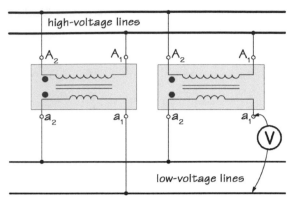

high-voltage lines

low-voltage lines

for correct connection, voltmeter reads 'zero'

Figure 11.18

the transformers have been installed incorrectly, then the voltmeter will read twice the value of the secondary voltage, and the transformers *cannot* be paralleled until they have been reconnected properly.

Paralleling three-phase transformers and banks

Now let's turn our attention to how **three-phase transformers** (and **three-phase transformer banks**) can be connected in parallel.

Before we can parallel two or more **three-phase transformers**, the requirements necessary for paralleling single-phase transformers still apply, together with the following *additional* requirements must be satisfied.

Same phase sequence

The '**phase sequence**' of a three-phase alternating-current system describes *the order in which the potential of each line terminal reaches its peak, positive, value.* The 'normal', or 'positive', phase sequence occurs in the order **A-B-C**. If, instead, the phase sequence should be **A-C-B**, then it is described as having a 'negative' phase sequence.

Although unusual, different distribution systems *can* have opposite phase sequences for a variety of reasons that are beyond the scope of this book. For this reason, whenever two three-phase transformers are supplied from separate voltages sourced from *different* distribution systems, the phase sequence of those two sources *must* be confirmed *before* any attempt is made to parallel the transformers.

If the primary windings of one three-phase transformer are supplied from a distribution system operating with a *positive* phase sequence, and the primary windings of a second three-phase transformer are supplied from a different system operating with a *negative* phase sequence then, during every cycle, each pair of phase windings will be short-circuited. A situation that must, obviously, be avoided at all cost!

Fortunately, this situation is comparatively rare and can only occur when the primary windings are either incorrectly connected, or connected to separate source voltages that originate from different distribution systems having different phase sequences. It *cannot* occur if the primary windings of both transformers are connected correctly to the *same* distribution system.

Same angular displacement

'**Angular displacement**' (also known as '**phase displacement**', which is a little misleading as it describes the relationship between line voltages, *not* phase voltages!)

is, to a *three-phase* transformer, what '**polarity**' is to a *single-phase* transformer. Angular displacement determines *whether* and, if so, *how*, two or more three-phase transformers may be connected in parallel with each other.

> **Angular displacement** is defined as *'the angle by which the secondary line-to-line voltages **lag** the corresponding primary line-to-line voltages'*.

The angular displacement for any particular connection can be determined directly from a phasor diagram representing that connection, by measuring (for a step-down transformer) the (*clockwise*) angle by which the *secondary* line voltages (e.g. $\bar{E}_{ab}$) lag their corresponding *primary* line voltages (e.g. $\bar{E}_{AB}$).

Angular displacement

As explained above, '**angular displacement**' is, to a *three-phase* transformer, what '**polarity**' is to a *single-phase* transformer.

The '**polarity**' of a single-phase transformer's secondary winding can only be either *in phase*, or 180° *out of phase*, with its primary winding. The '**angular displacement**' of a three-phase transformer's secondary line voltages, on the other hand, *can vary in increments of 30°*, measured relative to its primary line voltages.

Regardless of their polarities, two single-phase transformers can *always* be paralleled with each other, *provided* the appropriate terminals are interconnected correctly and the other requirements discussed earlier are met.

This, however, is *not* true for angular displacement; three-phase transformers having *different* angular displacements simply *cannot be connected in parallel with each other*.

Angular displacement, therefore, determines *whether* and, if so, *how*, two or more three-phase transformers may be paralleled with each other.

As we'll learn, shortly, angular displacement can be measured directly from a phasor diagram, by measuring (for a step-down transformer) the *clockwise* angle by which a secondary line voltage (e.g. $\bar{E}_{ab}$) is displaced from its corresponding primary line voltage (e.g. $\bar{E}_{AB}$).

A **three-phase transformer** is internally wired at the factory by its manufacturer, and its angular displacement is shown on its nameplate. In practice, national and international standards have limited the range of possible angular displacements to just *four* (0°, 180°, +30°, and −30°), with those transformer connections having the *same* angular displacement being assigned into one of *four* different '**vector groupings**', defined according to standards, such as **BS 171/EN 60076**.

This system of 'vector grouping' is fully explained towards the end of this chapter but, for now, all we need to understand is that the three-phase transformers within the *same vector grouping* all have exactly the *same angular displacement* and, so, can be paralleled with each other (providing all other requirements are met, of course). But transformer connections from one grouping *cannot* be paralleled with transformer connections from a different grouping.

Problems arise, however, with three-phase transformer *banks*. A **three-phase transformer bank** comprises three identical single-phase transformers connected together, often on site, to provide three-phase transformation. During installation, the way in which these single-phase transformers are connected together (delta/delta, wye/delta, etc.) determines the resulting angular displacement for that bank and, therefore, whether or not it can then be paralleled with another three-phase transformer bank.

So an understanding of *how* to connect a three-phase transformer bank in order to achieve a desired angular displacement is very useful.

Three-phase transformer bank connections

There are basically *two* types of situation which we encounter when dealing with the **angular displacement** of a three-phase transformer bank. These are:

1 For any given three-phase transformer bank connection, determine its angular displacement.
2 For any desired value of angular displacement, determine the transformer bank connection necessary to achieve that angular displacement.

For any given three-phase transformer bank connection, determine the angular displacement

In order to solve problems of this nature, we start by constructing a phasor diagram for the transformer bank connection, following the steps outlined below:

Step 1: Draw the primary phase voltages
Step 2: Draw the primary line voltages
Step 3: Draw the secondary phase voltages
Step 4: Draw the secondary line voltages
Step 5: Measure the angular displacement – i.e. the angle by which the secondary line voltage **lags** (i.e.: measured in the clockwise direction) the corresponding primary line-voltage.

The technique that follows, for constructing a three-phase phasor diagram, using **double-subscript notation** is fully described in the companion book, ***An Introduction to Electrical Science***.

Worked example 1

Given three identical, subtractive-polarity, step-down single-phase transformers, connected in wye/wye as shown below, determine the resulting angular displacement.

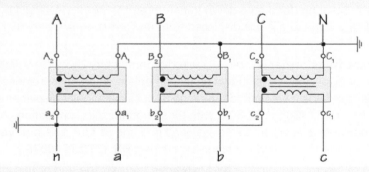

Figure 11.19

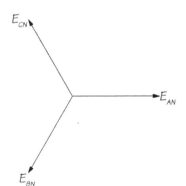

Figure 11.20

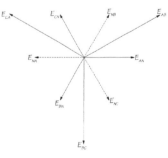

Figure 11.21

Solution

For clarity, each step will be drawn separately, to show how the final phasor diagram is built up. In practice, of course, only one phasor diagram need be constructed.

Step 1: Draw the primary phase voltages (Figure 11.20).

Step 2: Draw the primary line voltages (Figure 11.21), where:

$$\bar{E}_{AB} = \bar{E}_{AN} + \bar{E}_{NB}$$
$$\bar{E}_{BC} = \bar{E}_{BN} + \bar{E}_{NC}$$
$$\bar{E}_{CA} = \bar{E}_{CN} + \bar{E}_{NA}$$

Step 3: Draw the secondary phase voltages as shown in Figure 11.22 (for clarity, we have removed the 'construction phasors', shown in Figure 11.21).

Note Since $E_{a2\text{-}a1}$ is *always* in phase with $E_{A2\text{-}A1}$, then, in this case, $E_{n\text{-}a}$ must be in phase with $E_{A\text{-}N}$ ... etc. (Figure 11.22).

Step 4: Draw the secondary line voltages (Figure 11.23).

$$\bar{E}_{ab} = \bar{E}_{an} + \bar{E}_{nb}$$
$$\bar{E}_{bc} = \bar{E}_{bn} + \bar{E}_{nc}$$
$$\bar{E}_{ca} = \bar{E}_{cn} + \bar{E}_{na}$$

Step 5: Measure the clockwise angle between *any* secondary line voltage phasor, and its corresponding primary line voltage phasor (Figure 11.24) –e.g: between $\bar{E}_{ab}$ and $\bar{E}_{AB}$.

In this case, the angle by which the secondary line voltages lag the primary line voltages is 180°, so for this transformer connection, the **angular displacement is 180°**.

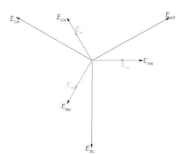

Figure 11.22

For any given angular displacement, determine the necessary transformer connection

In order to solve problems of this nature, we start by constructing a phasor diagram that gives us the specified angular displacement, following the steps outlined below:

Step 1: Draw the primary phase voltages

Step 2: Draw the primary line voltages

Step 3: Draw the secondary line voltages *at the specified angle* (i.e.: the angular displacement) *lagging the corresponding primary line voltages*

Step 4: Construct the secondary phase voltages, from the positions of the secondary line voltages

Step 5: Determine the polarity relationships between the secondary and primary phase voltages, and connect the secondary of the transformer-bank accordingly.

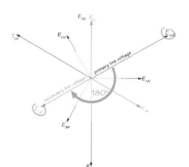

Figure 11.23

Worked example 2

Given three single-phase, subtractive-polarity, step-down single-phase transformers, with their primary high-voltage windings connected in **delta**, determine the necessary low-voltage connection in order to achieve an angular displacement of 210°. (Note that this is not a 'standard' angular displacement, listed in the vector groupings table, but we have deliberately chosen this figure to demonstrate that angular displacements with *any* multiple of 30° is possible.)

Solution

At this stage, it's always useful to make a sketch of the primary (high-voltage) connection; we cannot, though, draw the secondary (low-voltage) connection *until* it has been determined from the phasor diagram (Figure 11.25).

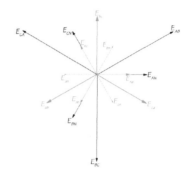

Figure 11.24

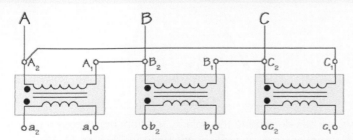

Figure 11.25

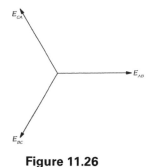

Figure 11.26

Note: The requirement is:

$$\bar{E}_{ab} \text{ is to lag } \bar{E}_{AB} \text{ by } 210°$$
$$\bar{E}_{bc} \text{ is to lag } \bar{E}_{BC} \text{ by } 210°$$
$$\bar{E}_{ca} \text{ is to lag } \bar{E}_{CA} \text{ by } 210°$$

Steps 1 and 2: Because the primary windings are connected in delta, the phase- and line voltages are, of course, the same. So, using $\bar{E}_{AB}$ as the reference phasor, construct the primary line voltage phasor diagram (Figure 11.26).

Step 3: Next, draw the secondary line voltages (Figure 11.27), at 210° *lagging* the primary line voltages (e.g. with $\bar{E}_{ab}$ lagging $\bar{E}_{AB}$ by 210°).

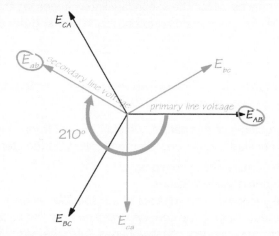

Figure 11.27

Step 4: Construct the secondary phase voltages from the positions of the secondary line voltages (Figure 11.28). As this is a 'reverse' step from the usual sequence (we normally construct the line voltages *from* the phase voltages, not the other way around!), it would be a good idea to sketch the secondary phasors on a separate piece of paper, then orientate that piece of paper until its secondary line voltage phasors line up with the main phasor diagram, and then transfer those phase voltages across onto the main phasor diagram:

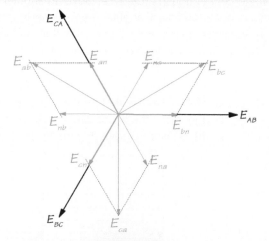

Figure 11.28

Step 5: Now, we can determine the polarity relationships between the primary (high voltage) and secondary (low voltage) phase voltages (Figure 11.29).

From the above phasor diagram, it is clear that:

$$\bar{E}_{bn} \text{ is in phase with } \bar{E}_{AB}$$

$$\bar{E}_{cn} \text{ is in phase with } \bar{E}_{BC}$$

$$\bar{E}_{an} \text{ is in phase with } \bar{E}_{CA}$$

Bearing in mind the 'Golden Rule' that $\bar{E}_{a2-a1}$ is *always* in phase with $\bar{E}_{A2-A1}$, the connection diagram can now be completed (Figure 11.29).

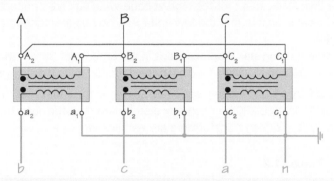

Figure 11.29

Practical implications of angular displacement

In the first of the two worked examples, above, we saw that the **wye/wye** transformer connection resulted in an angular displacement of **180°** while, in the second worked example, the **delta/wye** connection resulted in an angular displacement of **210°**. Having two completely different angular displacements means that *these two particular transformer banks can never be connected in parallel with each other*.

In fact, there are so many possible transformer connections, resulting in so many different values of angular displacement, that various national and international standards have evolved in order to reduce the numerous possible three-phase transformer connections into just **four** *standardised* groupings *based on their angular displacements*. These four groupings are generally known as '**vector groupings**', *and only those transformer connections within the same group can be paralleled with each other*.

The table illustrated in Figure 11.31, shows the **BS 171/EN 60076** standard's recommended **vector groupings** for various combinations of delta and star windings (less common connections, such as '**zig-zag**', have been removed from the table as they are outside the scope of this book).

As you can see, these connections are arranged into *four* main '**groups**', according to their resulting angular displacement.

Remember, **angular displacement** is defined as *'the angle by which the secondary line voltages **lag** the primary line voltages'*, and is measured in increments of 30°.

Despite the above definition, **vector group** tables usually express angular displacement in terms of the positive or negative angle measured between the secondary line voltages and their corresponding primary line voltages. For example, the angular displacement specified for **Group 4** connections is shown as '**+30°**', which is the angle measured counterclockwise from a primary line voltage – this corresponds to the secondary line voltage *lagging* the corresponding primary line voltage by 330°.

▶ **Group 1** comprises connections giving a **0°** angular displacement
▶ **Group 2** comprises connections giving a **180°** angular displacement
▶ **Group 3** comprises connections giving a **–30°** angular displacement
▶ **Group 4** comprises connections giving a **+30°** angular displacement.

Individual connections are identified, using a code such as '**41 Dy 11**' which, in this case, represents a **delta-star** connection in which:

▶ **4** – indicates the **main group** number.
▶ **1** – indicates the **first connection** within that group.
▶ **D** – (upper case) indicates that the high-voltage windings are connected in '**delta**'.
▶ **y** – (lower case) indicates that the low-voltage windings are connected in '**star**'('**wye**').
▶ **11** – indicates the **angular displacement** in increments of 30°, in this case equal to (11×30° =) 330° measured in a *clockwise* direction (or 30° *leading*).

In the UK, the **41 Dy 11** connection, specified above, is the most common designation for three-phase 11 kV/400-230 V secondary distribution transformers. Vector groupings for other primary distribution transformers are as follows:

Table 11.2

Transformation ratio:	Connection:
132/66 kV	32 Yd 1
132/33 kV	32 Yd 1
132/11 kV	32 Yd 1
66/11 kV	11 Yy 0
33/11 kV	11 Yy 0
11 kV/400-230 V	41 Dy 11

There are exceptions to the connections listed above, particularly with old installations.

Another way of explaining '**11**', in the above example, is by using the so-called '**clock convention**', as illustrated in Figure 11.30.

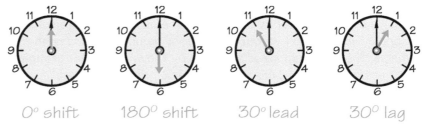

0° shift   180° shift   30° lead   30° lag

clock convention representing vector groups

Figure 11.30

The '**clock convention**' works as follows. Imagine a clock in which the **minute hand** represents the *primary line voltage* (i.e. the 'reference'), and the **hour hand** represents the *secondary line voltage*. The angle between two consecutive numbers on the clock represents an angle of 30°. So, if the hour hand is pointing at '11', then it is indicating (11×30°) = 330° lag, or a +30° lead.

The 'star' and 'delta' symbols shown in **columns 4** and **5** of Figure 11.31 indicate the relationship between the high-voltage and low-voltage winding connections. Although these symbols are shown in different columns, they can be more easily understood if

Vector Group Table for Star- and Delta- Connections

angular displace -ment	main group number	vector group ref. number & symbol	markings of line terminals and phasor diagram of induced voltages		winding connections and relative positions of terminals
			h.v. winding	l.v. winding	
0°	1	11 Yy 0	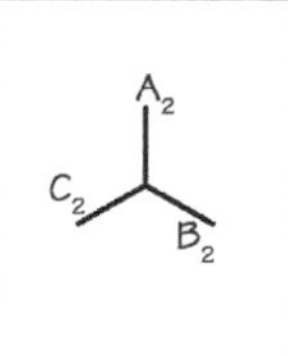	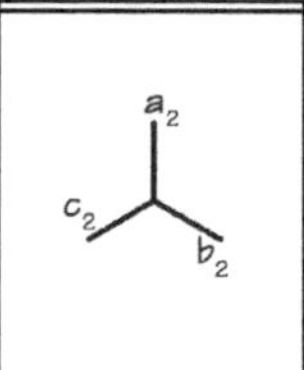	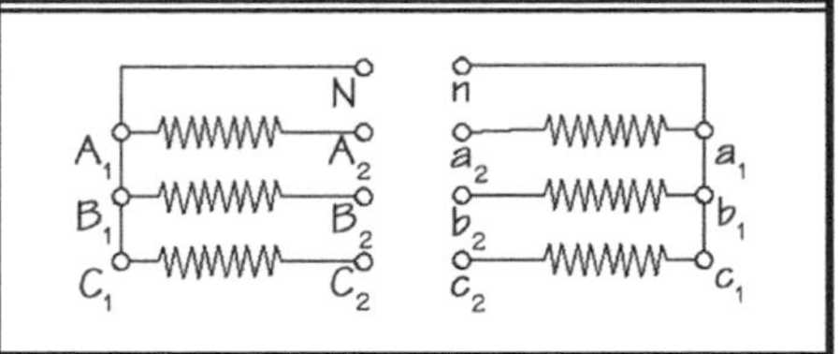
		12 Dd 0	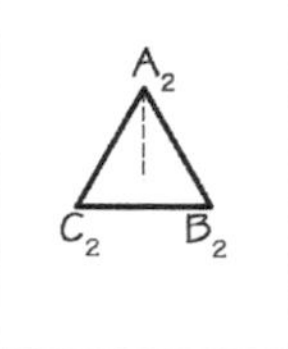	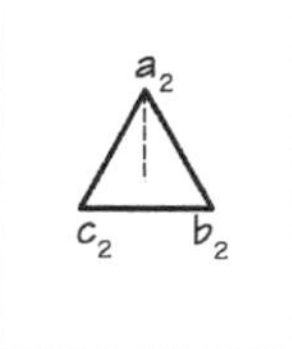	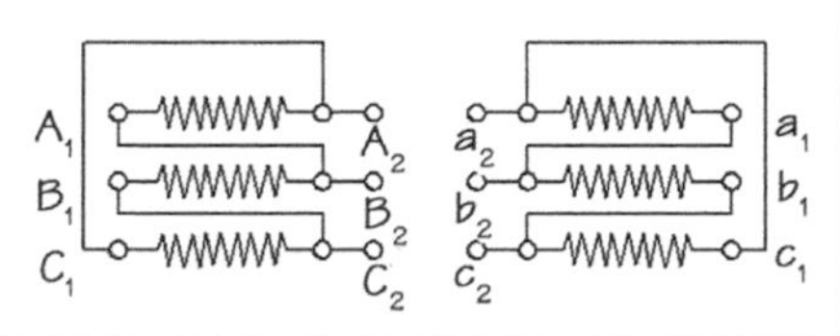
180°	2	21 Yy 6	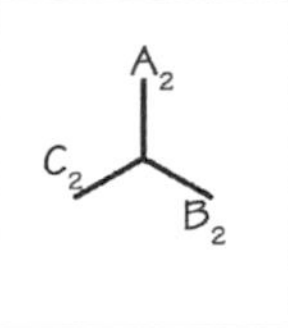	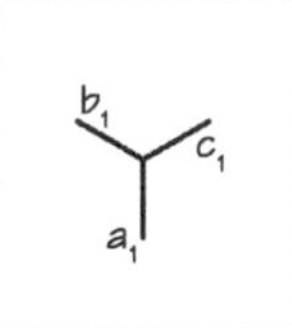	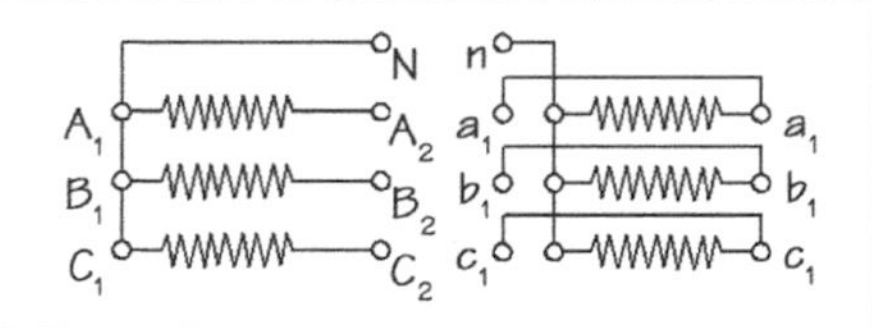
		22 Dd 6	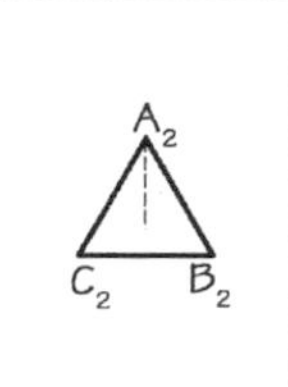	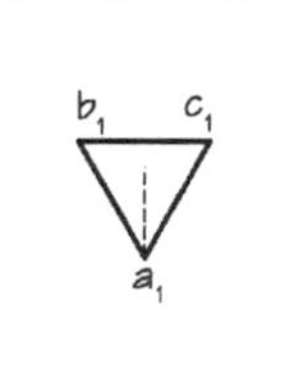	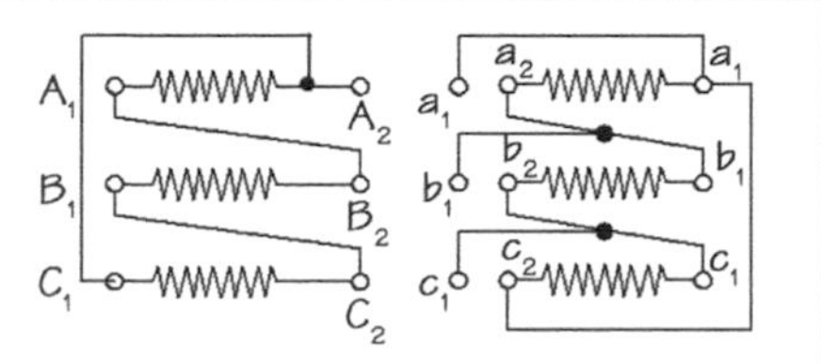
-30° (30° lag)	3	31 Dy 1	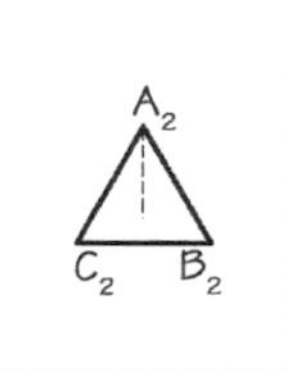	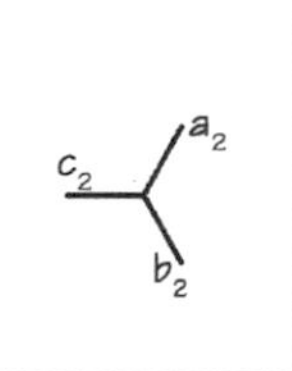	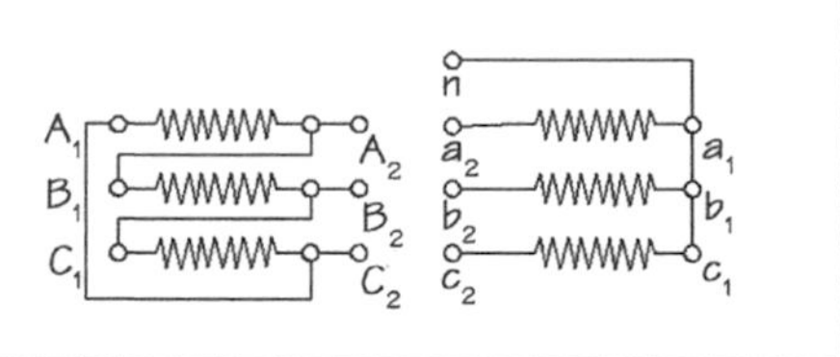
		32 Yd 1	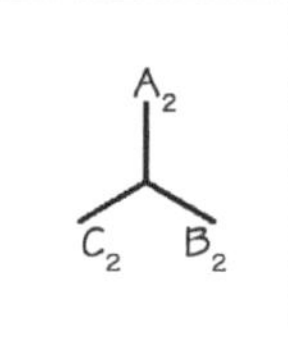	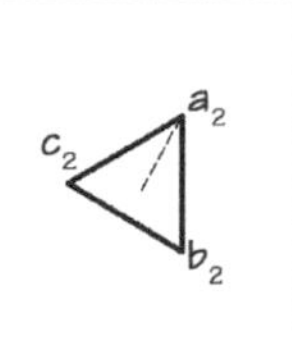	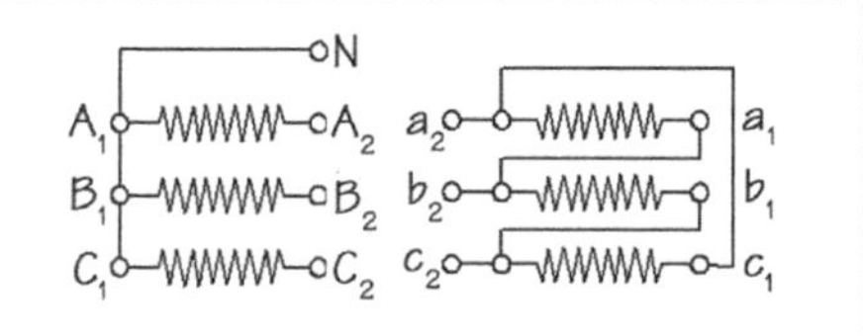
+30° (330° lag)	4	41 Dy 11	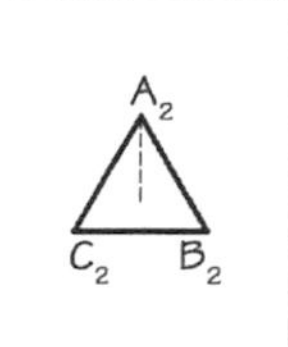	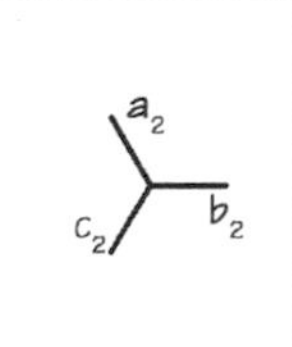	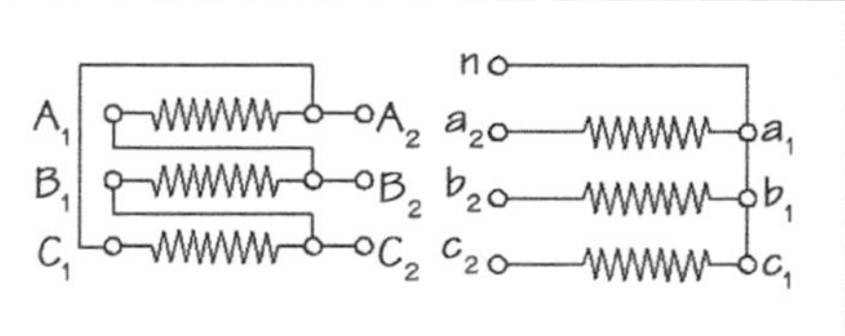
		42 Yd 11	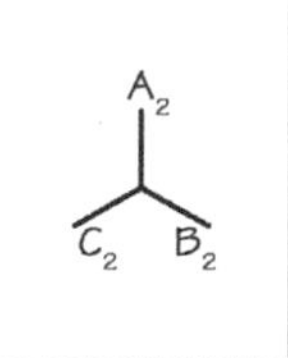	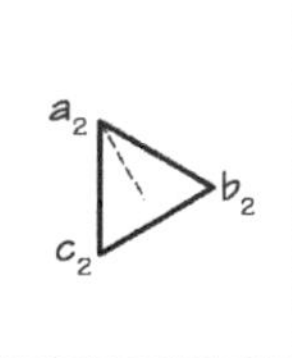	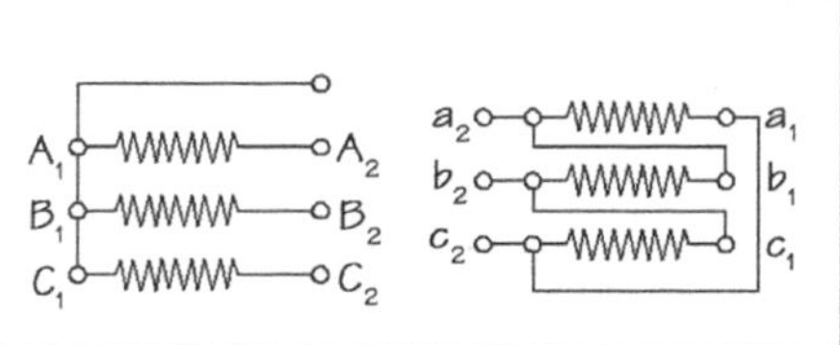

Figure 11.31

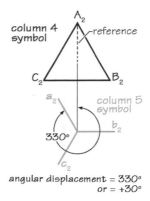

column 4 symbol

reference

column 5 symbol

330°

angular displacement = 330°
or = +30°

Figure 11.32

we were to superimpose *one above the other* – as illustrated in Figure 11.32 for the **41 Dy 11**:

The broken line superimposed over the 'delta' symbol represents the *reference* for the primary delta winding (in the above example, it extends from an 'artificial neutral point', at the centre of the triangle). The angle measured from this reference line, to the 'a_2' 'arm' of the 'star' symbol, then represents the angular displacement for that particular transformer connection. This example corresponds to '11 o'clock' using the clock convention, which represents the angle by which the secondary line voltage *lags* the primary line voltage – in this case is 330°, which corresponds to +30° (Group 4 in the vector grouping table).

Constructing three-phase transformer phasor diagrams

The construction of three-phase transformer phasor diagrams is explained in detail on this book's associated website; please go to: www.routledge.com/9781138849266.

Conclusion

Now that we've completed this chapter, we need to examine its **objectives** listed at its start. Placing a question mark at the end of each objective turns that objective into a **test item**. If we can answer those test items, then we've met the objectives of this chapter.

CHAPTER 12

D.C. generators

Objectives

On completion of this chapter you should be able to:

1. determine the magnitude and direction of a voltage induced into a conductor that is cutting a magnetic field.

2. describe the construction of a simple, single-loop, d.c. generator.

3. sketch the output waveform for a simple two-pole, four-pole, and six-pole d.c. generator.

4. compare the output waveforms of a simple single-loop generator and a multi-winding generator with a rotor.

5. explain 'action and reaction', as it applies to a d.c. generator.

6. Identify, and describe to functions of, the components of the stator and rotor of a practical d.c. generator.

7. explain what is meant by the 'magnetic circuit' of a d.c. generator.

8. explain the difference between 'concentrated-' and 'distributed-armature' windings.

9. explain the main features of, and method of installing the, armature windings onto the rotor of a practical d.c. generator.

10. recognise 'lap' and 'wave' armature windings, and explain how each is connected to the generator's commutator.

11. trace the 'current flow' through a schematic diagram of a lap- and wave-wound armature.

12. explain how the location of commutator brushes determine the number of parallel pathways through a generator's armature.

13. determine the number of parallel pathways through an armature for simplex/duplex/triplex lap and wave windings.

14. identify the components of a split-ring commutator, and describe its main functions.

15. solve simple problems relating to the e.m.f. equation for a d.c. generator.

16. explain what is meant by 'commutation', describe the difference between 'ideal' and practical commutation, and describe the problems with practical commutation.

17. summarise methods of improving commutation.

18. explain 'armature reaction', and discuss its consequences.

19. explain how armature reaction can be reduced using 'interpoles' and 'compensation windings'.

20. describe the schematic diagram, open-circuit characteristic, external characteristic, and typical applications for a (a) separately excited d.c. generator, (b) shunt generator, (c) series generator, and (d) compound generator.

21. describe the requirements for, and the process, of voltage build-up in a d.c. self-excited d.c. generator.

22. explain what is meant by the 'voltage regulation' of a d.c. generator.

23. discuss the losses that occur in a d.c. generator.

D.C. generator fundamentals

A **generator** is an electrical machine that does work by *converting mechanical energy into electrical energy*.

In this chapter, we are going to learn about **direct-current generators**. An historical term (familiar, no doubt, to those of us who are cyclists!) widely used as an alternative name for a d.c. generator is '**dynamo**' although, strictly speaking, this term can be applied to *any* type of electricity generator with a rotating armature: whether d.c. or a.c.

Factories producing aluminium, chlorine, etc., and those that carry out electroplating, use large quantities of direct current and, therefore, have traditionally used d.c. generators. D.C. generators, driven by a.c. motors, called 'motor-generator' sets were also widely used to convert alternating current into direct current.

However, thanks to the development of reliable and relatively inexpensive solid-state **rectifier** devices, direct-current generators have now been largely superseded by alternating-current generators ('alternators') whose outputs can be easily and efficiently changed from a.c. to d.c. Despite this, understanding the principles behind the operation of d.c. generators will provide us with an important insight into the operation of *all* types of rotating electrical machines: generators and motors; both d.c. and a.c. Furthermore, there is no difference between the constructional features of a d.c. generator and a d.c. motor, so there is a lot of common ground between these machines.

So let's start by examining the elementary theory behind what we call, '**generator action**'.

During the nineteenth century, the great English physicist, Michael Faraday (1791–1867), showed that whenever there is relative motion between a conductor and a magnetic field, *a voltage is induced into the conductor*.

For example, a conductor of length, *l*, cutting a magnetic field of flux density, *B*, perpendicularly at a velocity, *v*, as shown in Figure 12.1, *will develop a potential difference across its ends* as expressed by the following equation:

$$E = Blv$$

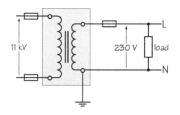

Figure 12.1

Important! By 'length of conductor', what we *really* mean is the length of the conductor that lies *within the magnetic field* – *not* the entire length of the conductor!

Actually, the above equation is *only* true when the conductor *cuts the flux at right angles*. For *any* other angle, the above equation needs to be modified as follows.

Assuming the conductor cuts the flux at an angle, θ (pronounced *'theta'*), at a velocity represented by vector **v**, then, in order to apply the above equation, we must find the *perpendicular component* of that velocity-vector, **v'** – as represented by the broken line in Figure 12.2.

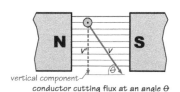

vertical component
conductor cutting flux at an angle θ

Figure 12.2

This can be determined from the sine ratio, as follows:

$$\sin\theta = \frac{\text{opposite}}{\text{hypotenuse}} = \frac{v'}{v}$$
$$v' = v\sin\theta$$

So we can now modify the original equation so that it can be used for conductors moving at *any* angle, θ, through the flux, as follows:

$$\boxed{E = B\,l\,v\sin\theta}$$

where:

E = induced voltage (volts)

B = flux density (teslas)

l = length of conductor in field (metres)

v = velocity of conductor (metres per second)

θ = angle at which the conductor cuts the flux (degrees)

From this equation, we should realise that, for a given velocity, a *maximum* voltage is induced into the conductor when it 'cuts' the magnetic flux at *right angles* whereas, if the conductor were to move, say, parallel with the flux, then *no* voltage would be induced into the conductor.

Worked example 1

A conductor moves through a permanent magnetic field of flux density 225 mT at a velocity of 50 m/s. If the length of conductor within the field is 150 mm, calculate the voltage induced into the conductor, if it cuts the flux (a) perpendicularly, and (b) at 30°.

Solution

$$E = Blv \sin\theta$$
$$= (225 \times 10^{-3}) \times (150 \times 10^{-3}) \times 50 \times \sin 90°$$
$$\simeq \textbf{1.69 V (Answer a.)}$$

$$E = Blv \sin\theta$$
$$= (225 \times 10^{-3}) \times (150 \times 10^{-3}) \times 50 \times \sin 30°$$
$$= 1.69 \times 0.5$$
$$= \textbf{0.845 V (Answer b.)}$$

> No potential difference is induced into a conductor that moves *parallel* to the lines of magnetic flux, because it will not be 'cutting' them.

Fleming's Right-Hand Rule

The **direction** of this induced voltage may be determined by using (for *conventional* current flow) **Fleming's Right-Hand Rule** for '**generator action**', credited to the engineer and academic, Sir John Ambrose Fleming (1849–1945), which works as follows:

If the *thumb, first finger (index finger)*, and *second finger* of the right hand are held at right angles to each other, as shown in Figure 12.3 . . .

. . . then:

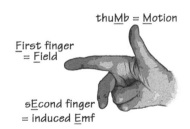

Figure 12.3

▶ the **thuMb** indicates the direction of **M**otion of the *conductor relative to the field*,

▶ the **First finger** (index finger) indicates the direction of the magnetic **F**ield (i.e. north to south),

▶ the **sEcond finger** indicates the direction of the induced **E**.m.f.

Remember, for Fleming's Right-Hand Rule to work, it's important to remember that the *direction of motion* always refers to the movement of the **conductor** relative to the magnetic field, *never* the other way around!

So if we apply **Fleming's Right-Hand Rule** to the downward moving conductor, shown previously, we will find that the resulting induced voltage will act *towards us* (i.e. *out* from the page). In other words, the nearer end of the conductor will be the positive end, and the far end will be the negative end (Figure 12.4).

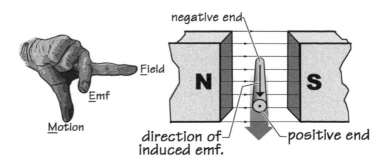

Figure 12.4

Now, if this conductor is connected to a closed *external circuit*, the induced voltage will cause a current to drift in the *same direction* as that induced voltage – i.e. in this case, *towards* us, as indicated by the 'dot' convention shown in the diagram.

Reminder: For conventional current, a **dot** represents current drifting *towards* us, and a **cross** represents current drifting *away* from us.

It's very important to remember that, whenever we talk about the 'direction of current' (i.e. in the sense of 'conventional flow' vs 'electron flow'), we are *always* referring to the *direction of current through the load; **never** through the voltage source itself* – so, in this particular case, the current will be leaving the moving conductor from its positive (nearer) end, drifting through the load, and re-entering the conductor at its negative (opposite) end.

The *plan* (downward) view of the above arrangement, shown in Figure 12.5, should help clarify this:

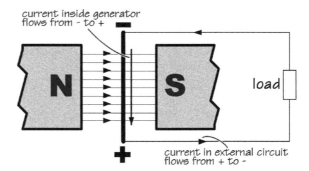

Figure 12.5

As you can see, with the downward moving conductor acting as a 'voltage source' for the external load, the resulting current drifts from positive to negative around the external circuit (while the current drift *within* that moving conductor is from negative to positive).

This, then, is the basic principle of operation of a **generator**, and is known as '**generator action**'.

Simple generator

A **simple generator** consists of a single, rectangular, **loop** of wire called an '**armature**', pivoted so that it can rotate within a magnetic field. When it is rotated by an external force, provided by a **prime mover**, a voltage is induced into each side (called the '**active conductors**') of the loop, and these voltages act in opposite directions to *reinforce* each other – in other words, if the voltage induced into one of the active conductors is **E** volts, then the voltage induced into the loop is **2E** volts.

When the rotating loop is connected to an external circuit, the resulting load current will flow in the direction shown in Figure 12.6. Again, it cannot be over-emphasised that when we talk about the 'direction of current', we are describing the direction of current flow *around an* **external** *circuit* and **not** within the armature.

Any device that drives a generator is called its **prime mover**. A typical prime mover for a d.c. generator is an internal combustion engine, often a diesel engine.

Split-ring commutator

In the case of a d.c. generator, the rotating armature loop is connected to its external load using a **split-ring commutator** and spring-loaded carbon **brushes**, as illustrated in Figure 12.7. The purpose of the springs is to firmly hold the carbon brushes against the rotating commutator. Carbon is used to manufacture the brushes because, as well as providing a relatively low resistance connection to the external circuit, it is self-lubricating.

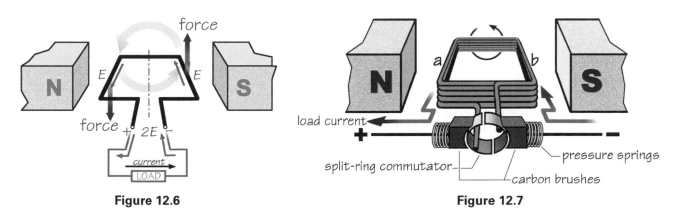

Figure 12.6   **Figure 12.7**

With the armature rotating, in this case, in a *counterclockwise* direction, the voltages induced into each side of the armature winding act *towards* us in the active conductors labelled '**a**', and *away* from us in the active conductors labelled '**b**'. So the resulting load current will *leave the armature via the left-hand commutator segment* towards the load, and *return to the armature via the right-hand commutator segment*. As the armature coil rotates past the vertical position, the voltages in the two active conductors will *reverse* direction, but the load current will *still* leave the left-hand commutator segment, and return via the right-hand segment.

In other words, thanks to the commutator, *the directions of the terminal voltage and load current will be unaffected by the rotation of the armature*.

Each end of the armature loop is connected to opposite halves, or 'segments', of the split-ring commutator. As well as providing a means of connecting the rotating armature loop to the fixed wiring of its external circuit, the commutator functions as a *rotary switch*, which *reverses the polarity of the output voltage after every 180° of rotation*. So, at the point when voltage induced into the armature naturally reverses, the commutator switches polarity, ensuring that output voltage always acts in the same direction as 'seen' by the external circuit. So, a commutator is essentially a **rectifier** whose output will consist of voltage pulses *acting in the same direction*, as shown (for a two-pole generator) in Figure 12.8.

12

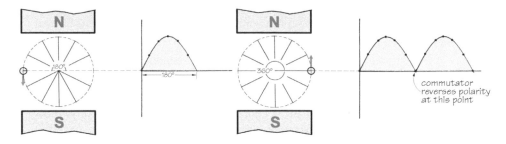

Figure 12.8

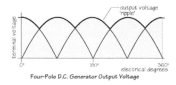

Figure 12.9

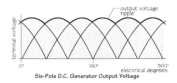

Figure 12.10

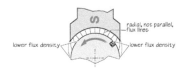

Figure 12.11

Figure 12.12

By using *multiple* poles and armature loops arranged at fixed physical angles to each other, a much smoother d.c. output can be obtained. For example, compare the *four-* and *six-pole* generator outputs with the two-pole machine in Figures 12.9 and 12.10.

Note that, *regardless* of the number of poles, a d.c. voltage will *always* produce some degree of 'ripple' – shown in bold, in Figures 12.9 and 12.10.

The waveforms illustrated in Figures 12.8, 12.9, and 12.10, are *sinusoidal* because the angle at which the conductor is cutting the parallel flux lines is constantly changing and, so, the output voltage is proportional to the sine of the angle by which each flux line is cut.

But, in a 'real' generator, *this isn't the case*. Because of the *cylindrical* shape of the faces of 'real' pole pieces, and the fact that they form a very narrow air gap with the cylindrical rotor, the air-gap flux lines are *not* parallel with each other, but radial.

As we can see from Figure 12.11, the flux lines that bridge the air gap are *not* parallel with each other but are *radial*. Furthermore, the flux density is roughly *constant* between the pole face and the rotor – but tends to *decrease* where the flux lines 'bulge' at the leading and trailing edges of the pole piece.

So, as the conductor moves through the field, the induced voltage will *build up* as the conductor *approaches* the leading edge of the pole piece, remain more-or-less *constant* as the conductor *passes in front of the pole piece* (where it continually cuts the flux at roughly the same angle), then *fall off* as the conductor *leaves* the trailing edge of the pole piece.

The resulting voltage waveform, therefore, bears little resemblance to a sine wave, but is very much 'flatter' – as shown in Figure 12.12.

Action and reaction

In the previous section, we learnt that if the conductor moving through a magnetic field is connected to a load, then the induced potential difference will cause a current to drift through any external circuit.

This current, of course, is capable of *expending energy* in any external load through which it passes – e.g. if the load were, say, a lamp, then the current would *heat* its filament. If the **Law of the Conservation of Energy** (*'energy can neither be created nor destroyed, but only changed from one form to another'*) is to be maintained, then this current can only expend energy *providing work has been done to produce it in the first place!*

The *work expended* in moving the conductor through a magnetic field is the product of the **force** applied to that conductor and the **distance** through which it is moved. And, according to **Newton's Third Law**, *this force **must** be opposed by an equal and opposite force* called a '**reaction**' (*'for every force, there is an equal and opposite reaction'*).

And this reaction *can **only** come from the magnetic field that the current itself produces!* That is, the magnetic field set up around the conductor by the load current flowing through it.

This reasoning led the Russian physicist, **Emil Lenz** (1804–1865), to conclude that . . .

> *'The current resulting from the induced potential difference, due to the motion of a conductor through a magnetic field, must act in such a direction that its own magnetic field will then oppose the motion that is causing that induced potential difference.'*

If we were to apply **Fleming's Left-Hand Rule** (for motor action) to our simple **generator** (try it – see Figure 12.13), where the 'current' is the **load current**, then we would find that the direction of the resulting reaction will be *opposite* to the direction in which the conductor is moving!

So when the generator is supplying current to the load, the armature loop behaves as *both as a generator **and** as a motor simultaneously!* The *greater* the load current, the *greater* will be the 'motor effect' that reacts against the movement of the armature loop.

This 'action' and 'reaction' effect plays a very important part in the operation of a generator, requiring it to ***react to changes in load*** – explaining why, if the load *increases*, the *load current increases*, causing an *increase in the torque opposing that of the prime mover*. This, in turn, causes the prime mover to tend *to slow down*.

On the other hand, if the load *decreases*, then the *load current decreases*, causing a *decrease in the torque opposing that of the prime mover*. This, in turn, causes the prime mover to tend *to speed up*.

However, the rated speed of a generator's prime mover is normally controlled by a **governor**. So, if the generator *does* tend to slow down or speed up, for the reasons described in the previous paragraph, *then the governor will adjust the speed of the prime mover to compensate*, and bring it back to its rated speed.

This action/reaction behaviour is a very important concept to understand and it explains why, when the generator has to supply more energy to its load, more energy is required from the prime mover to drive the generator.

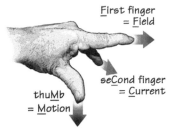

Figure 12.13

Construction features of a d.c. generator

Figure 12.14, below, shows an exploded view of the main components of a typical small two-pole **direct-current generator**.

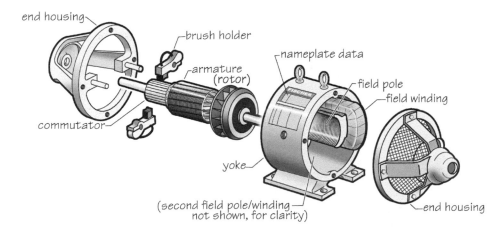

Figure 12.14

Its component parts are fairly representative of d.c. generators in general, regardless of their physical size, power rating, or number of poles. For the purpose of understanding its construction, we can divide this machine's component parts into *two* major categories: the **stator** and the **rotor**.

Stator

The **stator**, as the name suggests, is the *stationary* part of the machine, and comprises three parts: the **yoke**, the **poles**, and the **field windings** – as illustrated in Figure 12.16.

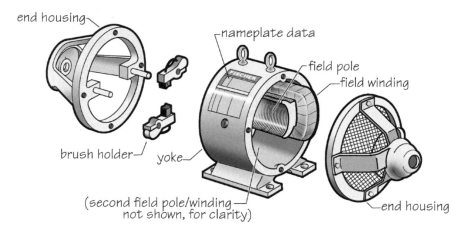

Figure 12.15

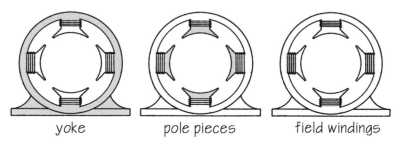

yoke   pole pieces   field windings

Figure 12.16

Yoke

This is the **outer frame** of the machine, comprising a ferromagnetic cylinder, which forms part of the magnetic circuit, and whose other functions are to support the field-pole assemblies and to enclose the machine. Bolted onto the front and back of the yoke are a pair of **end housings** that incorporate bearings which support the rotor's shaft, as well as supporting the **brush holders**. 'Brushes' are spring-loaded blocks of carbon, which 'ride' the commutator in order to connect the rotating armature with the fixed, external, wiring.

Pole pieces

Each **pole piece** is manufactured from a stack of lightly insulated silicon-steel laminations that are riveted or clipped together. Their purpose is to support the machine's *field windings, and their shape is such that they provide a relatively small cross-section* to accommodate the field windings while providing a significantly larger cross-section, and shaped to minimise the air gap between each pole piece and the rotor.

Field windings

Field windings provide the magnetomotive force necessary to set up the magnetic field within the air gap between the pole pieces and the rotor. These windings are pre-wound on formers, and slipped around individual pole pieces. The *direction* in which each field winding is wound and assembled on their pole pieces is such that, when energized, the resulting polarities of the poles are magnetised, alternately, *north-south-north-south-etc.* – as illustrated in Figure 12.17.

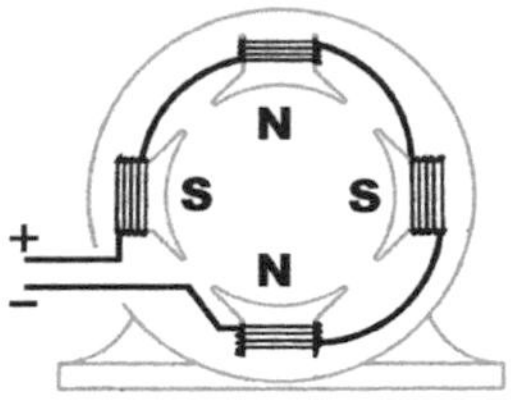

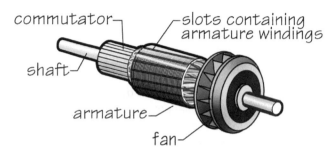

polarities of field windings

Figure 12.17

Rotor

The **rotor**, or '**armature**' (Figure 12.18), is the *rotating* part of the machine. As well as forming part of the machine's magnetic circuit, its function is to support the armature windings into which the generator's output voltage is induced. The main components are the **rotor** cylinder itself, **armature windings**, **commutator**, and cooling **fan**.

Figure 12.18

The **rotor** consists of a stack of disc-shaped, lightly insulated laminations, manufactured from silicon-steel, mounted on the machine's shaft to form an elongated cylinder. Each lamination has a number of rectangular holes punched out from around its circumference which, when the laminations are stacked together and lined up, the holes form longitudinal slots into which the armature windings are placed. In some cases, the slots are machined into the surface of a laminated rotor. The rotor is part of the machine's **magnetic circuit** and, together with the pole pieces, acts to concentrate the magnetic flux in the air gap.

In the example shown in Figure 12.18, the slots are 'skewed', rather than running parallel with the shaft, but both types are used.

Armature windings

The **armature windings** consist of insulated conductors, placed into the rotor's slots. In smaller machines, like the one illustrated, they are wound directly into the rotor's slots, by hand, from a long length of conductor. However, in most machines, they are machine-wound into numerous, individual pre-formed windings, which are then separately placed into the rotor slots and wedged into place.

Commutator

The voltage naturally induced into the armature windings, as they rotate within the magnetic field, is an alternating voltage, so it must be *rectified* and this is one of the *three* functions of a commutator. A **split-ring commutator** consists of a number of conducting segments surrounding the rotor's shaft, each insulated from the others as well as from the shaft, and its purpose is:

▶ to provide a means of connecting the *rotating* armature windings to the *fixed* external wiring supplying the load

12

> to act as a *rotating switch* which, together with the carbon brushes, functions as a **rectifier** by reversing the direction of the induced voltage every second half cycle, thus providing a direct-current output
> to provide a means of interconnecting the separate armature windings.

The **brushes** are roughly centred under the pole pieces because, in this position, the conductors to which they are attached are moving *parallel* with the field, so no voltage is being induced into them, so no voltage appears between the commutator segments being bridged by the brushes.

Cooling fan

Many motors incorporate a cooling fan on the rotor shaft. As we shall learn, ventilation is an important requirement in the design of a motor, which prevents excessively high temperatures from damaging the windings' insulation.

Magnetic circuit

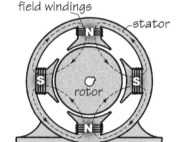

field windings

stator

rotor

Generator's Magnetic Circuit

Figure 12.19

The machine's **magnetic circuit** is a combination of the *field poles, yoke, air gap,* and *rotor*.

Figure 12.19 shows the magnetic circuit for a four-pole machine.

As explained in the companion book, ***An Introduction to Electrical Science***, the purpose of the magnetic circuit is to (a) provide a *low-reluctance/high-permeability* path that will (b) *concentrate* the magnetic flux, in order to ensure maximum flux density, and (c) to *direct* or *guide* that magnetic flux to where it is needed: in this case, to the air gap within which the armature's active conductors are located.

As the reluctance of silicon-steel is insignificant compared with even a very narrow air gap, *practically the entire magnetomotive force developed by the magnetic circuit's field windings will appear across the air gap*, in much the same way as practically *all* of the voltage applied to an electrical load will appear across that load because the voltage drops along the adjoining conductors are negligible due to their insignificant resistance.

Windings

A d.c. generator (in fact, *all* machines!) has *two* types of winding:

> *concentrated* field windings.
> *distributed* armature windings.

So what exactly do we mean by *'distributed'* or *'concentrated'* windings?

> A **'concentrated' winding** is the type of winding we would instantly recognise as a 'coil': i.e. a length of insulated conductor whose turns are tightly wound adjacent to each other, around a central ferromagnetic core. In the case of a d.c. generator, the **field windings** are 'concentrated'.
> A **'distributed' winding**, on the other hand, is one that is formed from conductors that are placed into longitudinal slots that have been machined into the surface of the rotor. In the case of a d.c. generator, the **armature windings** are 'distributed'.

How Armature windings are illustrator

Probably the most complicated and, therefore, most difficult to understand feature of *any* d.c. machine (motors as well as generators) is its **armature windings**. So, in this section, we are going to try our best to demystify this particular topic!

A d.c. generator's **armature windings** are *distributed* windings, which means they are laid into longitudinal '**slots**' that have been machined into the surface of the rotor. By installing them into *slots*, rather than securing them onto the *surface* of the rotor, the radial air gap between the surfaces of the pole pieces and the rotor can be minimised, thus concentrating the flux across that gap.

For most machines, the armature windings themselves are *not*, as one might expect, wound directly into the rotor's slots from a continuous length of wire but, instead, are manufactured as numerous *pre-formed* coils that are placed individually into the slots. Figure 12.20 illustrates schematic diagrams that represent *two* types of armature winding: a '**single turn**' **winding** and a '**multiple turn**' **winding**. In the case of the 'multiple turn' winding, two 'turns' are shown, but more turns are possible – *provided*, of course, there is enough space within the slots to accommodate them. The far-right image shows what a 'real' multiple turn winding looks like.

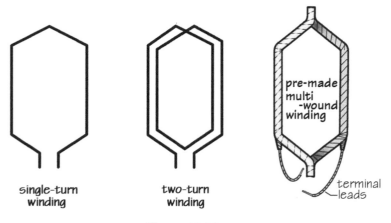

single-turn winding

two-turn winding

pre-made multi -wound winding

terminal leads

Figure 12.20

The parallel **sides** of a winding are termed the '**active**' **conductors** of that winding. By that, we mean those parts of the winding that lie within the rotor slots, cut the magnetic field, and have voltages induced into them. The distance between a winding's active conductors is such that they should be roughly centred over poles of opposite polarity, so that the voltages induced into the active conductors act in *opposite* directions and, thus, *reinforce* each other around the loop formed by the winding.

This is illustrated in Figure 12.21, in which each of the winding's active conductors are shown lying under the centres of pole pieces of opposite polarity.

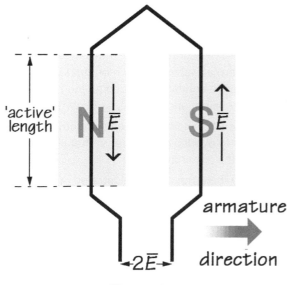

'active' length

N $\bar{E}$ S $\bar{E}$

$2\bar{E}$

armature

direction

Figure 12.21

The winding illustrated in Figure 12.21 consists of just a single turn but, of course, *most* windings will have several turns. If there are *Z* turns, then the voltage *(E)* induced into each winding will be 2 *ZE* volts.

For a simple, *two-pole*, machine the active conductors of the armature winding must be installed in slots which are 180° apart or *opposite* each other, as illustrated in Figure 12.22 (left).

But for, say, a *four-pole* machine, the active conductors of each armature winding must, ideally, be installed in slots that are located *one-quarter* (90°) of the way around the armature's circumference, as illustrated in Figure 12.22 (right).

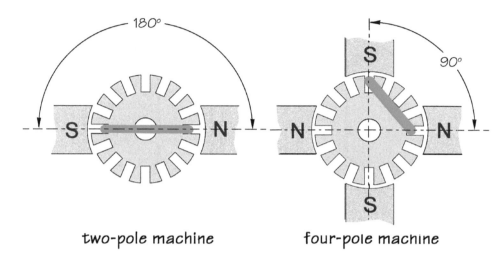

two-pole machine   four-pole machine

Figure 12.22

Whether it's actually *possible* to align the active conductors of the armature winding *precisely* with the *centre* of a pole piece of opposite polarity, depends on the number of slots cut into the circumference of the rotor. There is, of course, a practical limit to the number of slots, which depends upon (a) the circumference of the armature, and (b) the total cross-sectional area of the armature conductors/insulation they have to accommodate.

A less obvious reason is that if, in order to accommodate more conductors, the cross-sectional area of each of the remaining armature 'teeth' becomes too low, then they will be driven into saturation.

In practice, then, it may only be possible to align each side of an armature coil *approximately* under the centres of pole pieces of opposite polarity. But this presents no problem to the operation of the machine.

As already explained, each armature winding is manufactured in such a way that, when installed, one of its active conductors occupies the *outer*-half of one slot, while the other occupies the *inner*-half of another slot, as shown in Figure 12.23.

To achieve this in practice, each winding requires distinctive 'kinks' at opposite ends, which enable adjacent coils to overlap each other, rather like the **scales of a fish,** when they are installed on the armature.

For example, if we examine the armature coil illustrated in Figure 12.24, we can see that the left-hand side of the coil is higher than the right-hand side.

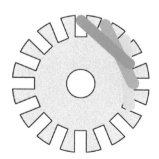

Figure 12.23

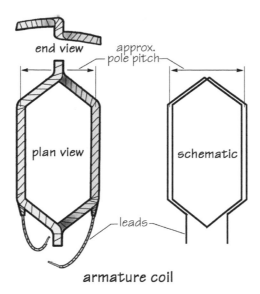

armature coil

Figure 12.24

This overlapping arrangement is illustrated in greater detail in Figure 12.25 below. Three coils are shown in the plan view but, for the purpose of clarity, only the *first* and *third* coils are shown in the end view. Note how the left-hand side (1) of the first coil occupies the *outer* half of one armature slot, while the right-hand side (1') occupies the *inner* half of the other armature slot. This 'fish scale' arrangement allows adjacent coils to overlap each other, as they are installed around the surface of the rotor.

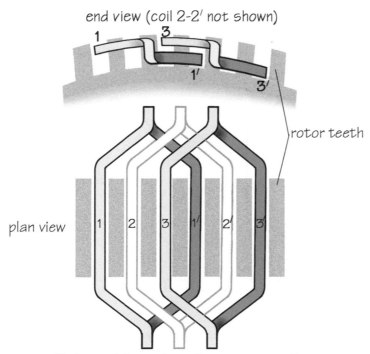

'fish-scale' overlap of armature coils

Figure 12.25

Once *all* the armature coils have been placed into the slots, they are then wedged to secure them against flying apart due to the rotor's centrifugal reaction.

12

Types of armature winding

Once installed into the rotor slots, the individual windings are connected together by soldering their terminal wires into small grooves cut into the commutator segments (see Figure 12.36). The way in which this is done determines the **type** of the resulting armature arrangement.

The two main types of armature winding are termed '**lap windings**' and '**wave windings**'. A third, less-common, type (and one beyond the scope of this book) is known as a '**frog's leg winding**', which is a combination of the simplex* lap- and duplex* wave-wound windings, and combines the attributes of both. These are illustrated in Figure 12.26.

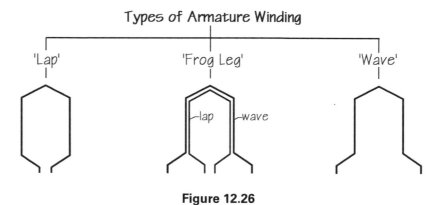

Figure 12.26

The difference between these types of armature winding is the way in which the individual winding's terminal leads are soldered onto the machine's commutator segments. As can be seen in Figure 12.27, in the case of a basic ('simplex') **lap winding**, the windings' leads are connected between *adjacent* segments whereas, for a **wave winding**, the windings' leads are connected between segments that are *displaced by approximately 360 electrical degrees.*

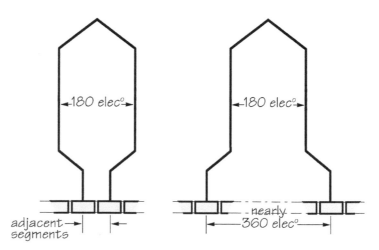

Figure 12.27

So, as well as functioning as a means of connecting the armature to the external circuit and rectifying the output voltage, the **commutator segments** also provide *the means by which the individual armature windings are connected to each other.*

To understand this, we must first appreciate how a machine's armature windings and poles are represented in the form of a schematic diagram.

---

*The terms 'simplex' and 'duplex' will be explained later in this chapter.

In fact, *it's very difficult* to illustrate how the individual armature windings actually physically relate to each other and to the field poles because, of course, they lie along the outer and inner surfaces of two concentric cylinders. To make matters even *more* complicated the armature coils are located on the outer surface of a *revolving* cylinder (the rotor), whereas the pole pieces are located on the inner surface of a *fixed* cylinder!

So, to simplify matters, an illustration technique has been devised that is not unlike the way we use the *Mercator Projection* to represent the earth's sphere in the form of a two-dimensional map.

Imagine cutting through the machine's cylindrical yoke, and flattening it out, as in Figure 12.28, so that the pole pieces are presented horizontally, alongside each other in plan view, and then doing *exactly* the same thing with the armature windings and commutator segments. We can then superimpose the armature windings and commutator segments *above* the pole pieces to show their relationship.

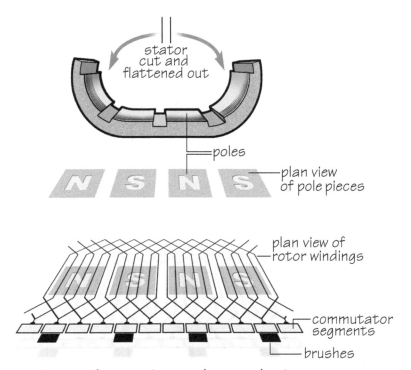

armature coils superimposed over pole pieces

Figure 12.28

The opposite ends of the flattened-out armature windings, of course, are actually connected to each other – this is where the cylindrical windings have been 'cut' and flattened out!

This allows us to easily relate the positions of the armature windings/commutator segments, not only to each other, but to the pole pieces below. Furthermore, it is easy to imagine 'sliding' the upper level horizontally over the lower level to represent the relative motion of the armature coils/commutator and the poles.

So, using this method of representing the armature and poles, let's start by examining the **lap winding** shown in Figure 12.29. At first glance, it's not in the least obvious that this *is* a lap winding! However, we *can* see that if the armature windings are moving to the right, applying Fleming's Right-Hand Rule to the conductors will result in voltages induced into them in the directions shown.

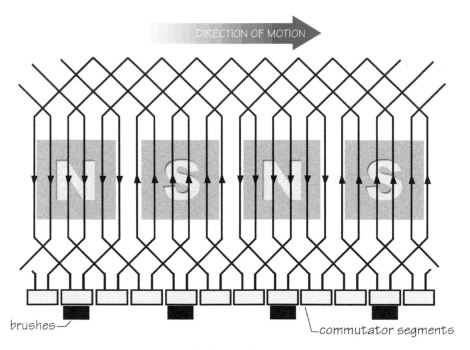

Figure 12.29

To confirm that this is, indeed, a lap winding, we simply need to trace adjacent windings *to see how they are interconnected*. This is what we have done in Figure 12.30. By highlighting a pair of adjacent windings, we can see that each one is terminated at *adjacent* commutator segments – confirming that it is indeed a **lap winding**.

Furthermore, by tracing the flow of current through a pair of adjacent windings, we can see that they are connected in *series* with each other, so that the voltages induced into each 'active' conductor are *cumulative*.

Figure 12.30

In Figure 12.31, on the other hand, we see that the individual windings we have bolded are terminated on commutator segments that are some distance (approximately 360 electrical degrees) apart, confirming that this is a **wave winding**. And, again, if we trace the flow of current through those windings, we can see that they too are connected in series with each other, so that the voltages induced into each 'active' conductor are, again, *cumulative*.

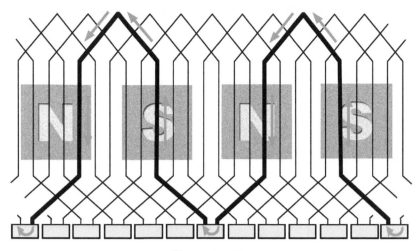

Figure 12.31

As should be obvious from Figures 12.30 and 12.31, the armature windings form *a continuous, closed, loop* around the circumference of the armature – essentially, *all* the windings are connected in *series* with each other – regardless of whether the armature is lap wound *or* wave wound – so the voltages induced into each winding are *cumulative* around the complete armature. In fact, as we shall learn next, if the machine wasn't fitted with **brushes**, there would be a large circulating current flowing around the armature (see the left-hand illustration in Figure 12.32), set up by these cumulative voltages!

However, if we take the locations of the commutator **brushes** (when connected to their external load) into account, we will see that two or more *parallel paths* are created: with each parallel path made up of an identical number of windings connected in series with each other.

For example, if *two* brushes were to be installed on *opposite* sides of the commutator, they would split the armature windings into *two* parallel pathways. Similarly, if *four* brushes were installed, they would split the armature windings into *four* parallel pathways. This should be made clear from Figure 12.32.

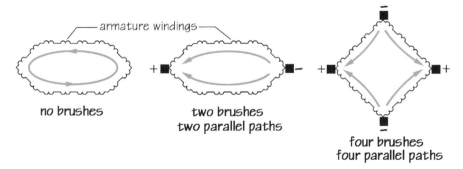

Figure 12.32

The *number of brushes* depends on the *number of poles*. For example, a *two*-pole machine would require *two* brushes, a *four*-pole machine would require *four* brushes, and so on. So we can say that, for a lap winding at least, *the number of parallel pathways through the armature windings must be equal to the number of poles*. And, of course, the brushes are located where they are centred opposite the pole pieces.

Brushes alternate in polarity around the commutator, and those of similar polarity are connected together in *parallel* with each other (see Figure 12.33).

Figure 12.33 shows a simplified circuit for lap winding for a *four-pole* machine. As explained, with the brushes located in the positions shown, there are *four* **parallel pathways** for current flow through the armature windings.

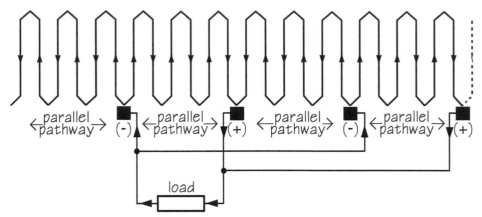

four-pole lap winding = four parallel pathways through windings

Figure 12.33

So, for a simple **lap winding**, the number of parallel paths *is always equal to the number of poles*. However, for a **wave winding**, the number of parallel paths is just *two*, *regardless of the number of poles* (we won't pursue the reason for this, as it is very difficult to draw as a schematic diagram!).

For example, in the case of a *six*-pole machine, a lap winding would form *six* parallel paths, while a wave winding would form just *two*. For this reason, a lap winding would be the better choice if the machine is required to supply a *large current*, whereas a wave winding would be the better choice if the machine is required to supply a *large voltage* (see **worked example 2**).

Sometimes, it's useful to *increase the number of parallel paths* in order to enable the armature to carry an even larger current, while *reducing the conductor current*. In order to do this, '**multiplex**' windings are used. The effect of this is summarised in Figure 12.34.

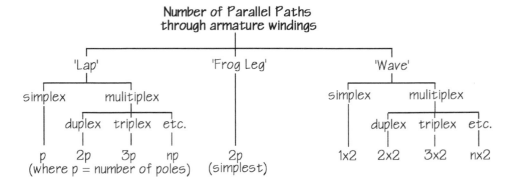

Figure 12.34

Multiplex windings are beyond the scope of this book but, essentially, they are a series of two, three, or (rarely) more, *independently connected* armature windings, whose active conductors are installed such that they occupy alternate slots around the circumference of the rotor. Figure 12.35 shows how a basic, simplex, lap winding is connected between commutator segments in order to accommodate duplex and triplex armature windings.

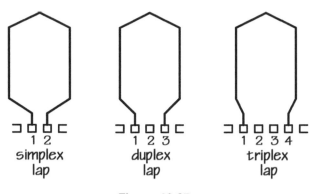

Figure 12.35

The purpose of multiplex windings is to *increase the number of parallel pathways* in order to enable the armature to supply an even larger load current, while *reducing the current through the individual conductors*. A **duplex winding** will *double* the number of parallel paths, while a **triplex winding** will *triple* the number of parallel paths, as demonstrated in worked example 2.

Worked example 2

Calculate the number of parallel paths through the armature of a six-pole d.c. generator, having (a) triplex lap winding, and (b) duplex wave winding.

Solution

(a) triplex lap winding $= 3p = 3 \times 6 =$ **18 paths (Answer a.)**

(b) duplex wave winding $= 2 \times 2 =$ **4 paths (Answer b.)**

Worked example 3

The armature of a six-pole d.c. machine comprises 498 conductors, each of which is capable of carrying a current of 120 A without overheating. The air gap flux density and speed of rotation of the armature is such that the average e.m.f. induced into each conductor is 2 V. Find (a) the total e.m.f. generated, (b) the maximum current the machine can supply, and (c) the maximum power generated, if the windings are simplex lap wound and simplex wave wound.

Solution

Simplex lap-wound machine:

(a) For a simplex lap-wound machine, the number of parallel paths is equal to the number of poles (six), so the e.m.f. generated must be:

$$E = \frac{\text{(voltage per conductor)} \times \text{(number of conductors)}}{\text{(number of parallel paths)}}$$

$$= \frac{2 \times 498}{6} = \textbf{166 V (Answer a.)}$$

(b) Because this simplex lap-wound machine has six parallel paths, the current it can supply must be:

$$I = \text{(number of parallel paths)} \times \text{(current rating of individual conductors)}$$

$$= 6 \times 120 = \textbf{720 A (Answer b.)}$$

(c) The power output of the machine is the product of its e.m.f. and the current it can supply:

$$P = EI = 166 \times 720 = \textbf{119.52 kW (Answer c.)}$$

Simplex wave-wound machine:

(a) For a simplex wave-wound machine, the number of parallel paths is 2, so the e.m.f. generated must be:

$$E = \frac{\text{(voltage per conductor)} \times \text{(number of conductors)}}{\text{(number of parallel paths)}}$$

$$= \frac{2 \times 498}{2} = \textbf{498 V (Answer a.)}$$

(b) Because this simplex wave-wound machine has two parallel paths, the current it can supply must be:

$$I = \text{(number of parallel paths)} \times \text{(current rating of individual conductors)}$$

$$= 2 \times 120 = \textbf{240 A (Answer b.)}$$

(c) The power output of the machine is the product of its e.m.f. and the current it can supply:

$$P = EI = 498 \times 250 = \textbf{119.52 kW (Answer c.)}$$

An important thing to note from worked example 2, is that the output power of the machine is the *same* regardless of whether it is connected in simplex lap or simplex wave.

Split-ring commutator

Figure 12.36 illustrates a typical split-ring commutator used with d.c. machines.

As we can see, it comprises a series of individual copper **segments**, separated by mica **insulation**, and clamped together to form a hollow cylinder. On the shaft side of the commutator, we can see the **slots** into which the armature winding leads are soldered.

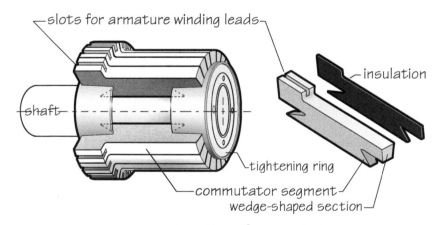

construction of commutator

Figure 12.36

We'll return to the subject of commutators and **commutation**, later in this chapter.

E.M.F. equation for a d.c. generator

We have already determined how to calculate the e.m.f. produced by a d.c. generator in worked example 3 by using good old-fashioned common sense.

Now let's see if we can develop a **general equation** of the e.m.f. developed by a d.c. generator that we can use in all circumstances.

Let's assume that the rotational speed of the rotor is N revolutions per minute.

We can express this rotational speed in terms of revolutions per second, in which case:

$$\text{rotational speed} = \frac{N}{60} \text{ rev/s}$$

So the time taken for one active conductor to complete one revolution must be:

$$\text{time} = \frac{60}{N} \text{ seconds per revolution}$$

In which case, the time taken for an active conductor to travel between the centres of adjacent poles (i.e. the pole pitch), must be:

$$\text{time between adjacent poles} = \frac{\text{time to complete one revolution}}{\text{number of poles}} = \frac{60}{Np} \text{ seconds}$$

If the useful flux, per pole, is Φ webers, then the rate at which the active conductor cuts flux as it passes each pole must be:

$$\text{rate of cutting flux} = \Phi \div \frac{60}{Np} = \frac{\Phi Np}{60} \text{ webers per second}$$

By **Faraday's Law of Electromagnetic Induction**, the average value of e.m.f. (E_{av}) induced into a single conductor passing through a uniform magnetic field is equal to the rate at which that field is cut by the conductor:

$$E_{av} = \frac{\Phi Np}{60} \text{ volts per conductor}$$

Let Z represent the total number of armature conductors, and c represent the number of parallel paths through the armature, then the total number of conductors in series within each path must be given by:

$$\text{total number of conductors in series within each parallel path} = \frac{Z}{c}$$

. . . where, for a **lap winding**, $c = p$ and, for a **wave winding**, $c = 2$.

Therefore, the total average e.m.f. appearing across the brushes is given by:

$$E_{av} \text{ between brushes} = \frac{\Phi Np}{60} \times \frac{Z}{c}$$

So the final equation for the average e.m.f. generated by a d.c. generator is given by:

$$\boxed{E_{av} = \frac{\Phi NpZ}{60c}}$$

where:
Φ = flux (webers)
N = rotational speed (rev/min)
p = number of poles
Z = number of active conductors
c = number of parallel paths

Worked example 4

A four-pole, wave-wound, armature has 51 slots with 12 conductors per slot, and is driven at a speed of 900 rev/min. If the useful flux per pole is 25 mWb, calculate the magnitude of the generated e.m.f.

Solution

As this is a wave-wound machine, the number of parallel paths, c, is 2 – regardless of the number of poles.

$$E_{av} = \frac{\Phi N p Z}{60 c} = \frac{25 \times 10^{-3} \times 900 \times 4 \times (51 \times 12)}{60 \times 2} = \frac{55\,080}{120}$$

$$= \textbf{459 V (Answer)}$$

Worked example 5

An eight-pole, lap-wound armature, driven at 350 rev/min, is required to generate a minimum e.m.f. of 260 V. The useful flux per pole is 0.05 Wb. If the armature has 120 slots, calculate the number of conductors required per slot.

Solution

As this is a lap-wound machine, the number of parallel paths, c, is equal to the number of poles, i.e. 8.

$$E_{av} = \frac{\Phi N p Z}{60 c}$$

$$Z = \frac{E_{av} 60 c}{\Phi N p} = \frac{260 \times 60 \times 8}{0.05 \times 350 \times 8} = \frac{124\,800}{140}$$

$$\simeq 891 \text{ conductors}$$

This, of course, is the total number of conductors, so the number of conductors *per slot* is given by:

$$\text{conductors per slot} = \frac{\text{total number of conductors}}{\text{number of slots}}$$

$$= \frac{891}{120} = 7.43$$

We obviously cannot have 7.43 conductors per slot and there cannot be an odd number of conductors, so we must increase the number to **8 (Answer)**.

Commutation

In this section, we are going to examine **commutation**, i.e. the way in which the armature supplies current to the generator's load, via its commutator and brushes.

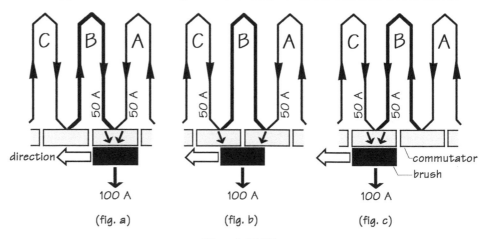

Figure 12.37

The three illustrations in Figure 12.37, represent a **carbon brush** (the black rectangle) moving between commutator segments. Of course, it is the *commutator* that's *actually* moving, *not* the brush, *but it's easier to illustrate the brush moving in the opposite direction!* So, although we are showing the brush moving to the left, what is *actually* happening, of course, is that the brush is stationary and the armature windings are moving to the right!

Remember, it's the position of the commutator's brushes (ideally centred opposite the pole pieces) that determine the location of the junctions of the parallel paths within the armature windings. For example, with the brush located as it is in Figure 12.37a, the junction between parallel paths occurs between windings **A** and **B** and, so, currents will approach the brush from *opposite* directions, as shown. In Figure 12.37b, the junction has moved to between windings **A** and **C**. And in Figure 12.37c, the junction has moved to between windings **B** and **C**.

With the brush located on the segment shown in Figure 12.37a, if we assume that windings **A** and **B** each supply, say, 50 A, then the total load current will be 100 A.

In Figure 12.37b, the brush has moved on, so that it now bridges two commutator segments, short-circuiting the centre winding (**B**) while windings **A** and **C** each supply 50 A – thus maintaining the load current of 100 A.

In Figure 12.37c, the brush has move further on to the next segment, where it draws 50 A each from windings **B** and **C** and, again, supplying 100 A to the load.

> Note how the current in coil **B**, falls to zero in Figure 12.37b , then reverses direction in Figure 12.37c, as the junction between the two parallel paths moves on.

The *reversal* of current direction in a winding as it passes a brush can be represented in the form of the graph shown in Figure 12.38.

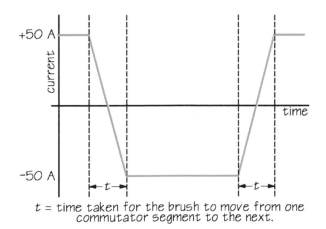

t = time taken for the brush to move from one commutator segment to the next.

Figure 12.38

Note, however, that this graph represents what is termed *'ideal'* commutation!

In practice, the armature winding has **inductance** and, therefore, any change in current will induce a voltage (called a '**reactance voltage**'), the direction of which, in accordance with **Lenz's Law**, will *oppose that change in current*. If you like, it will 'slow down' the rate of current change! So, returning to our graph, in the time *(t)* it takes the brush to move from one segment to the next, the winding's current, instead of reversing from +50 A to –50 A may only reverse from +50 A to, say, –25 A or thereabouts! This is illustrated in Figure 12.39.

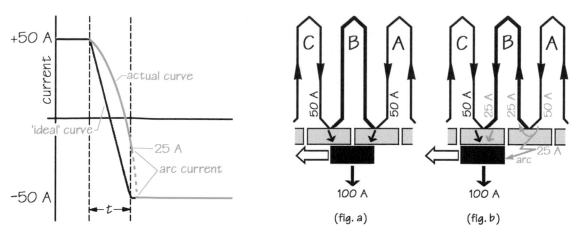

Figure 12.40

Figure 12.40

So, in *real life*, as the brush breaks contact with the commutator segment, there is only *one* way in which the remaining 25 A or so can possibly flow to the brush, and that is *through the air*, as a **spark**! This should be made clear in Figures 12.39 and 12.40.

So the 'reactance voltage' induced into a winding causes a spark to flash across between a commutator segment and the brush, and the heat thus generated can be high enough to *cause cumulative damage to the copper commutator segments*.

Methods of improving commutation

By '**improving commutation**', we mean *allowing commutation to take place with the minimum of arcing*.

There are *three* methods we can employ to reducing commutator arcing. These are:

(a) resistance commutation,
(b) e.m.f. commutation, and
(c) reducing the armature reactance.

Resistance commutation

In the companion book, **An Introduction to Electrical Science**, we learnt that the current lag in an inductive resistive circuit is determine by the **time constant** of that circuit, expressed in seconds, where the time constant is the ratio of the circuit's inductance to resistance.

Increasing a winding's *resistance*, therefore, reduces its time constant which, in turn, will reduce the 'current lag' which causes arcing.

In practice, it's *not* necessary to increase the resistance of the winding itself, as the same effect can be achieved by increasing the overall resistance of the winding/commuter/brush *circuit*. And this is achieved by increasing the brush/commutator **contact resistance**, by a combination of *using high resistivity carbon brushes* and *reducing the contact area* between the brush and the commutator segment.

E.M.F. commutation

The current lag that causes arcing is due to the 'reactance e.m.f.' induced into the winding, as the current rapidly falls towards zero.

If it was possible to 'inject' just enough voltage, in the correct direction, into the winding so as *to oppose and neutralise* this reactance e.m.f., then the falling current should more closely match the graph shown in Figure 12.38 and, so, reduce any arcing.

This is achieved using '**commutating poles**', also known as '**interpoles**'. These will be discussed later in this chapter.

Reduction of armature inductance

The only way of effectively reducing the armature winding's inductance is to reduce the number of its turns. Unfortunately, this may *not* be compatible with the design requirements of the machine, and is the least practical solution to the problem of arcing.

Armature reaction

Another potential cause of excess commutator sparking is something called '**armature reaction**', which we will briefly examine in this section.

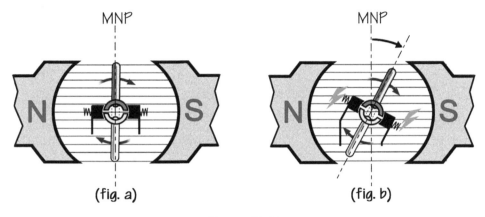

(fig. a)   (fig. b)

Figure 12.41

To start with, let's examine why the *position* of a d.c. generator's **brushes**, relative to the direction of the main magnetic field, is so important.

Figure 12.41a, which illustrates a simple, single-loop, d.c. generator, shows the position of what is known as the '**magnetic neutral plane**' (**MNP**). This, by definition, *lies at right angles to the main magnetic field*.

> A machine's **magnetic neutral plane** lies at right-angles to the main magnetic field.

Another important 'neutral plane' is the machine's '**geometric neutral plane**' (**GNP**), which *lies midway between, and perpendicular to, the pole pieces*. But we'll return to that later.

When the active conductors of the simple single-loop armature run parallel with the machine's main field, their positions lie along the magnetic neutral plane. This is illustrated in Figure 12.41a. In this position, no voltage is induced into the active conductors because they are *not* cutting the field and, therefore, no voltage appears across the commutator segments. This is why the brushes are centred opposite the pole pieces.

So, although the brushes are bridging the commutator segments, no current will flow between them and, so, no sparking can take place.

In Figure 12.41b, the brushes have been deliberately rotated clockwise *so they are no longer at right-angles to the magnetic neutral plane*. In this position, the loop's active conductors are now cutting the magnetic field and, so, an e.m.f. is being induced into the coil and, therefore, between the segments of the commutator. With the brushes located in this position, then, they are short-circuiting the commutator segments, and

12

a short-circuit current will pass between them resulting in unnecessary overheating and sparking.

> Clearly, therefore, *it is undesirable to locate the brushes away from their position at right-angles to the magnetic neutral plane*, if we wish to avoid damaging the commutator through sparking.

Let's now examine how all of this relates to '**armature reaction**'.

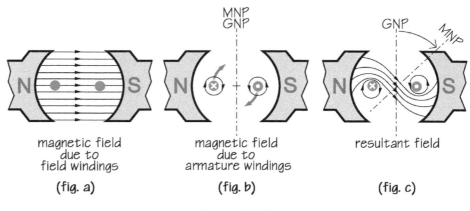

magnetic field
due to
field windings

(fig. a)

magnetic field
due to
armature windings

(fig. b)

resultant field

(fig. c)

Figure 12.42

In Figure 12.42a, we show the direction of the **main flux** (Φ_{main}) due to the machine's field windings.

In Figure 12.42b we have temporarily ignored the main flux, and show the direction of the **armature flux** ($\Phi_{armature}$) due to the armature currents set up as the armature loop rotates clockwise within the main field.

In Figure 12.42c, we show the resulting flux (Φ) due to the combination of Φ_{main} and $\Phi_{armature}$. As you can see, armature reaction has caused the main field to tilt in the *same direction in which the rotor turns*.

Remember, the magnetic neutral plane lies at right angles to the main field. So, as we can see, the resulting flux causes the magnetic neutral plane (MNP) to shift, *clockwise*, away from the geometric neutral plane (GNP) – *in the same direction as the rotation of the armature*.

The angle through which the magnetic neutral plane moves away from the GNP depends on the magnitude of the armature current; the *larger* the current, the *larger* this angle will be – so its position is entirely dependent upon the value of the **load current**.

In Figure 12.41, it was explained why, in order to avoid unnecessary sparking between commutator segments and the brushes, *it is important that the brushes should **always** lie at right angles to the magnetic neutral plane*.

At first consideration, therefore, it would seem reasonable to suppose that to avoid any commutation problems, all we have to do is to *advance the position of the brushes* so that they *always* correspond to the position of the magnetic neutral plane.

Unfortunately, however, the angle through which the MNP moves away from the GNP depends on the load current, and varies as the generator's load current varies. So *moving* the position of the brushes is not practical, *and is not the solution to the commutation problems caused by armature reaction!*

Before we move on to seek a solution to armature reaction, let's examine a further important problem caused by armature reaction. A problem that only becomes apparent when we examine a somewhat more realistic armature than the 'single-loop' armature described above. So, in Figure 12.43, we see an armature comprising eight 'active' conductors set around the circumference of a rotor that is rotating clockwise.

For the reasons already explained, we'll assume that armature reaction has caused the magnetic neutral plane to shift clockwise to the position shown.

By applying **Fleming's Right-Hand Rule** to the active conductors as they rotate clockwise, we can confirm that the load current flows through the conductors in the directions shown by the 'cross/dot' notation. By then applying the '**right-hand grip' rule** to each conductor, we can confirm the directions of the resulting flux surrounding those conductors.

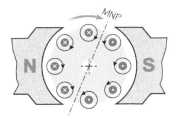

Figure 12.43

It should be obvious that the combined effect of the flux surrounding each of the conductors will result in a resultant field, due to the armature currents, that lie along the magnetic neutral plane. This 'armature flux', of course, will be significantly lower than the flux due to the main field, but we can determine the *resultant field* by plotting the two separate fields as a vector diagram, as shown in Figure 12.44:

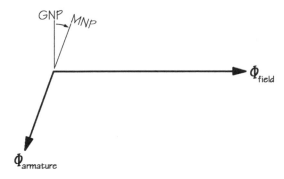

Figure 12.44

If we now (Figure 12.45) resolve the armature-field vector, $\Phi_{armature}$, into its *horizontal* (Φ_h) and *vertical* (Φ_v) components . . .

Figure 12.45

. . . it should be obvious (Figure 12.46) that the horizontal component (Φ_h) will act *to reduce the effective value of the field flux* (Φ_{field}).

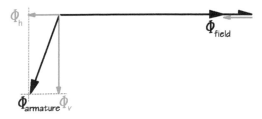

Figure 12.46

So it should now be clear that, as well as causing *commutation problems*, **armature reaction** *also acts to reduce the value of the field flux which, in turn, will act to reduce the value of the generated e.m.f.*

Let's finish this section by examining a *third problem* caused by armature reaction: **waveform distortion**.

Figure 12.47 illustrates how the flux would be evenly spaced around the air gap if the machine was not subject to armature reaction.

Figure 12.48 illustrates what happens to the flux when we take armature reaction into account. As we can see, the flux in the air gap is no longer evenly spaced, but becomes distorted and varies considerably around the air gap. In other words, the air-gap flux density *increases* in the direction of the magnetic neutral plane.

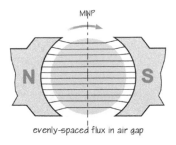

evenly-spaced flux in air gap

Figure 12.47

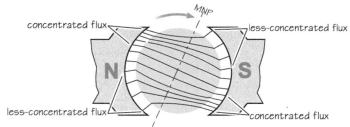

effect of armature reaction of flux distribution

Figure 12.48

As the voltage induced into the armature's active conductors is proportional to the *rate* at which they cut the flux then, at any given speed, the voltage will vary according to the concentration of the flux. And this will cause a distortion to the machine's output waveform, as illustrated in Figure 12.49!

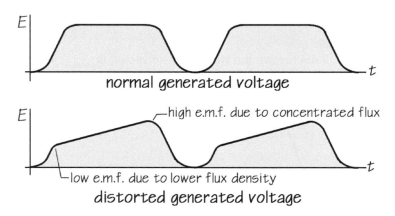

Figure 12.49

A *secondary* effect of the varying the concentration of the field, shown in Figure 12.48, is that the part of the field poles where the flux concentration is greatest can be driven into **saturation**, which will result in *a further reduction in the machine's output voltage*.

Let's end this section by summarising the problems associated with armature reaction.

▶ commutator sparking
▶ output voltage reduction
▶ distortion of output-voltage waveform
▶ partially saturated poles

Now that we know *what* armature reaction is, and the *problems* it can cause, it's time to move on to learn *how it can be reduced*.

Minimising armature reaction

As we have learnt, **armature reaction** is the *distortion* of a machine's main magnetic field due to the presence of a cross-field set up by the rotor's armature windings when they are delivering a load current. Its effect *is to move the magnetic neutral plane away from the geometric neutral plane, in the direction of rotation*, which results in excess commutator sparking, reduced voltage, distorted waveform, and partial pole saturation.

One widely used solution to this problem is to introduce small auxiliary pole pieces, called '**interpoles**' or '**compoles**' ('compensation poles'), between the main poles – as shown in Figure 12.50. Interpoles are wound with a relatively small number of turns of heavy conductor, and are connected *in series with the armature windings*. The direction (*opposite* to that of the cross-field) and magnitude of the field set up by these interpoles then acts to *counteract* and, effectively, *cancel the cross-field set up by the armature windings*. By counteracting the cross-field flux, armature reaction can be largely eliminated.

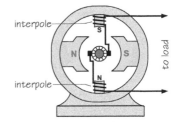

Figure 12.50

Because the interpole windings are connected in *series* with the armature winding, any variations in load current, which would normally shift the magnetic neutral plane, are automatically compensated for – as a result, the magnetic neutral plane will remain more or less fixed and, once the commutator brushes have been set correctly, they will *not* need to be adjusted any further.

An *alternative* method of reducing armature reaction is through the use of '**compensation windings**'. These are 'distributed' windings, set into longitudinal slots machined into the face of each pole piece, as illustrated in Figure 12.51, and connected in series with the load:

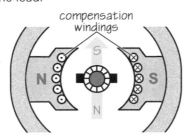

Figure 12.51

Just as with interpoles, when a load current passes through these windings (in the direction shown by the cross/dot symbols), a small field is created which *opposes* and cancels the field set up in the armature – thus eliminating, or reducing, any armature reaction.

Calculating the number of turns per interpole

To counter the effect of armature reaction, the total **magnetomotive force (m.m.f.)** of the interpole-pair must be equal and opposite that of the armature windings. The magnetomotive force of a winding is the product of the number of turns that form that winding and the current passing through it, expressed in amperes (although we normally speak this as 'ampere turns', to distinguish it from the unit of measurement for current).

So,

m.m.f. of armature windings = (number of turns per pole) × (armature current)

$$= \left(\frac{Z}{p}\right) \times \left(\frac{I_A}{c}\right)$$

$$= \frac{Z I_A}{p\,c}$$

... the requirement to eliminate armature reaction:

total m.m.f. of *pair* of interpoles = m.m.f. of armature windings

$$N_{interpole} \, X_A = \frac{Z X_A}{pc}$$

$$N_{interpole} = \frac{Z}{pc}$$

So the number of turns, *per interpole*, must be half this amount:

$$N_{per\ interpole} = \frac{1}{2} \times \frac{Z}{pc}$$

$$= \frac{Z}{2pc}$$

Sometimes, it's necessary to apply a *multiplication factor* (see worked example 6) to this to fully compensate for the armature reaction.

$$N_{per\ interpole} = \frac{Z}{2pc}$$

where: | N = number of turns per interpole
| Z = armature conductors
| p = number of poles
| c = number of parallel paths

Worked example 6

Calculate the number of turns required on each interpole for a four-pole, lap-wound, d.c. generator with 640 active armature conductors, so that the effect of armature reaction is neutralized in the region where the brushes are placed.

Solution

$$N_{per\ interpole} = \left(\frac{Z}{2pc}\right) = \left(\frac{640}{2 \times 4 \times 4}\right) = \textbf{20 turns (Answer)}$$

Worked example 7

A 150 kW, 250 V, six-pole d.c. generator has a lap-wound armature comprising 432 active conductors. Calculate the number of turns per pole required for the interpoles, assuming the brushes are placed at the geometric neutral axis, and that the interpole's m.m.f. per pole must be 1.3 times that of the armature poles in order to fully compensate for armature reaction.

Solution

$$N_{per\ interpole} = 1.3\left(\frac{Z}{2pc}\right) = 1.3 \times \left(\frac{432}{2 \times 6 \times 6}\right) = 7.8$$

The nearest whole number to this is **8 turns (Answer)**

Types of d.c. generator

Let's move on, now, to consider how d.c. generators are **classified**, and to examine the behaviour of different types of d.c. generator.

D.C. generators are classified according to their *method of excitation*.

The term, '**excitation**', refers *to the way in which a generator's field is produced*. In this chapter, we will consider the following types of d.c. generator.

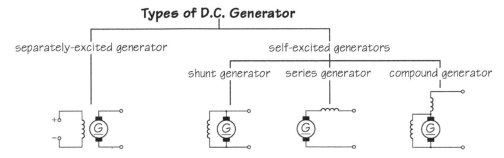

Figure 12.52

We can compare the *performance characteristics* of these types of d.c. generator by examining their **characteristic curves**. These are the generators'

▶ **open-circuit** characteristic curve, and their
▶ **external** characteristic curve.

The **open-circuit characteristic curve** (Figure 12.53) shows the change in a d.c. generator's **terminal voltage (E)** resulting from variations in its **field current (I_F)**, at no load, while driven at constant speed. This curve is also known as the 'magnetic characteristic' or 'no-load saturation' curve.

The **external characteristic curve** (Figure 12.54) shows the change in a d.c. generator's **terminal voltage (U)** resulting from variations in **load current (I_L)**, while driven at constant speed. The terminal voltage, when the machine supplies a load, will always be *less* than the no-load terminal voltage *(E)* because of a voltage drop occurring within the armature circuit. This curve is very important, because *it determines the suitability of a generator for a particular purpose.*

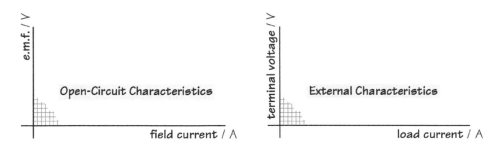

Figure 12.53   Figure 12.54

Separately excited d.c. generator

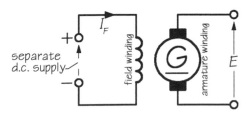

Figure 12.55

A **separately excited d.c. generator** is *not* normally used as a supply generator because it depends entirely on the availability of an *existing* d.c. supply, which rather defeats the point of needing a d.c. generator to start with! Instead, this type of generator is used for special applications, such as being part of the *'Ward-Leonard System'* for controlling the speed of d.c. motors. This control system is outside the scope of this chapter, but will be dealt with in the next.

The '**Ward-Leonard**' control system was introduced in 1891, and named after its designer, American engineer Harry Ward-Leonard (1861–1915), and is outside the scope of this book. It is a control system for smoothly varying the speed of a heavy-duty d.c. motor including changing its direction. Its applications include the drive systems for railway locomotives, naval gun turrets, military radar scanners, etc.

From Figure 12.55, it can be seen that the field current must be supplied from an *existing independent d.c. source* – hence the term, 'separately excited'.

The *flux density* of the machine's main field depends upon the magnitude of the field current provided, up to the point where that current drives the field poles into saturation. The *polarity* of the generator's terminal voltage may be reversed, by reversing the direction of the field current.

Self-excited d.c. generators

A **self-excited d.c. generator** is one *that generates its own field*.

At first glance, this seems to be a contradiction! A sort of 'chicken and egg' situation that beggars the question: *'If the generator requires a field in order to generate a terminal voltage, where does it get the necessary field current from to set up the field in the first place?'*

Well, the answer is to be found in the magnetic characteristics of the field poles. As we already know, the field poles are manufactured from silicon-steel (or, more accurately, 'silicon-iron') laminations, and these normally exhibit *a small amount of* **residual magnetism**. This residual magnetism is a small remnant of the flux generated from when the machine was previously operated.

So, with the rotor being driven by its prime mover, this residual magnetism is just sufficient to induce a small voltage into the armature windings, and this voltage, *provided it can overcome the resistance of the field circuit*, will then cause a small field current to flow. This small field current then *reinforces* the residual magnetism already in the poles, which then *increases* the voltage induced into the armature windings, which *increases* the value of the field current . . . and so on!

This effect is *cumulative* until, eventually, the field poles reach saturation and the flux density can increase no further, at which point the terminal voltage reaches its final constant value!

When a *new* generator is first installed, or when an older generator simply 'fails to self-excite', it is likely the poles have no, or insufficient, residual magnetism to enable the terminal voltage to build up. So, sometimes, it is necessary to *create* some residual magnetism in the field poles. This is done by temporarily disconnecting the field winding, and using an external d.c. supply to partially magnetise the poles. This is known as '*flashing the field*', and care must be taken to ensure that the *direction* of the resulting field is correct in order to ensure that the machine's terminal polarity is correct.

If we refer back to Figure 12.52, we can see that there are *three* types of self-excited d.c. generator: the **shunt generator**, the **series generator**, and the **compound generator**. And, as we shall learn, the terms, '**shunt**', '**series**', and '**compound**' each describe the way in which a machine's field windings are connected, *relative to the armature*.

Shunt generator

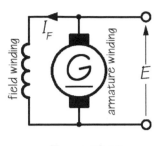

Figure 12.56

As we can see from Figure 12.56, a **d.c. shunt generator** is one *whose field winding is connected in parallel with its armature*. 'Shunt', we should recall, is simply a somewhat archaic term, which describes something that's connected 'in parallel' with something else. For the purpose of clarity, we are *not* showing any interpole windings which, as we learnt earlier, would normally be connected in *series* with the armature windings to reduce armature reaction.

As already explained, the field poles of a d.c. generator normally retain a small amount of residual magnetism from when the machine was previously run. So, with the rotor turning at its rated speed, a small voltage is initially induced into the armature thanks to this residual magnetism. With the field winding connected in parallel ('shunt') with the armature, this small terminal voltage, provided that it can overcome the resistance of the field circuit, will provide just sufficient field current (I_F) to *reinforce* the residual magnetism in the field poles which, in turn, increases the output voltage. This action is *cumulative*, and continues to build up until the poles reach saturation, at which point, the machine's final e.m.f. *(E)* is reached.

Before proceeding any further, let's remind ourselves about 'residual flux density' and 'saturation', by examining the **B-H curve** for silicon-steel. Remember, although the term 'silicon-steel' is the term widely used to describe the material from which the field poles are manufactured, that material is actually silicon-*iron*.

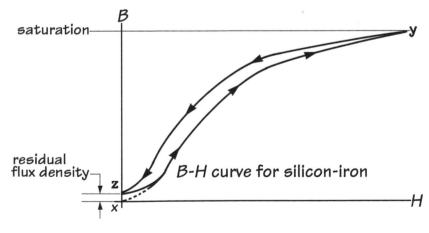

Figure 12.57

Figure 12.57 shows a typical **B-H curve** for silicon-steel. These curves were covered in great detail in the companion book, **An Introduction to Electrical Science**, so we'll only describe their main features here.

A quick reminder, before we start, that the symbol, **H**, represents **magnetic field strength**, which is defined as 'magnetomotive force per unit length', expressed in amperes per metre (normally spoken as 'ampere turns per metre') where magnetomotive force is the product of the number field winding turns *(N)* and the current *(I_F)* through them, i.e.:

$$H = \frac{I_F N}{l} \quad \text{therefore} \quad H \propto I_F$$

So the poles' magnetic field strength is proportional to the field current.

The symbol, **B**, represents flux density, which is defined as 'flux per unit area', expressed in teslas.

So, if we were to start with a completely unmagnetised sample of silicon-steel, and gradually *increase* the magnetic field strength, by increasing the field current, the flux density *(B)* will start to *increase* from zero (point **x** on the graph), and it will follow the lower curve until the sample reaches **saturation** (point **y** on the graph). Any further increase in magnetic field strength will have no effect on increasing the flux density once saturation is reached.

If we were now to *reduce* the magnetic field strength, by reducing the field current, the flux density will also reduce but, this time, following the *upper* curve. When the magnetic field strength reaches zero, however, the flux density falls to point **z** on the graph, *not zero!*

The difference between points **x** and **z**, along the vertical axis, represents the **residual flux density** for that sample of silicon-steel.

If we were now to repeat the increase/decrease cycle of the magnetic field strength, the resulting change in flux density would follow the solid line – that is, the **z-y-z** curve. We should recall that this behaviour is due to a phenomenon called **'hysteresis'**, from the Greek word meaning 'to lag'.

Now, the flux (Φ) is proportional to flux density, and m.m.f. is proportional to the field current and, since the machine's average e.m.f. is proportional to the field's flux and its field current is proportional to the field windings' m.m.f., we can completely re-label the graph as shown in Figure 12.58, where the horizontal axis now represents the field current and the vertical axis now represents the generated e.m.f.

$$\left[{}^*_{,}E_{av} = \frac{\Phi N p Z}{60c} \quad \text{therefore} \quad E \propto \Phi \right]$$

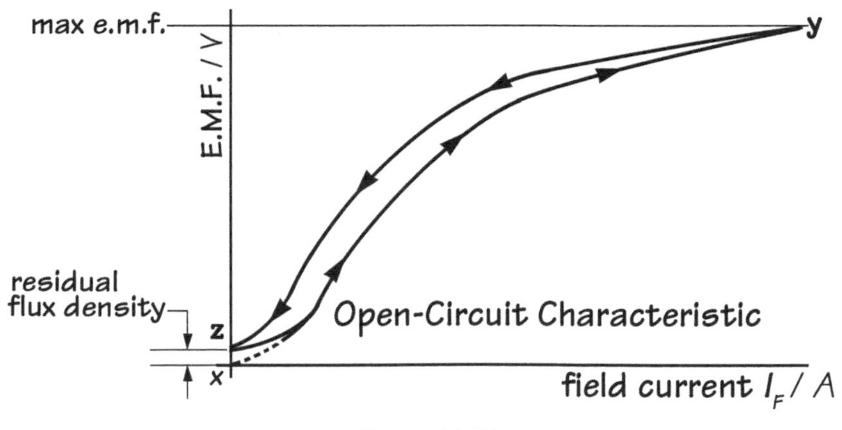

Figure 12.58

From this, we can confirm that the e.m.f. appearing across the output terminals of a shunt generator will follow the curves shown in Figure 12.58 for increasing/decreasing variations in field current.

Open-circuit characteristic

So, a shunt-generator's **open-circuit characteristic** for an increasing field current is as shown in Figure 12.59.

Let's now move on to consider the behaviour of the shunt generator *when it is connected to a load*, as shown in Figure 12.60.

When a **load** is connected to the terminals of a shunt generator, a load current (I_L) will flow. The field current (I_F) is unaffected by the load, as it is in parallel with the armature, but the armature current (I_A) must now *increase* to continue supplying the field winding *as well as* supplying the new load current, because:

$$I_A = I_F + I_L$$

As we learnt from our study of batteries and transformers, a load current will always act to *reduce* the terminal voltage of those devices, and the same is true for generators.

So the load current results in *a decrease in the generator's no-load voltage (E)*; an *increase* in load current will cause the no-load voltage *to fall even further*.

To understand this, we must examine the equivalent circuit for the armature, shown in Figure 12.61.

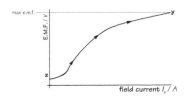

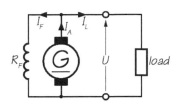

Figure 12.59

Figure 12.60

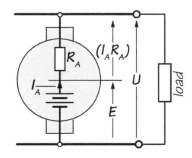

$$E = U + I_A R_A \quad so \quad U = E - I_A R_A$$

Equivalent Circuit of Armature

Figure 12.61

In this equivalent circuit (in which we are ignoring any interpoles), the 'battery' represents the **no-load voltage**, or **e.m.f.** *(E)* of the machine, and the resistor represents the **(internal) resistance** *(R_A)* of the armature windings. Because of the way in which the voltages are represented in the equivalent circuit, we might be forgiven for thinking that *U* is the largest voltage but, remember, *it's the no-load voltage, E, that is the largest!* So, applying Kirchhoff's Voltage Law around the armature/load circuit, we have:

$$E = U + (I_A R_A)$$

The **terminal voltage** *(U)*, therefore, must be the *difference* between the no-load voltage *(E)* and the internal voltage drop:

$$U = E - (I_A R_A)$$

This lower terminal voltage, of course, is not only applied to the load *but to* the field winding as well (they are, after all, in *parallel* with each other). This results in a reduction in the field current, which reduces the field's flux – *causing a further reduction in the terminal voltage!*

Earlier, we learnt that **armature reaction** causes the generator's terminal voltage to fall, so armature reaction – if uncorrected – will *also* contribute to the reduction in the generator's terminal voltage.

12

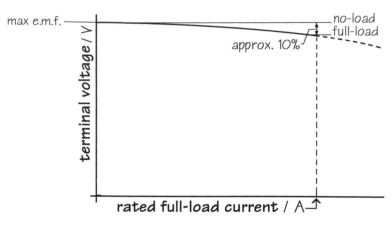

Figure 12.62

Despite this, the fall in terminal voltage, from no load to full load, is typically less than 10% for a shunt generator. This is perfectly acceptable and, for this reason, a shunt generator is considered to be a '**constant voltage**' generator, which makes it ideal as a **d.c. mains-supply generator**.

Series generator

As we shall learn, the terminal voltage of a **series generator** varies considerably as its load current changes, so it is *not* really suitable for use as a mains-supply generator and, so, is not widely used. So the *only* reason that we are going to discuss it, here, is because it will help us understand the operation of a **compound generator**.

From Figure 12.63 we can see that, for a series generator, the field windings are in *series* with the armature windings, so *its excitation depends entirely upon the load current*. Again, for the purpose of clarity, we are *not* showing any interpoles, which would normally be connected in series with the armature windings in order to reduce armature reaction.

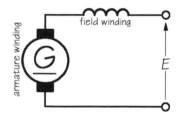

Figure 12.63

Without any load connected to the machine, no current can flow through the field windings and, so, it would seem reasonable to suppose that there can be no excitation and, therefore, no terminal voltage.

However, this *isn't* the case, because – as with a shunt generator – its poles exhibit a certain amount of **residual magnetism** from the previous time the generator was operated. So, even when it is *not* supplying a load, the generator *will* normally produce a very small e.m.f., but one which *cannot* increase in value because *there is no field current available to build up the field*.

A series generator's **open-circuit characteristic**, therefore, will look like that reproduced in Figure 12.64.

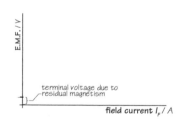

Figure 12.64

When a series generator is connected to a load, the resulting load current is common to both the armature and field windings – i.e. *they're all one and the same current*, as can be seen from Figure 12.65.

$$I_A = I_F = I_L$$

Because the load current tends to be much larger than the usual value of field current, a series generator's field windings tend to require *fewer turns*, compared with a shunt generator's, in order to produce the amount of m.m.f. required to produce the necessary field.

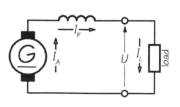

Figure 12.65

The **external characteristic** of a series generator is illustrated in Figure 12.66. As we can see, the terminal voltage is completely dependent upon the amount of load current up until the point where the poles become saturated.

When the generator is supplying current to its load, a situation arises that is more complicated than it might first appear because an *increasing* load current brings about *two* effects:

(a) it *increases* the field flux and, therefore, its generated e.m.f., and

(b) it *increases* the armature and field voltage drops, which tend to *reduce* the generated e.m.f.

So, while the build-up curve for a series generator might *appear* to identical to that of a shunt generator, it is *not* the same!

If the load current continues to increase beyond the point where the maximum value of terminal voltage occurs, then a sudden, sharp, collapse in that voltage will occur (Figure 12.67).

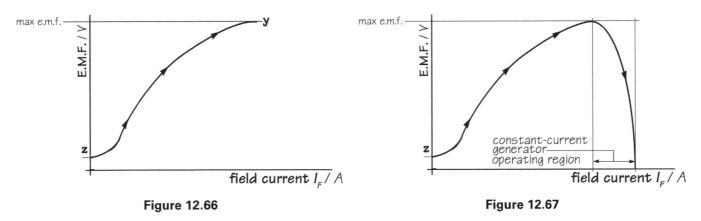

Figure 12.66	Figure 12.67

This collapsing portion of the curve indicates that, between points **x–y**, the machine may be operated as a '**constant-current**' **generator**. The variation in current between points **x–y**, isn't *literally* 'constant', of course, but is small enough to be considered as being 'relatively constant'.

Constant-current generators are used, for example, as **welding generators**, where a relatively constant current over wide variations in arc-voltages produce a relatively consistent welding temperature.

Compound generator

A **compound generator** has both shunt *and* series field windings, *wound together on the same pole pieces*.

A compound generator may be connected in '**long shunt**' or '**short shunt**' configurations, as illustrated in Figure 12.68.

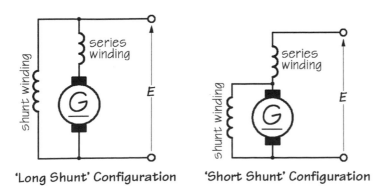

'Long Shunt' Configuration   'Short Shunt' Configuration

Figure 12.68

285

Furthermore, the **shunt** and **field windings** may be *cumulatively* wound or *differentially* wound.

In both cases, the *shunt* field always predominates. But, in the case of a '**cumulative-compound**' machine, the series field *aids* the shunt field whereas, in the case of a '**differential-compound**' machine, the series field *opposes* the shunt field. This is illustrated in Figure 12.69.

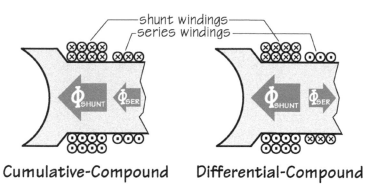

Cumulative-Compound   Differential-Compound

Figure 12.69

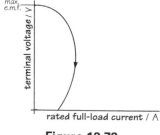

Figure 12.70

A '**cumulative-compound**' generator is used whenever it is necessary to avoid a drop in the terminal voltage, which is a characteristic of a shunt generator. As we have already learnt, whenever rated load is applied to a shunt generator, its terminal voltage can fall by up to 10%. However, an increasing load current through a series winding, wound cumulatively, acts to *boost* the terminal voltage, restoring it to, or even *above*, its original value.

The precise amount of boosting, or '**compounding**', necessary to restore the generator's terminal voltage, is controlled by means of a **series diverter**: a variable resistor, connected in parallel with the series winding (Figure 12.70).

The effect on a compound generator's external voltage characteristic curve, of adjusting the series diverter is illustrated in Figure 12.71.

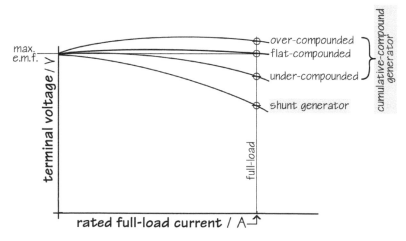

Figure 12.71

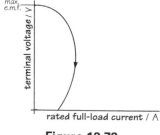

Figure 12.72

Note that, in the case of a '**differential-compound**' **generator**, the series field tends to *reduce* the machine's shunt field, and can result in *a really severe drop in terminal voltage* for increasing values of load current (Figure 12.72).

This curve indicates that, like a series generator, a **differential-compound** generator can, just like a series generator, be used as a '**constant-current**' **generator**, and its applications would be the same: e.g. welding generator.

Worked example 8

A 5 kW, 250 V, short-shunt compound d.c. generator has an armature winding, shunt field winding, and series field winding of resistances 0.8 Ω, 45 Ω, and 0.6 Ω respectively. Calculate the e.m.f. generated by the armature at full load.

Solution

Let's start by sketching a schematic diagram of the machine (Figure 12.73).

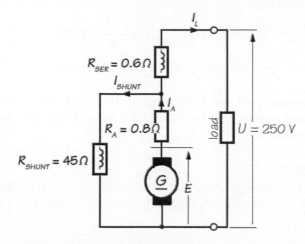

Figure 12.73

The *most important* thing to understand, here, is that the **no-load voltage, E**, is the *voltage source* in this circuit (see Figure 12.61 to remind ourselves of the equivalent circuit); *all* other voltages are voltage drops. In other words: $E = \left(I_A R_A + I_L R_{SER} + U\right)$.

So, the first step is to determine the load current *(IL)*:

$$I_L = \frac{P}{U} = \frac{5000}{250} = 20 \text{ A}$$

. . . which then lets us find the voltage drop across the series winding:

$$U_{SER} = I_L R_{SER} = 20 \times 0.6 = 12 \text{ V}$$

So the voltage drop across the shunt winding must be:

$$U_{SHUNT} = E_T - U_{SER} = 250 + 12 = 262 \text{ V}$$

. . . therefore, the current *(I_{SHUNT})* through the shunt winding must be:

$$I_{SHUNT} = \frac{U_{SHUNT}}{R_{SHUNT}} = \frac{262}{45} = 5.82 \text{ A}$$

So we can now determine the armature current *(I_A)*:

$$I_A = I_L + I_{SHUNT} = 20 + 5.82 = 25.82 \text{ A}$$

So the voltage drop across the armature will be:

$$U_{ARM} = I_A R_A = 25.82 \times 0.8 = 20.66 \text{ V}$$

Finally, we can determine the generator's no-load voltage:

$$E = U_A + U_{SER} + U = 20.66 + 12 + 250 = \textbf{282.66 V (Answer)}$$

12

Voltage regulation

Voltage regulation is defined as, *'the change in terminal voltage from no load to full load'*. This is normally expressed as a percentage of *either* the rated **no-load voltage** ('percentage voltage regulation down') *or* the rated **full-load voltage** ('percentage voltage regulation up').

British practice is to use the '**percentage voltage regulation down**', which appears to be more intuitive, whereas North American practice is to use 'percentage voltage regulation up'.

$$(\text{voltage regulation})_{down} = \frac{E-U}{E} \times 100\%$$

$$(\text{voltage regulation})_{up} = \frac{E-U}{U} \times 100\%$$

Clearly, 'ideal' regulation will occur when there is no change whatsoever in a generator's terminal voltage from no load to full load, in which case the voltage regulation will be *zero*.

Worked example 9

The no-load terminal voltage of a shunt generator is 200 V, and its full-load terminal voltage is 185 V. Calculate its (a) voltage regulation down, and (b) its voltage regulation up.

Solution

$$(\text{voltage regulation})_{down} = \frac{E-U}{E} \times 100\%$$

$$= \frac{200-185}{200} \times 100\% = \frac{15}{200} \times 100\%$$

$$= \textbf{7.5\% (Answer a.)}$$

$$(\text{voltage regulation})_{up} = \frac{E-U}{U} \times 100\%$$

$$= \frac{200-185}{185} \times 100\% = \frac{15}{185} \times 100\%$$

$$= \textbf{8.1\% (Answer b.)}$$

Energy losses

Energy losses are important because they determine the overall efficiency of a d.c. generator. For generators designed to operate at, or near, full load, manufacturers design their machines so that maximum efficiency occurs at full load.

Some generators, however, may operate for a large proportion of their life at *considerably less* than full load, and for only occasionally at full load. In these cases, the manufacturer may design the generator to operate at maximum efficiency at some point *below* full load.

So the design efficiency of a d.c. generator, therefore, is dependent not only upon the full-load output required, but also upon the service which the machine is expected to provide.

Energy is lost due to the:

▶ iron losses, consisting of hysteresis and eddy-current losses in the magnetic circuit.
▶ armature current passing through the resistance of the armature windings, through the contact resistance between the commutator and its brushes, and through the resistance of any interpoles or compensating windings where installed.

▶ field currents passing through the resistances of the field windings.

▶ frictional losses in the machine's bearings, and between the commutator and its brushes.

▶ windage* losses, due to the rotor having to overcome air resistance.

Some of these losses remain *approximately constant*, regardless of the machine's load current, providing its speed is more-or-less constant – e.g. iron losses, friction, windage, and excitation current (in the case of a shunt-generator).

> It can be shown that **maximum efficiency** occurs approximately whenever the variable losses are equal to the fixed losses.

The calculation of the actual efficiency of a d.c. generator is particularly difficult because of the problem of determining realistic values of frictional and windage losses, and is far beyond the scope of this book.

All the energy losses that occur in a d.c. generator must be dissipated by heat transfer away from the machine in order to limit the rise in the machine's temperature. Most of this heat is conducted from the external surface of the machine, via a thin layer of stagnant air, to a stream of circulating air. The machine must be constructed, therefore, so that this circulating air has free access to all those parts from which heat transfer must take place.

Conclusion

Now that we've completed this chapter, we need to examine its **objectives** listed at its start. Placing a question mark at the end of each objective turns that objective into a **test item**. If we can answer those test items, then we've met the objectives of this chapter.

12

---

*'Windage' is pronounced 'wind' as in 'breeze', not as in 'to wind up'.

D.C. motors

Objectives

On completion of this chapter you should be able to:

1. identify the factors that affect the force acting on a current-carrying conductor lying within a magnetic field.

2. identify a d.c. motor's magnetic circuit.

3. state the relationship between the torque developed by a d.c. motor and its volume.

4. derive an equation for power developed by a simple loop d.c. motor, in terms of its torque and rotational speed.

5. state the relationship between the power output of a d.c. motor and its volume and speed.

6. explain how a de Prony brake can be used to determine the torque, power, and efficiency of a d.c. motor.

7. derive an equation for the torque developed by a practical d.c. motor.

8. explain the difference between the armature reaction of a d.c. motor and a d.c. generator.

9. sketch schematic diagrams, showing the differences between a shunt, a series, and a compound d.c. motor.

10. sketch the torque characteristic, speed characteristic, and torque/speed characteristic for a shunt, series, and compound d.c. motor.

11. summarise the differences in behaviour of shunt, series, and compound d.c. motors.

12. provide examples of applications for shunt, series, and compound d.c. motors.

13. explain the need for d.c. motor starters.

14. determine the value of starting resistance for a simple d.c. motor.

15. provide examples of different types of d.c. motor starters.

16. explain the need for 'no volt' protection with d.c. motor starters.

17. summarise the main methods of d.c. motor speed control.

18. summarise the effect on d.c. motors of the advent of 'power electronics'.

Introduction

Electric motors are so common that we hardly give them any thought unless, of course, they happen to stop working! If we carried out an inventory of the motors we have in our cars, for example, we would probably be very surprised. For we rely on a motor to start our car, to cool its radiator, to run its ventilation/air-conditioning system, to operate the windshield wipers, to wind the windows up and down, perhaps to raise and lower the radio aerial and. if we are fortunate enough to own a luxury model, to adjust the position of the seats. And, in the case of motor vehicles, of course, these are *all* **d.c. motors**!

During the late eighteenth and early nineteenth centuries, many physicists began to wonder whether there was a relationship between **electricity** and **magnetism**. Until then, they had been considered to be two quite separate phenomena.

Then, one evening in 1820, apparently while preparing an experiment for his students, the Danish physicist, Hans Christian Ørsted (1777–1851), noticed that whenever he passed an electric current through a wire, it caused the needle of a nearby compass to deflect. As Ørsted was one of those physicists who believed there *was* a relationship between electricity and magnetism, it seems unlikely that his discovery was quite as 'accidental' as this story suggests but, nevertheless, his discovery was of great importance because it confirmed that a relationship did indeed exist between the two.

The publication of Ørsted's discovery triggered a great deal of further experiments by numerous other physicists, engineers, and inventors worldwide, all of whom realised that *if an electric current could produce a force* then, perhaps, *this force could then be used to produce useful motion*.

One of these physicists was the Englishman, Michael Faraday (1791–1867), a bookbinder by training, but whose great interest with science (derived, in part, from the books that he was tasked with binding) was recognised by *the* eminent physicist of the time, Sir Humphry Davy, who, in 1812, appointed him to the Royal Institute as a chemistry assistant. Nine years after his appointment, and within a year following Ørsted's discovery, Faraday created an experiment in which he was able to demonstrate the extremely important concept of **magnetic rotation**.

An illustration of Faraday's experimental apparatus is shown in Figure 13.1. It consists of a corked glass tube, sealed at the bottom with pitch, and containing some mercury into which a magnet had been placed, vertically. The free end of a straight wire or rod, hanging from a hook in the cork, was immersed in the mercury and, when a current was passed along the rod from the hook into the mercury (which is a conductor), the resulting electromagnetic field set up around that rod reacted with that of the permanent magnet, causing it to continuously rotate around the magnet.

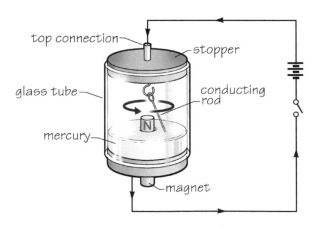

Figure 13.1

Although this experiment certainly *demonstrated* rotational 'motor action', *it clearly was of no practical use*. It was simply a laboratory demonstration of the results of Faraday's theoretical research, and was quite incapable of driving any type of load. But, nevertheless, it was a very important milestone in the quest to create a machine capable of converting electricity into useful motion.

A year later, another British physicist, Peter Barlow (1776–1862) was able to demonstrate a spinning star-shaped wheel known, these days, as '**Barlow's Wheel**', and is the first known example of a *rotating* machine to be driven by an electric current.

As can be seen in Figure 13.2, 'Barlow's Wheel' consisted of a simple star-shaped wheel, pivoted at its centre, which allowed it to rotate. With the tip of one of its pointed 'spokes' immersed in a slot containing mercury, current is able to pass along that particular spoke in the direction shown. By applying Fleming's Left-Hand Rule (for 'conventional current' flow) to that spoke, we see that the resulting force will push the spoke backward, causing the wheel to rotate (in a counterclockwise direction as viewed in Figure 13.2). This action is repeated for each spoke as its tip makes contact with the mercury, causing the wheel to rotate continuously.

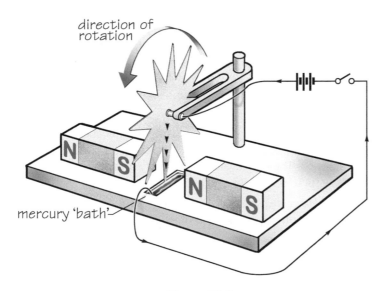

direction of rotation

mercury 'bath'

Figure 13.2

13

Again, although Barlow's spinning wheel certainly represented a major step forward in the quest to produce useful rotational motion from magnetism, just like Faraday's apparatus, the torque developed by the wheel was simply far too weak to do any useful work by which we mean driving a mechanical load! Nevertheless, it *was* a practical example of an important principle: that is *a current-carrying conductor, when placed within a magnetic field, will experience a force that will act to push that conductor out of the field* and, in this case, produce *rotational motion*.

In the companion book, ***An Introduction to Electrical Science***, we learnt that this force is expressed by the equation, $F = BIl$, which informs us that the force *(F)* acting on a conductor within a magnetic field is the product of the field's **flux density** *(B)*, the **current** *(I)* passing through the conductor, and the **length** *(l)* of the conductor within the field.

Although none of the nineteenth-century physicists would have recognised this equation, these three requirements would have been instinctively obvious to them. So, during the years following the publication of the discoveries made by Ørsted, Faraday, and Barlow, physicists and inventors all over the world began working quite independently of each other to produce a practical '**electromagnetic engine**', as 'motors' were known in those days, capable of producing sufficient force or torque to do useful work.

A major step forward in this direction was the invention of the **electromagnet** by William Sturgeon (1783–1850) in 1825. Up to this point, only weak permanent magnets were available to provide the magnetic field necessary to produce motion. But electromagnets were capable of producing a significantly higher flux density than any permanent magnet.

Some based their designs on *rotational* machines while others, influenced by the operating principles behind the reciprocating **steam engines** in use at that time, based their designs on *reciprocating* machines. But it was the *rotational* machines that ultimately proved to be the better designs.

The first practical rotating electric motor, i.e. one with a useful output power, was developed in Russia, in 1834, by a German engineer, Moritz Hermann von Jacobi (1801–1874), from which, four years later, he developed one that was capable of driving a 28-foot boat, carrying up to 14 passengers, on the River Neva. However, *none* of the motors developed by von Jacobi and his contemporaries led to the electric motors that we would recognize today.

In fact, the modern motor emerged, perhaps rather surprisingly, *not* from the development of motors by people like von Jacobi, but from the development of power *generators* and the invention of the **commutator**. This was predicted by the Russian, Emil Lenz (1804–1865), who, in 1833, wrote of *'the reversibility of electric generators and motors'*, describing experiments in which he was able to use a generator as a motor.

D.C. motors have, traditionally, scored over a.c. motors, because of their versatility, offering a *high starting torque*, *variable speed characteristics*, and the ability to *reverse their direction of rotation*. They are available to supply an enormous range of power outputs: from milliwatts to hundreds of kilowatts.

These characteristics have always made d.c. motors ideal for applications such as lifts (elevators), railway locomotives, cranes and drag lines, rolling mills, and so on.

As an example of the d.c. motor's versatility, let's consider the characteristics required of, say, a **lift (elevator) motor**. It must be able to start and stop with various loads, ranging from the weight of the empty 'car', to that of one carrying its full complement of passengers. The motor must then be able to accelerate and decelerate the car smoothly, and without any sudden changes that may cause the passengers to fall over! It must be also able to change direction, raising or lowering the car quickly along the lift shaft while being able to make stops at precise locations along the way. This is asking a great deal from *any* electric motor, but tasks like this are handled with ease by the appropriate d.c. motor.

Like the d.c. generators described in the previous chapter, d.c. motors have always been categorised as being '**separately excited**' or '**self-excited**' and, within the second category, as being '**shunt-**', '**series-**', or '**compound-wound**' – according to the way in which their field windings are connected relative to the armature windings. However, *these descriptions all date back to the days before the introduction of 'power electronics', which now enable a separately excited d.c. motor to perform just like a shunt, series, or compound motor!*

Nevertheless, in this chapter, we will retaining these traditional categories, and will be examining the characteristics of each of them. Finally, we will briefly examine how power electronics has changed things.

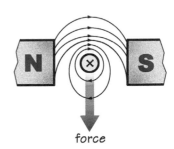

force

Figure 13.3

Principle of operation of a d.c. motor

We have already learnt that whenever a current-carrying conductor is placed within a magnetic field, a **force** will act on that conductor, which will try and *push the conductor out of the field*. This force results from the interaction of the magnetic field surrounding the conductor due to the current, and the permanent magnetic field.

The way in which these two magnetic fields react with each other to produce that force, is illustrated in Figure 13.3.

We have also learnt that a very useful method for determining the **direction** of this force, if we already know the direction of the *conventional* current and the direction of the magnetic field (always north to south), is by applying **Fleming's Left-Hand Rule for 'Motor Action'**.

We start by extending our first finger (index finger), second finger, and thumb at right-angles to each other. If the **f**irst finger is then pointed in the direction of the magnetic **f**ield (*always* north to south), and the se**c**ond finger pointed in the direction of the conventional **c**urrent, then the thu**m**b will indicate the direction of the force which results in the conductor's **m**otion:

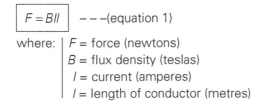
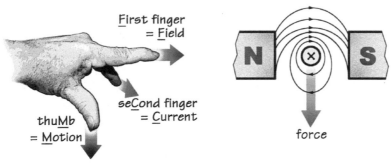

First finger = Field

seCond finger = Current

thuMb = Motion

force

Fleming's Left-Hand Rule

Figure 13.4

The **magnitude** of the resulting force is directly proportional to the values of the flux density of the **field**, the **current**, and the **length** of the conductor *within that field*:

$$F = BIl \quad ---\text{(equation 1)}$$

where:
F = force (newtons)
B = flux density (teslas)
I = current (amperes)
l = length of conductor (metres)

As we shall learn, while the magnetic field, of course, is *essential* for motor action to take place, there is a limit to the flux density that can be achieved even with electromagnets and, so, *nearly all of the output energy supplied by a motor to its mechanical load is in fact derived from the electrical energy supplied to the armature conductors*.

The force acting on a single conductor within a magnetic field set up by an ordinary permanent magnet will be very small indeed. Typically, the flux density near the pole of a typical magnet will be just a fraction of a tesla – even the strongest permanent magnet (e.g. a neodymium-iron-boron type) produces a flux density of hardly more than around 1.25 T! So a current of, say, 1 A, will result in a force of around only 1.25 N *per metre length* of conductor. If we then consider that the length of the conductor actually *within the field* (i.e. across the face of the magnet) might amount

13

to just 20 mm or so, then the *actual* force acting on that conductor will be only around 25 mN – i.e. equivalent to the force produced by a mass of a little more than a couple of grams!

This force is *far too small* to be of any practical use, particularly if we want that force to be able to drive a mechanical load.

So the question that we must ask ourselves is *how would it be possible to increase the value of this force?*

Well, if we return to equation (1), it should be obvious that we need to be able to:

(a) create the highest **flux density** possible through the use of an efficient 'magnetic circuit', using *electromagnets* rather than permanent magnets,

(b) increase the effective **length** of the conductor within the magnetic field, by using the sides of a *coil* wound with numerous turns, rather than a single straight conductor, and

(c) allow the conductor to carry as high a value of **current** as possible without causing the machine to overheat.

So, let's move on and see how we can achieve these different requirements.

Magnetic circuit for a d.c. motor

Magnetic circuits were discussed in depth in the companion book, ***An Introduction to Electrical Science***, and we discussed them again in the previous chapter, so it will be assumed that we are sufficiently familiar with this topic to understand what follows.

The purpose of a magnetic circuit is to (a) provide a *low-reluctance/high-permeability* path, which will (b) *concentrate* the magnetic flux, in order to ensure maximum flux density, and (c) to *direct* or *guide* that magnetic flux to where it is needed: in the case of a motor, to the air gap within which the current-carrying conductors are located. The broken line in Figure 13.5 shows the magnetic circuit for a simple four-pole d.c. motor.

The **magnetic flux** (symbol: Φ), which is expressed in webers (Wb), is produced by a **magnetomotive force** (symbol: F) set up by a field winding wound around part of the magnetic circuit, and whose magnitude is the product of the number of turns *(N)* of the magnitude of the current *(I)* flowing through the coil:

$$F = IN \ \text{--- (equation 2)}$$

In SI, magnetomotive force is expressed in amperes (A), although it is usually 'spoken' as *'ampere turns'* in order to distinguish this unit from that of current.

The resulting magnetic flux not only depends upon the magnetomotive force that creates it, but upon the *opposition* of the magnetic circuit's material to the formation of that flux. We call this opposition '**reluctance**' (symbol: R_m), and it is expressed in amperes per weber (but usually spoken as *'ampere-turns per weber'*). The relationship between magnetic flux (Φ), magnetomotive force *(F)*, and reluctance *(Rm)*, is sometimes termed the *'Ohm's Law of magnetic circuits'*, and is given by:

$$\Phi = \frac{F}{R_m} \ \text{--- (equation 3)}$$

As the reluctance of silicon-steel is insignificant compared with that of even the narrowest of air gaps, *practically the entire magnetomotive force developed by the magnetic circuit's excitation winding will appear across the air gap*, in much the same way as practically *all* of the voltage applied to an electrical load will appear across that load because the voltage drops along the adjoining conductors are negligible due to their insignificant resistance.

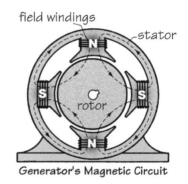

field windings

stator

rotor

Generator's Magnetic Circuit

Figure 13.5

And, just like resistance, reluctance of the **air gap** is *directly proportional* to its axial length (symbol: *l*), and *inversely proportional* to its cross-sectional area (symbol: *A*), that is:

$$R_M \propto \frac{l}{A}$$

To change the 'proportional' sign into an 'equals' sign we must, of course, introduce a *constant*. This constant is called 'magnetic reluctivity', but it's more usual to use its *reciprocal* instead which, for air, we call the **'permeability of free space'** (symbol: μ_o):

$$R_M = \left(\frac{1}{\mu_o}\right) \times \frac{l}{A}$$

$$R_M = \frac{l}{\mu_o A} \quad ---\text{(equation 4)}$$

Where: $\mu_o = (4\pi \times 10^{-7})$ henrys per metre.

We can now rewrite equation (3), as follows:

$$\Phi = \frac{F}{R_m} = \frac{IN}{R_m} = IN \times \frac{\mu_o A}{l}$$

$$\Phi = \mu_o \frac{INA}{l} \quad ---\text{(equation 5)}$$

. . . from which we can finally derive an expression for the *flux density* in the air gap:

$$B_{\text{air gap}} = \frac{\Phi}{A} = \mu_o \frac{IN\cancel{A}}{l\cancel{A}}$$

$$\boxed{B_{\text{air gap}} = \mu_o \frac{IN}{l}} \quad ---\text{(equation 6)}$$

From equation (6), it should be clear that the flux density in the air gap depends upon *the radial length (l) of that air gap* together with the *magnetomotive force* appearing across it and, so, we can safely ignore its cross-sectional area!

The only *variable* in equation (6), of course, is the **exciting current** (*I*) responsible for creating the magnetomotive force in the field winding. But it would be a *mistake* to assume that, by *increasing* the current, a *proportional increase* in flux density would result.

This is because silicon-steel, in common with *all* ferromagnetic materials, is subject to magnetic **saturation**. Saturation occurs when *all* or *most* of the domains in the steel have fully aligned with the field: at which point any further increase in magnetomotive force will result in absolutely no increase whatsoever in its flux density.

A *secondary* effect of saturation is that the reluctance of the steel increases significantly, which then requires significantly more magnetomotive force to 'drive' the flux through the steel part of the magnetic circuit, leaving less to drive the flux across the air gap. This is equivalent to the resistance of the connecting wires in an electric circuit increasing to a point where they start to cause a significant voltage drops along their lengths, resulting in a much lower voltage appearing across the load.

The effect of saturation depends upon the type of steel used but, typically *restricts the maximum achievable air gap flux density in practical motors to around just 1.5 T.*

Armature conductors and rotor

From equation (6) we can see that, in order to achieve maximum flux density within the air gap of the motor's magnetic circuit, the axial length ($l_{\text{air gap}}$) of that air gap should be as small as possible. This is achieved by installing the conductors not *on* the surface of the rotor but, instead, *into* longitudinal **slots** machined along its axial length. Conductors installed in this way are called *'distributed windings'*.

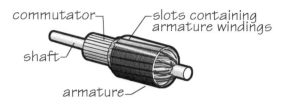

Figure 13.6

From equation (1), for any achievable value of flux density in the air gap, the force acting on each conductor must be proportional to the *length* of that conductor within the field which, in practice, corresponds approximately to the axial length of the rotor, as well as to the magnitude of *current* it is carrying.

In order to maximise the force (and, therefore, the **torque**) acting on each conductor, then, it's necessary to pass as much current as possible through them without causing the motor to overheat. This is determined by (a) the 'temperature class' (i.e. the maximum operating temperature) of the insulation selected, and (b) the method of cooling used by the machine. For any given class of insulation, the more effective the cooling system used, the higher the current and, therefore, the higher the resulting torque.

Unfortunately, if we try to increase the current by increasing the cross-sectional area of the conductors, it will be necessary to *increase* the width of the slots to accommodate them, thus *narrowing* the width of the 'teeth' which, in turn, could result in them saturating and reducing the effectiveness of the magnetic circuit! Alternatively, increasing the *circumference* of the rotor would accommodate slots of greater cross-sectional area.

There is, therefore, a direct link between the achievable overall **torque** on the rotor and its *diameter* and *axial length*; in other words, between the achievable torque the *volume* of the rotor. And, of course, the *volume of the rotor* has a direct effect on the *overall volume of the complete motor*.

> The desired torque developed by a motor generally determines the *volume* of that motor.

So, as a general rule, we can conclude that *if we want a motor to develop a lot of torque, then that motor must be physically large!* And 'large', of course, usually equates to 'expensive'!

Relationship between torque, speed, and power output

Let's now consider the relationship between a d.c. motor's **torque**, **speed**, and **power output**. We'll start by considering a single conductor, subject to a constant force *(F)* moving along a circular path, of radius *(r)*, as illustrated in Figure 13.7.

Work is usually defined as *'force multiplied by the distance moved in the direction of that force'*.

So, the work done by this force upon the conductor, *during one completion revolution* of that conductor, is the product of that force and the distance *(s)* travelled by the conductor around that circular path, that is:

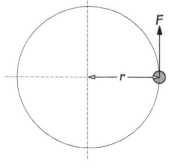

Figure 13.7

$$W = Fs \; - - - \text{(equation 7)}$$

The distance moved by the conductor, of course, is the *circumference* of the circular path along which it is moving, so we can rewrite the above equation, as follows:

$$W = F(2\pi r)$$

The product of force *(F)* and radius *(r)* is termed **torque** *(T)*, so we can rewrite the above equation as:

$$W = 2\pi T$$

The above equation, then, represents the work done during *one* completion revolution of the conductor. If the conductor were to rotate *'n'* times, then the equation can be rewritten as:

$$\boldsymbol{W = 2\pi nT} \; - - - \text{(equation 8)}$$

Now, **power** is defined as the 'work done per unit time', so if we divide the above equation by time *(t)*, we can obtain an equation for power:

$$P = \frac{W}{t} = \frac{2\pi nT}{t}$$

The expression, $\left(\dfrac{2\pi n}{t}\right)$, is known as '**angular velocity**' (symbol: . ω) . . .

. . . so we can rewrite the previous equation as:

$$P = \omega T$$

In SI, angular velocity (ω) is expressed in 'radians per second', but this is usually an inconvenient unit for the practical measurement of the speed of a rotating machine. Normally, we measure the rotational speed of a machine in '**revolutions per minute**' (**rev/min**), so let's modify the earlier equation as follows:

$$\boxed{P = \frac{2\pi NT}{60}} \; - - - \text{(equation 9)}$$

where: $\quad P$ = power output (watts)
$\quad N$ = rotational speed (rev/min)
$\quad T$ = torque (newton metres)

This equation, of course, is for the **output power**, of a motor, expressed in watts. In North America, the **horsepower** is still generally used to measure the output power, or '**brake horsepower**', of a motor.

We can determine the output power of a motor, expressed in horsepower, by dividing the output power, expressed in watts, by 746. For example, a motor with an output power of 8 kW, is equivalent to:

$$\text{output power, in horsepower} = \frac{8000}{746} = \textbf{10.72 h.p.}$$

The '**power rating**' of a motor is *always* its **output power**, *never* its input power, because it's the **output power** that drives its mechanical load. When selecting a motor for a specific application, knowing its *output* power is far more relevant than knowing its *input* (electrical) power.

Worked example 1

If an electric motor develops an output power of 8 kW at a rotational speed of 1200 rev/min, calculate (a) the work done over a period of 45 min, and (b) the torque developed by the machine.

13

Solution

Since $P = \dfrac{W}{t}$, then

$$W = Pt = 8000 \times (45 \times 60) = \textbf{21.6 MJ (Answer a.)}$$

$$P = \frac{2\pi N T}{60}$$

$$T = \frac{60P}{2\pi N} = \frac{60 \times 8000}{2\pi \times 1200} = \textbf{63.67 N·m (Answer b.)}$$

From equation (9), it is clear that the output power of a motor is proportional to its *speed* as well as to its *torque*:

$$P \propto NT$$

And, since torque depends upon the *volume* of the motor, we can conclude that *for any desired power output, we have a choice between using a small, high-speed motor or a large, low-speed motor!*

> For any desired **power output**, we have a choice between using a *small, high-speed* motor or a *large, low-speed* motor!

In practice, the **output power** of a motor, at any particular rotational speed, may be determined by measuring the amount of torque it develops at that speed. This can be done using a '**dynamometer**', which is a device for measuring the *force, torque,* or *power* of a motor. A simple example of a dynamometer is the '**de Prony brake**', named in honour of its inventor, a French mathematician and hydraulics engineer with the impressive name of Gaspard Clair François Marie Riche de Prony (1755–1825).

> The **de Prony brake** is sometimes, incorrectly, referred to in the field as a '**pony brake**' or, worse still, a '**pony break**'!

Actually, the de Prony brake is rarely used these days, as much more accurate dynamometers are available. However, it is a useful learning tool and is widely used in college laboratories because its principle of operation is very understandable without the need for a lengthy explanation, and it is obvious *how* and *what* it is measuring.

There are *two* versions of de Prony's dynamometer, as illustrated in Figure 13.8, and *either* version can be used to determine the torque developed by a motor.

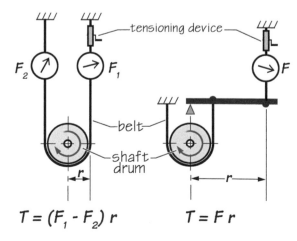

$$T = (F_1 - F_2)\, r \qquad\qquad T = F\, r$$

Figure 13.8

In its simplest form (Figure 13.8 left), the de Prony brake consists of *two* **spring balances**, F_1 and F_2, calibrated in newtons, connected together by a **leather strap** that is passed beneath a **brake drum** attached to the motor's drive shaft. With the motor running at its rated speed, which can be confirmed and monitored using a hand-held **tachometer**, the belt is gradually tightened until the motor can no longer maintain that rated speed, at which point the spring-balance readings are recorded.

The resultant force is the *difference* between the two spring-balance readings, that is:

$$F = (F_1 - F_2)$$

The slightly more complicated version, shown in Figure 13.8 (right) uses a **lever arm** that is supported by a single spring balance *(F)* calibrated in newtons. In this case, the spring balance indicates the total force applied by the lever.

For the de Prony brake shown in Figure 13.8 (left) the **torque** developed is the *difference between the two spring-balance readings*, multiplied by the *radius of the brake drum*.

For the de Prony brake shown in Figure 13.9 (right), the **torque** developed is the *reading of the spring balance*, multiplied by the *length of the lever arm*.

Worked example 2

In a 'de Prony brake' test on a d.c. motor, the effective load on the brake drum was found to be 222.7 N, the effective diameter of the brake drum was 0.5 m, and the rotational speed was 960 rev/min. Under these conditions, the input to the motor was 30 A and 230 V. Calculate (a) the torque, (b) the power output, (c) the power input, and (d) the efficiency.

Solution

(a) torque $=$ force $\times$ radius

$$= f\, r$$

$$= 222.7 \times \frac{0.5}{2}$$

$$= \mathbf{55.68\ N \cdot m\ (Answer\ a.)}$$

(b) Power Output $= \dfrac{2\pi NT}{60}$

$$= \frac{2\pi \times 960 \times 55.68}{60}$$

$$= \mathbf{5598\ W\ (Answer\ b.)}$$

(c) Power Input $= EI = 230 \times 30 = \mathbf{6900\ W\ (Answer\ c.)}$

(d) Efficiency $= \dfrac{\text{output power}}{\text{input power}} \times 100\%$

$$= \frac{5598}{6900} \times 100\%$$

$$= \mathbf{81.13\%\ (Answer\ d.)}$$

Torque developed by a practical d.c. motor

As we already know, a current-carrying conductor placed within a magnetic field, will be subjected to a force that acts to push the conductor out of that field. For conventional current flow, the *direction* of that force can be determined using

Fleming's Left-Hand Rule, and its magnitude can be determined from the following general equation:

$$F = BIl \qquad --- \text{(equation 1)}$$

where: F = force (newtons)

B = flux density (teslas)

I = current (amperes)

l = length of conductor (metres)

For a simple, **single-loop motor** . . .

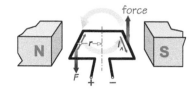

Figure 13.9

. . . when an armature current (I_A) flows in the direction shown in Figure 13.9, opposing forces will act on each conductor, and subject one loop to a counterclockwise torque (T), where:

$$T = Fr \qquad --- \text{(equation 10)}$$

where: T = torque (newton metres)

F = force (newtons)

r = radius of loop (metres)

For a **multiple pole motor** with **distributed armature windings**, the torque will depend upon the number of 'active' conductors that are *within the magnetic field*. However, not *all* the active conductors will necessarily lie within the magnetic field, because *some* of them must lie midway *between* the pole pieces, where there is no (or very little) flux. So, if the total number of active conductors is Z, and x represents the 'per unit' (as opposed to 'percentage') of those conductors within the field, then:

$$T = FrZx$$

Since the current flowing through each active conductor is determined by the total armature current (I_A), *divided* by the number of **parallel paths** (c) through the armature then, if we substitute $(BI_A l)$ for F in the above equation,

$$T = B\left(\frac{I_A}{c}\right)l\,rZx$$

$$\boxed{T = \frac{BI_A\,l\,rZx}{c}}$$

where: T = total torque (N·m)

B = flux density (T)

I_A = armature current (A)

l = length of active conductor (m)

r = radius of armature (m)

Z = number of active conductors

x = per unit of conductor in field

c = number of parallel paths

Worked example 3

A four-pole, lap-wound, d.c. motor draws a line current of 135 A, while its armature rotates at 2550 rev/min. The field pole flux density is 0.81 T, and 72% of the active conductors are within the field. The rotor carries 96 active armature conductors, has an effective diameter of 187.3 mm, and an effective length of 152.4 mm. Determine (a) the maximum torque developed, and (b) the output power of the machine.

Solution

Remember, for a lap-wound machine, the number of parallel paths *(c)* is equal to the number of poles, in this case, 4:

(a) $T = \dfrac{BI_A lrZx}{c}$

$= \dfrac{0.81 \times 135 \times (152.4 \times 10^{-3}) \times \left(\dfrac{187.3}{2} \times 10^{-3}\right) \times 96 \times 0.72}{4}$

$= \textbf{26.97 N·m (Answer a.)}$

(b) Power Output $= \dfrac{2\pi NT}{60}$

$= \dfrac{2\pi \times 2550 \times 26.99}{60}$

$= \textbf{7.2 kW (Answer b.)}$

Construction of a d.c. motor

There are *no significant differences* between a **d.c. motor** and a **d.c. generator** in terms of their construction. In fact, a d.c. motor can function perfectly well as a d.c. generator, and a d.c. generator can function perfectly well as a d.c. motor!

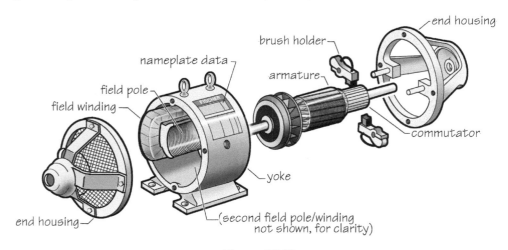

brush holder

nameplate data

field pole

field winding

armature

end housing

commutator

yoke

end housing

(second field pole/winding not shown, for clarity)

Figure 13.10

So, *everything* we've already learnt regarding the construction of a *d.c. generator* applies equally to a *d.c. motor*. The *only* difference is that a d.c. motor *drives* a mechanical load, whereas a d.c. generator *is driven* by a prime mover.

Everything we've learnt about d.c. generators regarding, for example, their major components, their armature windings and whether they are lap- or wave-wound, simplex or multiplex, and the number of parallel paths, etc., *applies equally to the d.c. motor*.

As with d.c. generators, **armature reaction** can be a problem, but we should be aware, however, that resulting distortion to the field caused by armature reaction is *opposite* to that of a generator – i.e. the magnetic neutral plane (MNP) *lags behind* the direction of rotation, rather than *moves ahead* of it. The reason for this should be obvious from the combined armature and field fluxes illustrated in Figure 13.11.

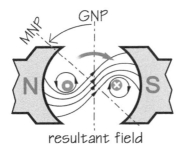

Figure 13.11

Armature reaction causes exactly the same sort of problems regarding commutator sparking, partial saturation of the pole pieces, etc., as we discussed in the previous chapter, and the solution is exactly the same too: the addition of **interpoles** or **compensation windings**, connected in *series* with the armature windings.

Types of d.c. motor

Let's move on, now, to consider how d.c. motors are **classified**, and to examine the behaviour of different types of d.c. motor.

If we ignore small, permanent magnet motors, of the sort used to run toys, etc., **d.c. motors** are classified in *exactly* the same way as d.c. generators – i.e. according to their *method of excitation*.

A reminder that the term, '**excitation**', refers *to the production of a generator's field*. In this chapter, we will consider the following types of d.c. generator.

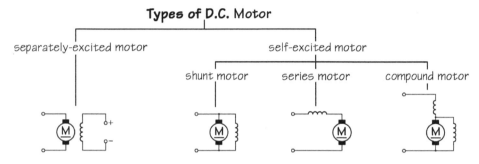

Figure 13.12

Just as with d.c. generators, we can compare the *performance characteristics* of these types of d.c. generator by examining their **characteristic curves**. For d.c. motors, these are the:

▶ **torque/armature current** characteristic curve,
▶ **speed/armature current** characteristic curve, and their
▶ **torque/speed** characteristic curve.

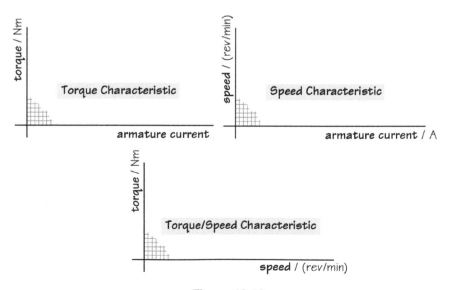

Figure 13.13

Separately excited motor

The torque developed by a d.c. motor is due to the interaction of the magnetic fields set up by the currents flowing through the armature winding and the field winding.

Shunt motor

A **shunt motor**'s field winding is connected in parallel with both the supply *and* the armature winding, as shown in Figure 13.14.

Let's compare this with the schematic diagram shown in Figure 13.15.

From this schematic diagram, we can see that the **shunt field winding** is connected *directly across (in parallel with) the supply voltage*. Therefore, the resulting **field current** (I_F) must be a *constant* value, and quite independent of any variations in the armature current (I_A).

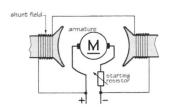

Figure 13.14

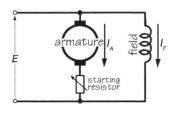

Figure 13.15

Torque characteristics

With a *constant* field current, the flux-density of the field is also *constant*. So the torque produced by the motor must be *directly proportional to the armature current* – producing the straight-line graph until saturation is reached, as shown in Figure 13.16.

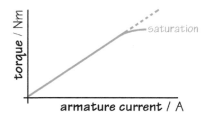

Figure 13.16

Starting torque. On starting, because there is no back-e.m.f., the armature current is deliberately restricted using a starting resistor, which prevents it from becoming excessively high. Because, on starting, the armature current is deliberately kept low, the starting torque will *also* be low.

Load increasing. When the motor is running, any *increase* in load will cause the armature to slow down, *reducing* its back-e.m.f. This causes an increase in the armature current, which then provides the larger torque necessary to drive the increasing load.

Load decreasing. When the load *decreases*, the armature will speed up, *increasing* its back-e.m.f. This increase in back-e.m.f. results in a decrease in armature current, which causes a reduction in torque.

To summarise, *whenever a shunt motor's load changes, the torque always adjusts to meet the new load*.

Speed characteristics

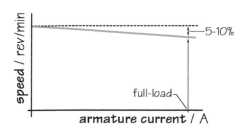

Figure 13.17

Figure 13.17 shows the variation of speed from no load to rated full load is between 5–10% of its no-load speed. As this does not represent a great difference in its operating speed, *the shunt motor is generally considered to be a **constant-speed motor***.

As shunt motors are classified as 'constant-speed motors', they are used wherever a device must operate at approximately constant speed between no load and full load. Applications, therefore, include: rollers in steel rolling-mills, lathes, drilling machines, fans, etc.

Series motors

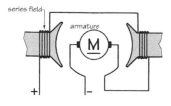

Figure 13.18

A **series motor**'s field winding is connected in series with the armature winding, as illustrated in Figure 13.18.

Let's compare this with its schematic diagram shown in Figure 13.19.

From the schematic diagram, we can be see that the **series field winding** and **armature winding** are in series. Therefore, the **field current** (symbol: I_f) and the armature current (symbol: I_a) are *one and the same current*. The flux-density of the field, therefore, depends upon the magnitude of the armature current.

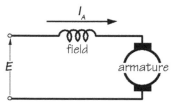

Figure 13.19

Torque characteristics

The torque developed by any d.c. motor depends upon flux densities produced by both the field *and* armature windings. In the case of a series motor, these flux densities (up to saturation) are *both* proportional to the armature current, therefore:

$$\text{torque} \propto (\text{armature current})^2$$

So, *doubling* the armature current will *quadruple* the torque; *tripling* the armature current will increase the torque by a factor of *nine*, and so on. This relationship produces a curve, as shown in Figure 13.20, which continues until the fields become saturated (i.e. no further increase in flux density):

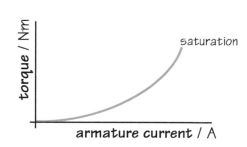

Figure 13.20

Starting torque. On starting, the motor's rotor is initially stationary and, so, its back-e.m.f. will be non-existent, resulting in a very high armature current. As the rotor then begins to move and its speed begins to increase, its back-e.m.f. will also start to increase, reducing its armature current. Therefore, upon starting, the motor's torque will be very high – so, a series motor will have *a high starting torque*.

Increasing load. When the motor is running, if the load *increases*, it will cause the motor to slow down. This will result in a reduction in the back-e.m.f., allowing the armature current to increase causing an increase in torque. So, an increase in load is compensated for by a corresponding increase in torque.

Decreasing load decreasing. When the load *decreases*, the armature will speed up, increasing its back-e.m.f. This increase in back-e.m.f. results in a decrease in armature current, which causes a reduction in torque – allowing the torque to match the new load.

To summarise, *whenever a shunt motor's load changes, the torque adjusts itself to meet the new load.*

Speed characteristics

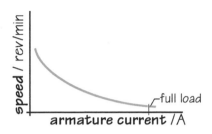

Figure 13.21

The speed characteristic curve shown in Figure 13.21 indicates a substantial difference in speed variation with changes in its load, so *a series motor is considered to be a **variable-speed motor**.*

So a series motor runs very slowly with heavy loads, and very fast with light loads. If the load is suddenly removed from a series motor, it will dangerously over-speed to the point of self-destruction! Accordingly, *series motors must **never** be run without a load.*

Applications. As series motors have a very high starting-torque, they are ideal for traction motors. Applications, therefore, include: railway locomotives, drag line excavators, cranes, etc. Series motors are also ideal for driving loads having a torque that increases with speed, such as heavy-duty fans.

Drag-Line Excavator

Figure 13.22

A **drag-line excavator** is a piece of heavy civil-engineering equipment, used for open-cast strip-mining of coal fields, tar-sands, etc. A drag-line consists of a bucket, suspended from a boom, which is manoeuvred by a number of wire ropes and chains, and dragged horizontally across the surface of a strip mine to harvest the coal, oil-shale, etc.

Drag-line excavators are amongst the very largest land machines ever constructed, particularly in North America where they can weigh up to 13 000 t (tonnes) and stand 22 stories high! These larger machines have, historically, used d.c. electric drive systems – including the series d.c. motors described in this chapter – to provide the high torque they require. Newer drag-line excavators, thanks to industrial electronics, are being built with a.c. drive systems and older excavators are being converted to arc drives.

13

307

Compound motors

A **compound motor** has *two* sets of field winding. As with d.c. generators, there are two types of compound motor: **long-shunt compound** and **short-shunt compound**. Figure 13.23 represents a long-shunt compound motor:

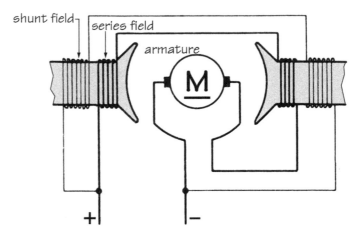

Figure 13.23

Compare this with its schematic diagram, shown in Figure 13.24 (left).

A **compound motor** combines the advantages of both the shunt and the series motor.

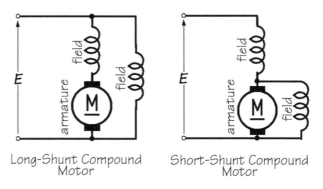

Long-Shunt Compound Motor Short-Shunt Compound Motor

Figure 13.24

The left-hand machine is called a **long-shunt compound motor**, and has *two* field windings: one *in parallel with the supply*, and the other *in series with the armature winding*.

The right-hand machine is called a **short-shunt compound motor**, and has *two* field windings: one *in parallel with the armature winding*, and the other *in series with the supply*.

There are *two* sub-categories of compound motor:

Differentially compounded motor – in which the series winding is wound in such a way that its flux *opposes* the flux produced by the shunt winding. This causes the effective field to *weaken* as the load increases. This type of compound motor has a very low value of starting-torque, and is seldom used.

Cumulatively compounded motor – in which the series winding is wound in such a way that its flux *reinforces* the flux produced by the shunt winding. There are two types of cumulatively compounded motor:

Predominant shunt – weak series – this gives the motor a characteristic that is similar to a shunt motor, but has a greater drop in speed from no-load to full-load (between 10% and 15%).

Predominant series – weak shunt – this gives the motor a characteristic that is similar to a series motor, but will limit its no-load speed to a safe value (in this case, the shunt-winding is interlocked with the starter so that, should the shunt-winding become open-circuited, then the motor will be made to stop).

Comparison of the characteristics of d.c. motors

Let's compare the **speed/torque** and **torque/armature-current** characteristics of these d.c. motors, as shown in Figure 13.25.

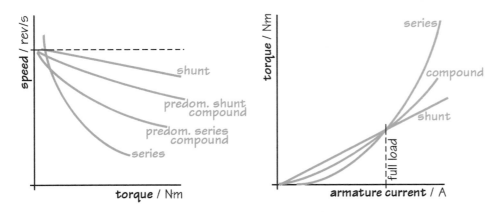

Figure 13.25

Starting with the **speed/torque characteristics**, we notice that the **shunt** motor's speed drops the *least* for an increase in torque, whereas the **series motor**'s speed drops *significantly* for an increase in torque. As far as the **compound motors** are concerned, they have characteristic curves that lie between these two extremes with, as we might expect, the **predominant-shunt compound** characteristic *closer to the shunt characteristic*, and the **predominant-series compound** characteristic *closer to the series characteristic*.

For the **torque/armature current characteristics** for motors with the same output power, we see that the **shunt motor**'s characteristic is a *straight line* because the torque is directly proportional to the armature current. For the **series motor**, the curve reflects the fact that the torque is proportional to the *square* of the armature current confirming that, beyond their full-load current, the torque becomes significantly greater than that of a series motor. Again, as we might expect, the compound motor's characteristic curve lies somewhere between these two extremes.

The effect of power electronics on d.c. motors

'**Power electronics**' is a branch of industrial electronics concerned with the application of **solid-state electronics** *to the control and conversion of electrical energy*. The development in this particular field of electrical engineering has had a profound effect on motors in general, and upon d.c. motors in particular.

In the days *before* the introduction and development of power electronics made variable voltage supplies readily available, d.c. motors had to operate from constant-voltage d.c. supplies, and their characteristics then depended entirely on whether their armature and field windings were connected in parallel ('shunt') or in series: with 'shunt motors' good for *constant-speed* applications, while 'series motors' were good for *high-torque* applications.

13

However, with the introduction of variable voltage supplies, a separately excited d.c. motor can now be made to do practically *anything* that a shunt, series, or compound motor can do.

So, in a way, *the need for shunt, series, and compound motors has now become redundant* (although series motors are still widely used for heavy-traction applications, such as drag-line excavators, etc.).

Perhaps of even greater importance is the fact that 'inverter-fed'* a.c. motors, thanks to their simpler construction, are significantly less expensive than equivalent d.c. motors, and can now match most of the characteristics previously only achievable with d.c. motors! So inverter-fed a.c. motors are now *replacing* d.c. motors and, eventually, may render them obsolete.

*An 'inverter' (or, more specifically, a 'voltage-source inverter' or 'VSI') is a solid-state electronic device whose output voltage can be controlled to match the required operating conditions of the motor.

Because an understanding of power electronics requires a thorough understanding of fundamental electronics, variable voltage supplies are well beyond the scope of this book and we will *not* be pursuing this topic any further.

Starting d.c. motors

Immediately a d.c. motor's armature starts to rotate, its windings will begin to cut the field, and an e.m.f. will be induced into them – just like a generator!

In accordance with **Lenz's Law**, the *direction* of this induced e.m.f. must *oppose* the rotation that is causing it. As the rotation is brought about by the armature current supplied to the motor, this means that the induced e.m.f. must act *to oppose and reduce the value of this current.*

Passing a current through the motor's armature winding causes it to rotate ('motor action'), but the movement of that winding through the field causes it to generate a voltage ('generator action') that opposes the applied voltage!

So, the armature winding behaves both as a motor *and* as a generator *simultaneously!*

The magnitude of the back-e.m.f. will be *exactly* the same as if the machine was operating purely as a generator which, if we refer back to the voltage equation for a d.c. generator, means that *it must be proportional to the speed (N) of the machine.* That is:

$$E_{back} = \frac{\Phi N p Z}{60\,c} \quad \text{so} \quad E_{back} \propto N$$

So, the *actual* current flowing through the armature winding will be:

$$I_A = \frac{(E - E_{back})}{R} \quad ---\text{(equation 12)}$$

where: I_A = armature current (amperes)

E = supply voltage (volts)

E_{back} = back e.m.f. (volts)

R = resistance of armature winding (ohms)

The 'action' and 'reaction' effect, described above, plays a **very important** part in causing d.c. machines (both motors *and* generators) to ***automatically react to changes in load***.

In the case of a d.c. motor, if its mechanical load *increases*, the motor tends to *slow down*. As it *slows down*, the back-e.m.f. *decreases* and the current *increases*: creating more torque to compensate for the increasing load.

Similarly, if the motor's mechanical load *decreases*, the motor's speed will *increase*. As its speed *increases*, its back-e.m.f. also *increases*: reducing the current and reducing the torque to match the reduced load.

While a d.c. motor is running, then, it is generating a **back-e.m.f.** through *generator action*. As we have learnt, a 'back-e.m.f.' is so called because it *opposes the supply voltage applied to the motor*.

When a *stationary* d.c. motor is initially connected to its supply voltage, there is no back-e.m.f. generated – so the armature current is restricted only by the relatively low resistance of the armature winding. The **starting current**, therefore, can be *very* high. However, as the motor starts to turn, the back-e.m.f. starts to increase, causing the armature current to fall towards a safer value.

So, on start-up, a large current (up to 10× the running current) is drawn by the motor. With large d.c. motors, this starting current may be high enough, and last long enough, to overheat the armature windings and damage the insulation, *unless it is restricted in some way*. For this reason, d.c. motors generally have their starting current restricted by means of an **external stepped resistance**, which can be gradually reduced in value as the motor's speed increases and its back-e.m.f. takes effect.

Worked example 4

A 15 kW shunt motor runs off a 230 V d.c. supply. Calculate (a) the value of the total starting resistance necessary to limit the starting current to 1.5 times the full-load current, if the machine's full-load efficiency is 88% and the resistance of the armature circuit is 0.2 Ω. Also, calculate (b) the back-e.m.f. generated by the motor when the current has fallen to its full-load value – assuming the whole of the starting resistance is still connected. Ignore the field current.

Solution

Let's start by sketching a schematic diagram of the motor:

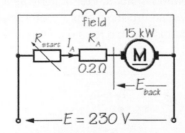

Figure 13.26

Next, we need to determine its input power:

$$P_{input} = \frac{P_{output}}{\text{efficiency}} = \frac{15\,000}{0.88} = 17\,045 \text{ W}$$

From this, we can determine the full-load armature current:

$$I_A = \frac{P_{input}}{E} = \frac{17045}{230} = 74.11 \text{ A}$$

From the schematic diagram,

$$E = E_{back} + I_A(R_A + R_{start})$$

We are told that we need to restrict the armature current to 1.5 times the full-load current, so we need to modify this equation as follows:

$$E = E_{back} + 1.5\,I_A(R_A + R_{start})$$

Rearranging, to make **R_{start}** the subject of the equation:

$$E = E_{back} + 1.5\,I_A(R_A + R_{start})$$

$$R_{start} = \left(\frac{E - E_{back}}{1.5\,I_A}\right) - R_A$$

$$= \left(\frac{230 - 0}{1.5 \times 74.11}\right) - 0.2$$

$$= \left(\frac{230}{111.17}\right) - 0.2 \approx \textbf{1.87 } \Omega\,\textbf{(Answer a.)}$$

Moving on to the *second* part of the question, we have already established that the full-load armature current, I_A, is 74.11 A, so:

$$E = E_{back} + I_A(R_A + R_{start})$$

$$\text{so } E_{back} = E - I_A(R_A + R_{start})$$

$$= 230 - 74.11(0.2 + 1.87)$$

$$= 230 - 153.41$$

$$= \textbf{76.59 V (Answer b.)}$$

Motor starters

The device that performs this function is called a '**motor starter**'. Essentially, this is a variable resistor or, more commonly, a series of individual resistors, placed in series with the armature winding. On start-up, the starter is set to *maximum* resistance; as the motor runs up to speed, the starter's resistance is *gradually reduced* until, by the time the motor reaches it full speed, the external resistance has been removed altogether.

The schematic diagram, shown in Figure 13.27, represents the basic principle of operation of a manually operated motor starter, known as a '**faceplate starter**'. In this particular example, it's shown controlling a shunt motor, but it can be adapted to start *any* motor. This type of starter is suitable for a d.c. motor with an output power of up to around 5 kW.

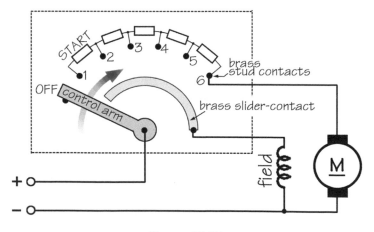

Figure 13.27

When the **control arm** is moved from its 'off' position to the 'start' position (stud 1), the five resistors are all inserted in series with the motor's armature winding. At the same time, the field winding is connected directly to the supply voltage, via the **brass slider-contact**.

As the motor starts to pick up speed, the control arm is gradually moved clockwise, one stud at a time – with the operator pausing at each stud to allow the motor's speed to catch up. At position 2, the first resistor has been removed from the armature circuit. At position 3, the first two resistors have been removed. Eventually, at position 6, the machine has reached its full speed and *all* the resistors have been removed from the armature circuit.

In practice, all motor starters *must* have what is termed '**no-volts**' **protection**, which prevents the machine from self-starting upon the restoration of supply following a temporary loss of the mains supply for some reason. Obviously, any motor that can start by itself without any warning can be *very* dangerous indeed!

So *all* practical motor starters *must* incorporate some form of 'no-volts' protection, and this feature is shown in Figure 13.28.

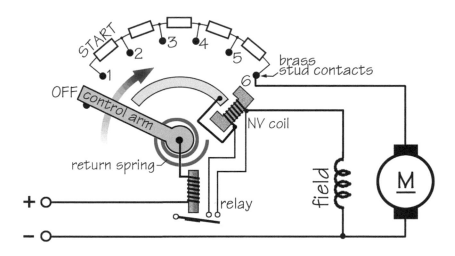

Figure 13.28

In this example, a 'no-volt' protection feature has been added to the basic starter. It takes the form of a **spring-loaded control arm**, a **relay**, and a **NV** ('no volts') **electromagnet**.

When the arm is moved from the 'off' position onto stud 1, as well as starting the motor, the relay's operating coil is energised which closes its normally open contact and energises the operating coil of the NV electromagnet.

When the control arm is eventually moved fully clockwise (onto stud number 6), against the tension in the spring, the now energised NV electromagnet *holds the arm in that position* against the restoration force of the spring. In the event of a loss of supply, the NV electromagnet is immediately de-energised, and the spring *returns the control arm, counterclockwise, to its 'off' position*. This ensures that, when the supply is eventually restored, the motor *cannot* restart by itself, and can *only* be restarted by hand.

The faceplate starter, described above, has been included in this chapter because it provides us with a simple example of the fundamental principle behind *most* d.c. motor starters. However, this type of starter is considered to be obsolete these days, and it has been superseded by **automatic starters** which can be used for both small and large machines. With automatic starters, the operator needs only to press a 'start button', and the starter takes care of the rest: i.e. bringing the motor safely up to speed while controlling its start-up current.

There are various types of automatic starters available, including '**time-delay**', '**current-limit**', and '**counter-e.m.f.**' types.

▶ **Time-delay starter**. This type of starter cuts out each starting resistor, in sequence, after predetermined time intervals – in rather the same way as a manually operated faceplate starter.

▶ **Current-limit starter**. This type of starter cuts out each starting resistor, in sequence, according to the level of armature current flowing.

▶ **Counter-e.m.f. starter**. This type of starter cuts out each starting resistor at successively increasing values of generated e.m.f.

As this can be a lengthy and complicated subject in its own right, let's look at an example of just *one* of these: a basic **time-delay automatic starter**, as illustrated in Figure 13.29.

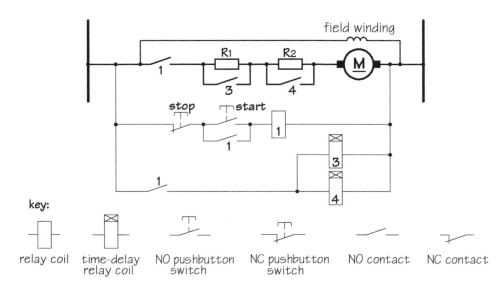

Figure 13.29

For the sake of simplicity, only *two* starting resistors (**R1** and **R2**) are illustrated. In practice, of course, there would be *more* than just two resistors. In order to understand the operation of this starter circuit we must realise that, for all schematic diagrams:

(a) *all* switches and contacts are shown in their *de-energised* positions.

(b) relay *contacts* are not necessarily shown anywhere near their *operating coils* although, in reality, they are individual, self-contained, units.

(c) relay contacts are identified by numbers and are operated by the relay operating coil identified with the *same* number.

When the normally open (**NO**) '**start**' pushbutton is pressed, **relay coil 1** is energised. This closes its normally open (**NO**) **contact 1**, connected in parallel with the start button, which acts to bypass the start button, and keep **relay coil 1** energised (this arrangement is called a *'hold-in'* or *'latching'* feature) when the pushbutton has been released. At the same time, the normally open (**NO**) **contact 1** in the top branch is closed, energising the motor's armature winding. Also at the same time, the normally open (**NO**) **contact 1** in the lower branch is closed, energizing both **time-delay relay coils 3** and **4**.

After a predetermined period of time, **time-delay relay coil 3** will close **contact 3**, in parallel with **resistor R1**, removing that resistor from the circuit. After a further predetermined period of time, **time-delay relay coil 4** will close the normally open (**NO**) **contact 4**, in parallel with **resistor R2**, removing that resistor from the circuit.

By the time both resistors have been removed from the circuit, the motor will be running at its normal speed (remember, that in a 'real' starter, there will be *more* than just two resistors).

To stop the motor, the normally closed (**NC**) 'stop' pushbutton switch is pressed. This de-energizes **relay coil 1**, which releases the 'latch-in' feature, and reopens the normally open (**NO**) **contact 1** in the top branch, disconnecting the armature. At the same time, the normally open (**NO**) **contact 1** in the bottom branch de-energizes the two **time-delay relay coils** (which reset themselves), and the two **resistors** are reinserted into the armature circuit.

Relay coil 1 also provides the circuit with its '**no-volts' protection**. If the supply should *fail*, this coil will de-energize, causing the 'latching' **contact 1** (in parallel with the 'start' pushbutton switch), to open and preventing the motor from restarting by itself when the supply is restored.

Further '**start**' and '**stop**' **stations** can be added by connecting additional 'start' NO pushbutton switches in *parallel* with the existing one, and by connecting additional 'stop' NC pushbutton switches in *series* with the existing one.

D.C. motor speed control

Earlier, we learnt that the back-e.m.f. generated by a motor was exactly the same as the e.m.f. produced by a generator running at the same speed and, so, the general voltage equation for a generator applied, that is :

$$E_{back} = \frac{\Phi N p Z}{60 c}$$

In this equation, the only variables are E_{back}, Φ, and N – the rest being *constants* for any given motor, so we can rewrite the above equation as follows:

$$N = k \frac{E_{back}}{\Phi}$$

And, since the back-e.m.f. generated by the motor is also given by . . .

$$E_{back} = E - I_A R_A$$

. . . we can substitute for E_{back} in the first equation, giving us:

$$\boxed{N = k \frac{(E - I_A R_A)}{\Phi}} \qquad \text{(equation 13)}$$

where:
N = rotor speed (rev/min)
k = constant for the motor
E = line voltage (V)
I_A = armature current (A)
Φ = field flux (Wb)

This equation tells us that there are *three* ways of varying the speed of the motor:

▶ varying the *field flux* (Φ) by varying the *field current*)
▶ varying the *line voltage* (*E*)
▶ varying the *armature current* (I_A).

Of these three methods, we can dismiss the third as being impractical. So, speed control is normally achieved by either *varying the field current* or by *varying the line voltage*.

But *which* of these is the better choice?

This may be determined by re-examining the fundamental equations for **torque** and for **power**:

$$T \propto \Phi I_A$$

$$P = \frac{2\pi N T}{60}$$

$$\boxed{T = k\Phi I_A} \text{ —(equation 14)}$$

$$\boxed{P = k'NT} \text{ —(equation 15)}$$

Speed control by varying the armature voltage

From equation (14), we can see that varying the line voltage will have absolutely no effect on the torque. But if the torque remains constant then, from equation (15), the motor's power will be proportional to its speed.

So, by *varying the armature voltage*, the:

▶ **torque** *remains constant*, while
▶ **power** *is proportional to speed.*

Speed control by varying the field flux

From equation (14), the torque is proportional to the field flux. So, if we substitute for torque (equation 14) and speed (equation 13), in equation (15), we have:

$$P = k'TN$$
$$= k'(k\Phi I_A)\left(k\frac{E - I_A R_A}{\Phi}\right)$$
$$= k''\left(EI_A - I_A^2 R_A\right) \text{ ---(equation 16)}$$

We can now see that, by varying the field flux, the:

▶ **power** *remains constant*, while the
▶ **torque** *is proportional to the field flux.*

From the preceding, it should be clear that speed control *by varying the armature voltage* is a better choice than by varying the field current, because:

▶ maximum torque is available at all times, and
▶ the chances of the motor's speed 'running away' with itself due to the inadvertent reduction, or complete loss, of the field is eliminated.

The ease by which variable voltage supplies are now readily available, thanks to **power electronics**, means that *d.c. motor speed control is achieved almost exclusively by varying the armature voltage.*

So, thanks to power electronics, it's now possible to use the rectified output from a.c. generators (alternators) to achieve smoothly variable speed control.

Prior to this, when variable line voltages were *not* common, a method called the '**Ward-Leonard System**' was a widely used method of speed control. In fact, until well into the 1960/70s, the Ward-Leonard system was the *only practical way of obtaining the variable voltage d.c. supply* needed for the speed control of a d.c. motor (Figure 13.30).

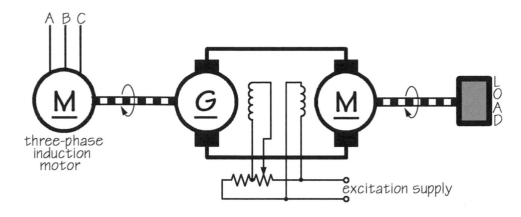

Figure 13.30

The 'Ward-Leonard System', which we touched upon briefly in the previous chapter, is named after its designer, the American engineer and inventor, Harry Ward-Leonard (1861–1915). It consists of a **three-phase induction motor**, which drives a **d.c. generator** at a constant speed, the output voltage of which then supplies the armature of a **d.c. motor**, as shown in Figure 13.30.

The constant-speed d.c. generator's output voltage was varied by adjusting the excitation voltage to its separately excited field winding, while the d.c. motor's field winding was also separately excited and fixed at its maximum field current. The excitation for both the generator and the motor was normally provided by an '**exciter**' – i.e. a small d.c. generator (not shown) also driven by the a.c. induction motor.

For naval applications, such as driving a warship's gun turrets, the d.c. generator's field would have been controlled by a **control-amplifier** whose input would have been provided by the ship's (gun) fire-control system.

Because the Ward-Leonard system's d.c. motor has a *fixed* field, but *variable* armature voltage, it essentially provides a **constant torque/variable power output** to its mechanical load.

The advantages of the Ward-Leonard system over other methods of d.c. motor control, includes:

▶ a wide-range of speeds from 'crawling' to high speed; the motor is started, accelerated, speed adjusted, and stopped by simply adjusting the generator's output voltage.

▶ reversing, achieved by adjusting the field of the generator, rather than by reversing the motor's field windings using contactors.

Conclusion

Now that we've completed this chapter, we need to examine its **objectives** listed at its start. Placing a question mark at the end of each objective turns that objective into a **test item**. If we can answer those test items, then we've met the objectives of this chapter.

13

CHAPTER 14

Synchronous generators

Objectives

On completion of this chapter you should be able to:

1. explain what is meant by the term, 'synchronous machine'.

2. recognise types of synchronous generator, including

   ▶ vehicle alternator
   ▶ portable alternator
   ▶ standby diesel alternator
   ▶ high-speed steam turbine alternator set
   ▶ hydroelectric turbine alternator set

3. explain the difference between 'distributed' and 'concentrated' windings.

4. explain the difference between 'rotating' and 'stationary' field systems, and explain the advantages of the 'stationary' type.

5. identify 'salient' and 'cylindrical' rotors, and compare the advantages/disadvantages of each type.

6. explain the relationship between the *speed* of the rotor, the *number of poles*, and the *frequency* of the e.m.f. induced into the armature.

7. describe the features of a synchronous generator's stator.

8. describe the armature windings of a single- and three-phase synchronous generator in general terms.

9. describe what is meant by an armature's 'active conductors'.

10. describe the relationship between the e.m.f. induced into a single active conductor and the total e.m.f. induced into a complete phase winding.

11. explain the reason for the truncated shape of the e.m.f. induced into a single active conductor, and how its shape can be improved towards that of a sine wave.

12. explain the effect of armature reaction on the terminal e.m.f./load current characteristics of a synchronous generator.

13. explain the effect that the power factor of the load has upon the terminal e.m.f./load current characteristics of a synchronous generator.

14. define what is meant by the 'percentage voltage regulation' of a synchronous generator.

15. list the factors that affect the e.m.f. induced into the armature windings of a synchronous generator, and identify the only practical method of controlling that e.m.f.

16. describe the basic field-excitation system of a typical synchronous generator.

17. describe the basic principle of operation of an automatic voltage regulator (AVR).

18. list the requirements necessary for synchronising a generator to an infinite bus system.

19. explain the principle of operation of the 'cross-connected lamp' synchroscope.

20. describe the steps necessary to synchronise a generator to an infinite bus system.

21. describe the behaviour of a synchronised generator that is supplying energy to an infinite bus system.

Historical background

During the nineteenth century, while much important and successful research was being conducted in Europe on designing practical alternating-current machines such as generators (alternators), motors, and transformers, much of the early work in designing practical means of *distributing* electrical energy was being conducted in the United States.

At the forefront of this development was the American inventor, **Thomas Edison** (1847–1931), who was a strong advocate of *direct current* as a means of distributing electrical energy. The main reason for this, of course, was that he held numerous patents for the d.c. generators and motors that his company was manufacturing and, therefore, had a huge financial interest in the success of his direct-current distribution system.

Emerging in direct competition with him was the American industrialist, **George Westinghouse** (1846–1914), who had been so impressed by the development of alternators and transformers in Europe that, against the advice of his engineers, he decided to invest his company's vast resources into *alternating-current* distribution systems.

Edison and Westinghouse therefore became the two leading protagonists in what came to be known as the *'War of the Currents'*, in which Edison used his considerable wealth to promote his direct-current system and to undermine and discredit Westinghouse's alternating-current system. Edison was not shy of making full use of any underhanded method he could employ to bring alternating current into disrepute!

His main argument was that alternating current distribution, which operated at higher voltages than his direct-current system (thus enabling it to distribute energy over greater distances at lower currents) was therefore far more dangerous and, to demonstrate this, he conducted numerous public meetings in which he used alternating current to electrocute, with various degrees of success, hundreds of stray dogs, horses and even, on one occasion, a disobedient circus elephant! This 'execution' of an elephant, incidentally, was not unprecedented; elsewhere in the United States, another circus elephant was actually publicly tried and *hanged*, using a railroad crane, for 'murdering' its trainer after he had tortured it with a pitchfork for being disobedient!

At about the same time, the US justice system was searching for 'more humane' methods for executing convicted murderers, and was attracted towards using electrocution as the means of doing so. Edison quickly leapt on this particular bandwagon, insisting that alternating current was the obvious choice because of its '*very dangerous nature*'. Typical of the man, was his efforts to popularise the expression '*to be Westinghoused*', meaning '*to be electrocuted*' – by alternating current, of course! – in order to undermine Westinghouse's technological advances!

> 'Rectify', in the sense of 'making things right', was introduced by Edison, intended to demean alternating current. Edison used this term to describe how it was necessary to rectify the 'problems' he associated with a.c., by changing it into d.c! It's a term that stuck, and we still use today to describe the process of changing alternating current to direct current!

In fact, alternating current provides a *significantly* better method of distributing electrical energy than by direct current. The reason for this is that alternating current allows a *single large power station* to supply energy *to numerous consumers over long distances*, whereas direct current requires *numerous small power stations* to supply energy *to fewer consumers over short distances*.

Consequently, it was Westinghouse's **alternating-current** transmission/distribution system that prevailed in the end and, ultimately, Edison was forced not only to adopt

Westinghouse's alternating-current distribution system, but to pay him royalties for the privilege of doing so!

Westinghouse had won *'the War of the Currents'!*

So why was Westinghouse's alternating-current system so much better than Edison's direct-current system? In one word: the **transformer**. As we know, the transformer is an *a.c. machine* that allows us to easily and efficiently *increase and decrease voltages*. And high transmission voltages are *essential* for energy distribution.

For any given load, the *higher* the transmission voltage, the *lower* the resulting transmission current. Low transmission currents mean:

▶ minimising the energy loss along the transmission line (meaning that most of the energy gets to where it's going, as opposed to be lost *en route*!)

▶ minimising the voltage drop along the length of the transmission line

▶ ability to use transmission conductors of manageable cross-sectional area and weight.

In this chapter, then, we will be learning about the *main source of alternating current*: the **synchronous a.c. generator** or '**alternator**'.

A.C. synchronous generator

An **a.c. generator**, also called an '**alternator**', is an example of a '**synchronous machine**'.

So let's start by learning what we mean by the term, '**synchronous machine**'.

A **synchronous machine** is a type of a.c. rotating machine whose *speed (N)* and the *frequency (f) of the current in its armature* are proportional to each other under steady-state conditions. That is,

$$f \propto N$$

A **synchronous machine** can operate both as an *a.c. generator* and as an *a.c. motor*.

In the case of a **synchronous generator**, the *frequency of its terminal e.m.f.* is proportional to *the speed at which its rotor is turning*.

And, in the case of a **synchronous motor**, the *speed of its rotor* is proportional to the *frequency of the mains supply*.

> A synchronous machine is an a.c. rotating machine whose speed and the frequency of the current in its armature are proportional to each other under steady-state conditions.

In this chapter, we will be concentrating on a synchronous machine used as an **alternator**, while leaving its use as an a.c. motor to another chapter.

This is because synchronous machines are principally used as *alternators* which supply the national grid: the main source of energy used by industry, commerce, and domestic consumers.

Alternators convert *kinetic energy* into *electrical energy*. The immediate source of this kinetic energy is termed a '**prime mover**', which may be an internal combustion engine, a gas- or steam-turbine, or a water turbine.

In the UK, the most common prime mover is the **steam turbine**. In other countries with large mountainous regions, such as in the province of British Columbia in western Canada, the most common prime mover is more likely to be a **water turbine**.

Prime movers such as **petrol** or **diesel engines** are primarily used to drive standby (known as 'non-spinning reserve') generators.

14

Typical alternators

In common with *all* electrical rotating machines, an alternator's output volt-ampere rating is proportional to its **volume**. So alternators are manufactured in a wide range of physical sizes, which reflect their volt-ampere ratings. Therefore, the physical size of an alternator ultimately determines its maximum output. At the time of writing this book (2014), the world's largest alternator is a Siemens/GE **700 MV·A** unit, with a mass of over 700 t (tonnes), installed as part of *Three Gorges* hydroelectric project in China.

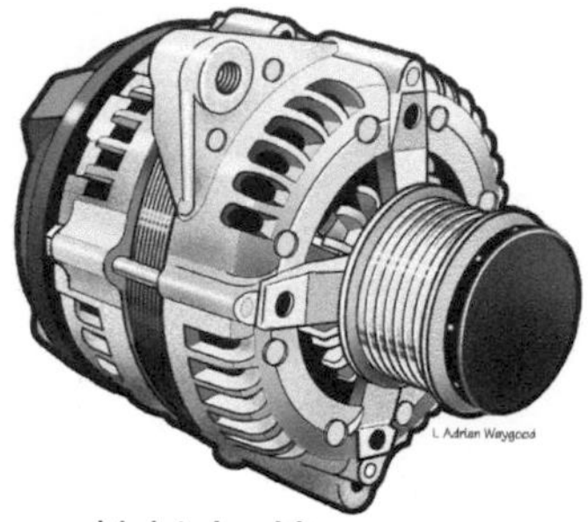

The *efficiency* of an alternator is related to its output volt amperes. For example, a 1000 V·A alternator might have an efficiency as low as just 50%, whereas the very largest alternators can have efficiencies approaching 99%.

Perhaps the most common alternator, and one with which we are all very familiar, is the type we see in our vehicles (Figure 14.1). This is a compact, three-phase alternator, with a rectifier board incorporated into its housing, so that it produces a relatively smooth *direct-current* output voltage of around 15 V for charging the vehicle's 12 V battery and running its electrical load. Its rectified output is typically between 100 and 200 W. Its prime mover, of course, is the vehicle's engine, and it's driven via the fan-belt.

Another type of alternator is the internal combustion engine-driven alternator (Figure 14.2). These alternators can be small two- or four-stroke petrol engine-driven, single-phase, portable types, used to provide electricity for driving power tools, for example, on building sites where no permanent mains power supply is available.

Alternatively, they can be large, diesel engine-driven three-phase units used to provide 'standby' power to supply hospitals, offices, factories, apartment buildings, etc., with a secure power source in the event of a mains' power failure (Figure 14.3).

In the UK, there is an agreement between the generating companies and many of the organisations that own standby alternators (known as *'non-spinning reserve'*) under which they can be connected to feed onto the national grid to help maintain system stability during emergency operating conditions or unforeseen load swings. This is a very expensive source of additional energy for the grid, and is intended only to be used in extreme conditions.

Vehicle Alternator

Figure 14.1

Portable Alternator

Figure 14.2

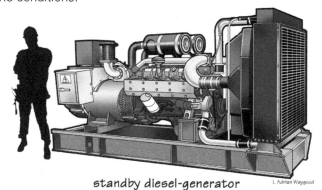

standby diesel-generator

Figure 14.3

Alternators used in power stations that normally provide power to the grid ('spinning reserve'), however, are *very* large units whose prime movers are typically high-speed steam turbines in thermal power stations or low-speed water turbines in hydroelectric power stations.

As will be explained later in this chapter, alternators driven by high-speed steam turbines (see Figure 14.4), have *small* diameters but *large* axial lengths whereas those driven by low-speed water turbines (see Figure 14.5), have very *large* diameters but *small* axial lengths.

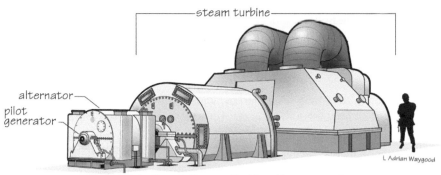

High-Speed Steam Turbine Alternator Set

Figure 14.4

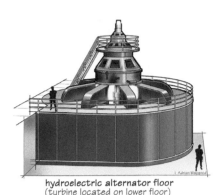

hydroelectric alternator floor
(turbine located on lower floor)

Figure 14.5

The combination of an alternator, together with its prime mover, is generally referred to as an '**alternator set**'. Most power stations will operate *several* alternator sets, allowing individual sets to be temporarily taken out of service for routine maintenance while the others continue to supply energy to the grid.

The axis of rotation of high-speed steam-turbine alternator sets is *horizontal*, with the alternator's axis being *in line* with that of the turbine. Rather like an iceberg, however, much of the turbine's bulk such as its condenser (where the expended steam is condensed into water) lies out of sight *below* the alternator floor.

The speed of a steam turbine is controlled by adjusting the steam-inlet valve to the turbine.

The type of water turbine used in hydroelectric power stations depends on the head of water available from the dam. **Reaction turbines**, such as the *Kaplan* turbine illustrated in Figure 14.6, are typically used with heads between 10 and 60 m, while **impulse turbines**, such as the *Pelton* wheel, are used with heads of 300 m or more.

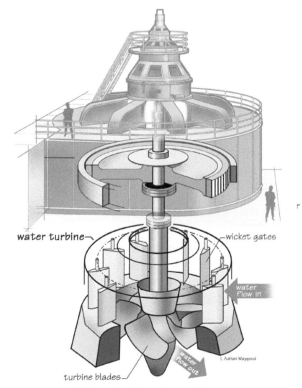

Water Turbine driven Alternator

Figure 14.6

323

We've not illustrated the Pelton wheel but it consists of a large wheel with a horizontal shaft, with numerous 'buckets' installed around its circumference. A water jet, directed at these buckets causes the wheel to rotate, driving an alternator attached to its shaft.

The speed of a reaction turbine is controlled by adjusting the angle of attack of its blades; the speed of an impulse turbine is controlled by adjusting the water flow.

The axis of rotation of reaction-type turbo-alternator set is *vertical*, with the turbine installed over separate levels, *below* the alternator floor. The axis of rotation of a Pelton wheel alternator is *horizontal*.

The very largest hydroelectric alternators are manufactured up to around 700 MV·A.

Reaction turbines, such as the Kaplan and Francis types, operate rather like propellers, with the water flowing past blades that rotate the turbine shaft. **Impulse turbines**, such as the Pelton wheel, use a jet of water to strike 'buckets', mounted around the circumference of a wheel, to drive the turbine.

Construction of synchronous machines

In common with *all* rotating electrical machines, a synchronous machine consists of two main groups: the **stator** and the **rotor**.

As we already know, the **stator** is the *stationary* part of the machine, while the **rotor** is the *rotating* part of the machine. Together, they form the machine's **magnetic circuit**, whose function it is to concentrate a strong magnetic field in the airgap between the stator and rotor.

Synchronous machines have *two* main groups of **windings**:

▶ *distributed* or *concentrated* windings, through which flow direct (excitation) currents.
▶ *distributed* three-phase windings, through which flow alternating currents.

So what exactly do we mean by '*distributed*' and '*concentrated*' windings?

▶ A '**concentrated**' **winding** is the type of winding we would instantly recognise as a 'coil': i.e. a length of insulated conductor whose turns are tightly wound adjacent to each other, around a central ferromagnetic core. As we shall learn, concentrated windings are associated with 'salient pole' rotors.
▶ A '**distributed**' **winding** is one that is formed from a conductor that is placed into longitudinal slots that have been machined into the inner surface of the machine's stator, or into the outer surface of its rotor.

Synchronous alternators are categorised according to (a) their type of **field system**, and (b) the **shape** of their rotor.

    Field system. An alternator's field system may be *stationary* or *rotating*.
    Rotor shape. Low-speed alternators, such as those driven by water turbines and internal-combustion engines, use '*salient-pole*' rotors. High-speed alternators, such as those driven by steam turbines, use smooth '*cylindrical*' (or '*non-salient*') rotors.

Field system

Synchronous alternators *below* around 5 kV·A are often '**stator fed**' or '**rotating armature**' types, which means that the magnetic field is provided by a d.c. excitation current flowing through **field windings** installed on the *stator*, while an alternating voltage is induced into the **armature windings** mounted on the *rotor*.

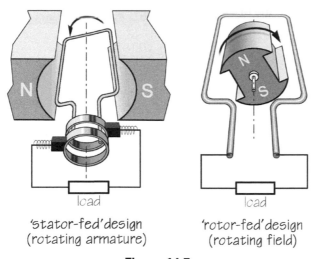

'stator-fed' design
(rotating armature)

'rotor-fed' design
(rotating field)

Figure 14.7

Synchronous machines *above* around 5 kV·A, on the other hand, are *always* '**rotor fed**' or '**rotating field**' types, which means that the magnetic field is provided by a d.c. excitation current flowing through *field windings* mounted on the *rotor*, while an alternating voltage is induced into *armature windings* installed on the *stator*.

In *both* cases, the direct current that is responsible for establishing the magnetic field is known as the '**excitation current**' and, for that reason, another name for the 'field system' is the '**excitation system**'. In this context, the terms, 'field' and 'excitation', can be used interchangeably.

Theoretically, it doesn't really matter whether the field or the armature rotates. However, in practice, *all* large alternators are of the '**rotating field**' type. And there are very good reasons for this.

The main advantage of using a rotating field design is straightforward. Suppose, for example, we have an 11 kV, 50 MV·A, three-phase alternator. This machine is capable of supplying a line current of:

$$I_L = \frac{S}{\sqrt{3}E_L} = \frac{50 \times 10^6}{\sqrt{3} \times 11 \times 10^3} \simeq 2624 \text{ A}$$

If this machine was a 'rotating armature' type, then *three* slip rings and brush assemblies would be required, one per line, and, as well as being capable of handling a line current of up to 1312 A, they must be insulated to withstand a phase voltage of $(11 \div \sqrt{3} \simeq)$ 6350 kV to earth (ground). As three-phase armatures are generally 'star' ('wye') connected, with the star point connected to earth, there would also be a requirement for a *fourth* slip ring and brush set to enable this connection to be made.

Handling such high currents and voltages in this way would *not* be practicable, and would also necessitate frequent maintenance shutdowns due to the extensive arcing that would take place between the slip rings and their brushes.

The d.c. excitation supply voltage, however, is typically 400 V, and its power rating is a small percentage of the output rating of the alternator's rated apparent power output: varying from around 2.5% for a 1 MV·A alternator, to around just 0.5% in the case of a 500 MV·A alternator.

So, we would expect the power of the excitation system of the 50 MV·A machine, referred to earlier, to be around 200 kW (remember, volt amperes only apply

14

to *alternating* current; the excitation is *direct* current so we must use watts) or thereabouts. In which case:

$$I_{excitation} = \frac{P_{excitation}}{E_{excitation}} = \frac{200 \times 10^3}{400} = 500 \text{ A}$$

Only *two* slip rings, insulated to just 400 V, would be necessary to supply a rotating field and these can easily cope with such a low excitation current without requiring excessive maintenance. Furthermore, *this arrangement would make the construction of a rotor-fed field system machine much simpler than one with a stator-fed field system.*

Rotor shape

Rotors may be either '**salient pole**' or '**cylindrical**' types.

The term, 'salient', simply means to *project* or to *stick out*, so a **salient-pole rotor** is one whose pole pieces *project from the shaft of the rotor*, with each pole piece carrying a *concentrated* winding. Figure 14.8 shows an example of a six-pole salient-pole rotor, with the individual pole pieces secured to a 'spider' which is keyed onto the shaft. For clarity, only one field winding is shown.

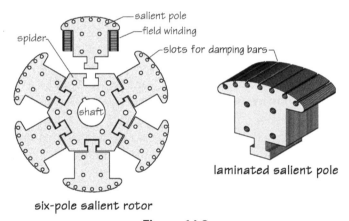

Figure 14.8

'**Cylindrical**' (also known as '**non-salient**') rotors, on the other hand, have *distributed* windings, set into longitudinal 'slots' that have been machined into their surface, leaving that surface smooth. Figure 14.9 shows an example of a two-pole **cylindrical rotor** manufactured from a single piece of cast steel with its distributed windings

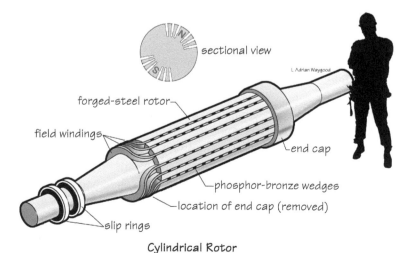

Cylindrical Rotor

Figure 14.9

secured into the slots using phosphor-bronze wedges. End caps or rings secure the looped ends of the windings against high centrifugal forces.

The choice between these two types of rotor design is determined by the magnitude of the centrifugal reactions that act on the rotor resulting from its speed of rotation: the higher the speed, of course, the higher the resulting centrifugal reaction.

Salient pole rotors are simply unable to cope with high centrifugal reactions. Accordingly, they are used with *low-speed* prime movers, such as internal combustion engines and water turbines, while cylindrical rotors are used with *high-speed* prime movers, such as steam turbines.

So why not make *all* rotors 'cylindrical'?

Well, in order to withstand the enormous centrifugal reactions to which it can be subjected, a cylindrical rotor's diameter is limited to less than 3 m. Furthermore, cylindrical rotors are usually manufactured from a solid piece of cast steel, so there is also a limit to their diameters imposed by their weight.

This means that cylindrical rotors typically have only sufficient circumference to accommodate enough distributed windings to form *four* poles at most; in fact, *most* cylindrical rotors have only *two* poles.

Now, the equation that relates rotational speed *(N)* and generated frequency *(f)*, to the number of poles *(p)* is:

$$N = \frac{120f}{p}$$

. . . so, for a two-pole machine to generate a voltage at 50 Hz, it must run at:

$$N = \frac{120f}{p} = \frac{120 \times 50}{2} = 3000 \text{ rev/min.}$$

This speed is typical of prime movers such as steam turbines. Practically all other prime movers run at much lower speeds than this: a water turbine, for example, typically runs at 500 rev/min or less. To generate a voltage at 50 Hz at this speed, would require the rotor to have:

$$p = \frac{120f}{N} = \frac{120 \times 50}{500} = 12 \text{ poles.}$$

This number of poles can *only* be accommodated on a rotor with *a large circumference*, and this means using a salient pole design.

Many of the large alternators used in hydroelectric power stations actually run much *slower* than 500 rev/min and, therefore, *require far more poles* - e.g. in Canada, Hydro Québec's *La Grande-3* hydroelectric power station uses water turbines that rotate at just 112.5 rev/min, requiring alternators with . . .

$$p = \frac{120f}{N} = \frac{120 \times 60}{112.5} = \textbf{64 poles}$$

. . . in order to generate a terminal voltage at 60 Hz (this being the standard frequency in North America).

Figure 14.10 illustrates a large-diameter, multi-pole, **salient-pole** rotor typical of those driven by water turbines in a hydroelectric power station. To put this in perspective, it is not unusual for each pole piece to have a mass of around 10 t (tonnes)! Despite this, their slow speeds result in much lower centrifugal forces than those generated by steam turbine-driven cylindrical rotors.

As we learnt from the chapters on d.c. generators and motors, it can be shown that *the output power of an alternator is proportional to its volume*. So, for any given power output, we can say that a:

▶ **salient-pole rotor** will have a *large diameter*, but a *small axial length*, whereas a
▶ **cylindrical-pole rotor** will have a *small diameter*, but a *large axial length*.

14

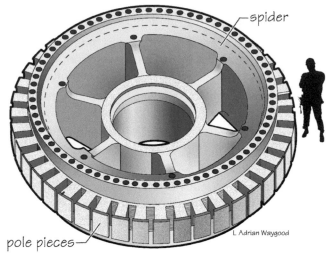

Multi-pole Salient Rotor
(water-turbine driven)

Figure 14.10

Typically, cylindrical rotors can have diameters up to 3 m with axial lengths of up to 12 m, whereas salient-pole rotors can have diameters up to 15 m with axial lengths of up to 2 m.

Stator windings

As already explained, *all* large synchronous alternators are '**rotor fed**': that is, the field is produced by a direct current flowing through the *rotor* windings, while the armature is installed on the *stator*. This arrangement eliminates the need to connect the armature to its load via slip rings and brushes. But it also has the secondary advantage of providing greater room in which to accommodate the high-voltage armature windings and their necessary insulation.

Figure 14.11 illustrates the stator for a high-speed cylindrical-rotor machine. It's manufactured from a stack of silicon-steel laminations to reduce eddy-current losses, with longitudinal slots machined into its inside surface to accommodate the armature windings.

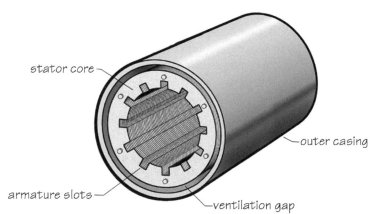

Stator for Cylindrical-Rotor Alternator

Figure 14.11

The distribution of the armature winding's conductors in the stator's slots can be both complicated and extremely confusing, as there are various ways of distributing those conductors. As we are describing alternators in general terms, in the example that follows, we will be using what is known as a '*whole-coil concentric*' winding.

One of the greatest difficulties we face when describing an armature's winding arrangement is, *'How can it be clearly illustrated?'* That is, how do we illustrate the *three*-dimensional inside-surface of a cylinder in just *two* dimensions! We are, in fact, faced with a similar dilemma that faced early cartographers when they had to represent our spherical-shaped Earth as a two-dimensional map, and we adopt a similar solution.

So, let's see how we can represent a **single-phase armature winding** for an alternator having a six-pole (salient-pole) rotor – as illustrated in Figure 14.12.

Imagine that we are looking through the centre of the stator cylinder. The conductors inserted into the longitudinal slots running along the length of the stator are those parts of the winding that will be cut by the rotating field, so we call them the 'active' parts of the winding, or the '**active conductors**'. These are the parts of the winding into which voltages will be induced. Beyond the 'active' lengths of conductor, at the nearest end and at the furthest end of the stator cylinder, these conductors must be looped around to form individual *turns*. In Figure 14.12, we've labelled these 'non-active' parts of the winding as 'front-end' and 'rear-end connections'.

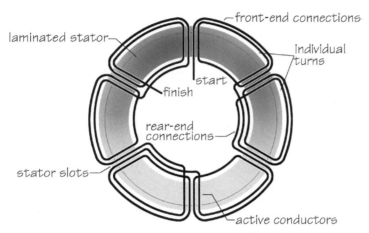

schematic representation of single-phase
six-pole armature winding

Figure 14.12

Now, if we remove the stator itself, what we are left with are the stator-winding's six turns, which resemble the petals of a flower – as illustrated in Figure 14.13.

You should notice *two* things about the way in which the armature is wound.

Firstly, the armature is wound from a *single length of conductor*, laid in the stator's slots, and looped around at their front and rear ends. It has a 'start' end and a 'finish' end, which are connected to the machine's external load.

Secondly, the conductor is arranged within the slots to form six sets of distributed turns, with adjacent turns wound in *opposite* directions. This ensures that, as the rotor poles pass the active conductors, the voltages induced into them are *cumulative,* or in *series* with each other – i.e. the voltage appearing across the start and finish of the armature conductor is *the phasor-sum of the voltages induced into each of the individual active conductors*. This is more clearly illustrated in Figure 14.13.

It would be a good idea, at this point, to pause from reading this text and spend some time tracing the path taken by the conductor around the armature, from which it should become obvious that the directions of the voltages induced into the active conductors are cumulative.

The reason that the final, cumulative, voltage is the *phasor-sum*, rather than the *algebraic-sum*, of the individual voltages induced into each of the active conductors is because the way in which they are distributed around the armature results in a small phase shift (α) between each voltage. As can be seen from the resulting phasor

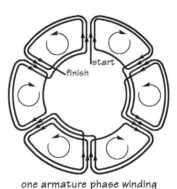

one armature phase winding
for six-pole alternator

Figure 14.13

14

diagram in Figure 14.14, the *resulting* voltage is somewhat *less* than the *algebraic-sum* of the individual voltages (for simplicity only eight induced voltages are shown; in reality, of course, there are far more).

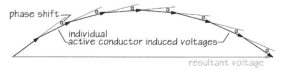

Figure 14.14

Returning, now, to Figure 14.13, as we can see there are *six* sets (of four) active conductors distributed around the inner surface of the stator – making a total of 24 active conductors. Each set of active conductors must be 60 'mechanical' degrees apart. As one of the rotor's north poles passes any one set of these active conductors, its south pole passes the adjacent set – *inducing one complete cycle* between those adjacent sets.

It follows, therefore, that for every 60 'mechanical' degrees through which the rotor turns must be equivalent to 360 'electrical' degrees. In other words, there must be 360 'electrical' degrees between adjacent sets of active conductors.

And, for a six-pole machine, 360 'electrical' degrees must correspond to 60 'mechanical' degrees – illustrated in Figure 14.15.

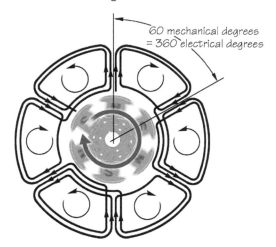

Figure 14.15

In other words, for this particular armature, *one* complete rotation of the rotor will generate *six* cycles of alternating voltage.

Most large alternators are not single-phase, of course, but *three-phase*. So a three-phase alternator's armature must be made up of *three*, completely separate, single-phase distributed windings, like the one illustrated above, but *with each phase-winding superimposed over each other*, and displaced from each other by 120 *electrical* degrees, which corresponds to 20 *mechanical* degrees in the case of our six-pole machine. This arrangement is illustrated in Figure 14.16.

As we can see, once assembled and installed in the stator slots, the windings' active conductors for all three phase windings are evenly displaced around the inside surface of the stator.

This leaves *three* pairs (S_1-F_1, S_2-F_2, and S_3-F_3) of 'start' and 'finish' terminals that must be connected to the machine's external load. As already mentioned, most alternators are '**star**' ('**wye**') connected and, again, there's a very good reason for choosing this configuration, which is because the individual phase windings need then only be insulated to withstand a maximum voltage of 57.7% (the reciprocal of $\sqrt{3}$) of the line voltage.

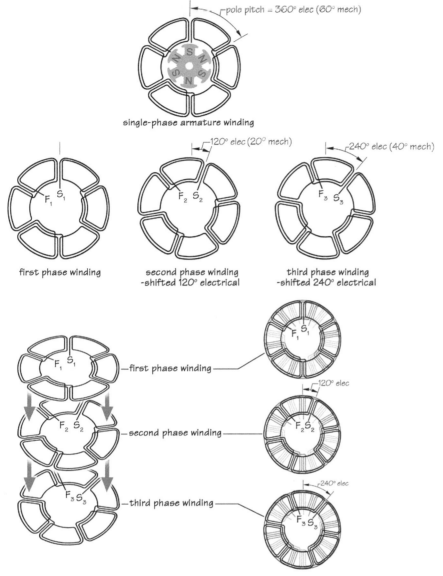

Schematic diagram showing three-phase armature winding 'layers'

Figure 14.16

As we can see in Figure 14.17, by connecting together the 'end' connections of each phase winding, it's possible to connect the armature windings to form a 'star' connection.

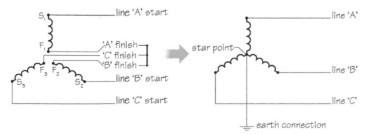

How the three individual phase windings are connected in 'star' ('wye')

Figure 14.17

The amount of space required by the armature conductors and their insulation limits an alternator's line voltage to around 30 kV. Beyond this value, the overall volume of an alternator will become impractically large.

In the UK, the standard line voltages for alternators used at power stations are typically 11 kV or 25 kV.

Generated waveform

In theory, we generally assume that an alternator always generates a voltage having a *sinusoidal* waveform, however this isn't actually true in practice. For example, if we were to simplify our earlier illustrations of the armature windings so that the rotor cuts only a *single* 'active' conductor, rather than several, then the resulting induced voltage's waveform would *not* be sinusoidal but, rather, one having the truncated shape illustrated in Figure 14.18.

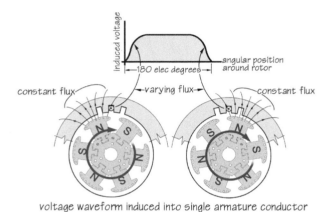

voltage waveform induced into single armature conductor

Figure 14.18

The reason for this is because the magnetic flux leaving the pole is evenly distributed *across its face*, only varying at its *leading* and *trailing* edges. As the pole face sweeps past the conductor, its evenly distributed flux means that the *rate* at which it cuts the conductor will be constant, and the resulting voltage induced into the conductor will also be constant (rather than sinusoidal).

However, if *two*, rather than just one, adjacent 'active' conductors were to be cut by the rotating pole, the following waveform would result. As we can see from Figure 14.19, the two induced voltages (**A** and **B**) are slightly displaced from each other, and their phasor-sum will start to approach the shape of a sinusoidal voltage.

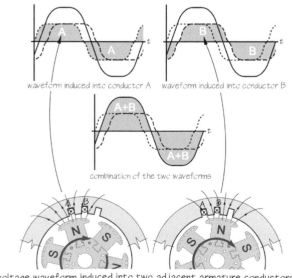

voltage waveform induced into two adjacent armature conductors
(closer to a sinusoidal waveform)

Figure 14.19

Additional pairs of active conductors, such as those shown in our earlier illustrations, would improve the waveform even more until, eventually, it becomes pretty close to the ideal sinusoidal waveform.

Achieving a true sinusoidal waveform (or as close as we can get to one) is important, because non-sinusoidal waveforms contain 'harmonics'. Harmonics are beyond the scope of this book, but are components of a waveform having multiples of the 'fundamental' frequency (i.e. in this case, 50 Hz). Harmonics are undesirable in electrical systems because they result in power-quality problems, such as overheating and torque pulsations in motors.

Effect of armature reaction

We discussed **armature reaction** at length, in the chapter on d.c. generators. Armature reaction describes the effect that the armature field has upon the excitation field and how it acts to *distort* the resultant, combined, field.

When the alternator is supplying a *balanced* three-phase load, the three armature (load) currents produce a combined magnetic field which *naturally rotates within the air gap between the stator and rotor at synchronous speed.*

The *reason* for this *naturally rotating field* will be fully explained in the following chapter on **three-phase induction motors** but, for now, it's sufficient to accept this is so.

So let's examine the behaviour of the alternator for extreme types of load – i.e. '**purely resistive**', '**purely inductive**', and '**purely capacitive**'. In each case, we'll assume that the field current remains constant.

Purely resistive load

Figure 14.20 (left) illustrates the pattern of the excitation flux surrounding the rotor's field winding *alone*. This is the pattern that would exist if we completely ignored the flux produced by the armature. We'll assume that the rotor is actually spinning in a *clockwise* direction, although we've illustrated it 'frozen' it in the upright position.

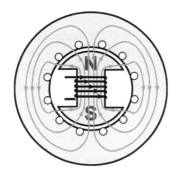
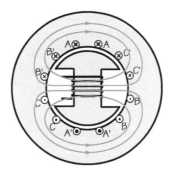

field due to rotor alone   field due to armature alone

Figure 14.20

Figure 14.20 (right) illustrates the pattern of the magnetic flux set up by the three-phase armature currents *alone*, at the instant when the current through armature phase-winding **A – A'** predominates. This is the pattern that would exist if we completely ignored the flux produced by the rotor. 120-electrical degrees, later, when the current through armature phase-winding **B – B'** predominates, the field would have rotated, clockwise, by 120 electrical degrees and, when the current through armature phase-winding **C – C'** predominates, the field would have rotated by yet another 120 degrees, and so on – in other words, the field will continuously rotate in a *clockwise* direction.

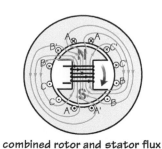

combined rotor and stator flux

Figure 14.21

In *both* cases, we can confirm the direction of the magnetic fields by applying the '**right-hand grip**' rule.

The *speed* at which the armature's magnetic field rotates *exactly matches that of the rotor*, which is why, of course, the machine is called a 'synchronous' alternator.

Now let's see what happens when these two, separate, fields *combine* to form their **resultant field**. Careful examination of the two diagrams shown in Figure 14.20 shows that the direction of the armature field's flux lines *oppose* those emerging from the *leading edge* of the rotor, but act in the *same direction* of those emerging from the *trailing edge*, resulting in the combined field, illustrated in Figure 14.21.

As we can see, thanks to the effect of **armature reaction**, the resultant field is *not* symmetrical, with the flux lines 'stretched' at both *trailing edges* of the rotor.

If we think of the flux lines behaving somewhat like 'rubber bands', then it should be obvious that the 'stretched' flux lines must act *to produce a 'rearward pull' on the rotor*. Note that this rearward pull *doesn't* slow down the rotor but, instead, causes the prime mover to compensate by exerting an *increased torque* compared with that which would be delivered when the generator *isn't* supplying a load.

Now, let's move on and consider what happens if the generator is supplying a (theoretical) **purely inductive** load.

Purely inductive load

Theoretically, when the generator supplies a **purely inductive load**, the armature (load) current will *lag* the generated e.m.f. by 90 electrical degrees.

So, when the e.m.f. induced into armature phase-winding **A – A'** is at its *peak* value, *the rotor will have turned 90 degrees clockwise by the time that the armature current through that same phase-winding will have reached its peak value.*

So, at this particular instant, the field due to the armature (load) current alone would be *identical* to that shown in Figure 14.20 (left), and acting from *right-to-left through the centre of the rotor.*

The field due to the rotor would produce *exactly the same pattern* as the armature current, *but it would be acting from left-to-right through the centre of the rotor!*

In other words, the armature flux and the rotor flux will be *opposing each other* – which would act to **weaken** the excitation field and, therefore, **reduce** the e.m.f. *induced into the armature windings!* As the excitation flux is greater than the armature flux, the resulting combined field will be as shown in Figure 14.22. As this combined field is symmetrical, the prime mover provides no torque (other than that necessary to overcome friction and windage).

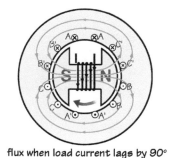

flux when load current lags by 90°

Figure 14.22

Purely capacitive load

Theoretically, when the generator supplies a **purely capacitive load**, the armature (load) current will *lead* the generated e.m.f. by 90 electrical degrees.

The resulting flux pattern would be exactly the same as that illustrated in Figure 14.22, except that *the magnetic polarity of the rotor would be reversed*, and its field would be acting from right-to-left, and be 'assisting' the armature field, **strengthening** *the combined field and, therefore,* **increasing** *the e.m.f. induced into the armature windings!*

Again, as this combined field is symmetrical, the prime mover provides no torque (other than that necessary to overcome friction and windage).

Terminal-voltage/load-current characteristics of an alternator

Of course, there are *no* such things as 'purely resistive', 'purely inductive', or 'purely capacitive' loads. 'Real loads' are a *combination* of resistance, inductance, and capacitance, but these 'theoretical' loads help us understand the **terminal-voltage/ load-current characteristics** of an alternator.

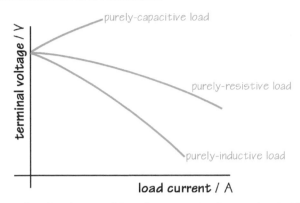

terminal voltage / load current characteristic

Figure 14.23

Assuming a constant value of excitation current, Figure 14.23 shows the influence of armature reaction for changes in terminal voltages resulting from variations in load current.

For a purely resistive load, the fall in terminal voltage for an increase in load current is relatively small compared to that for a purely inductive load. 'Real' loads would fall between these two extremes.

For example, the voltage/load current curve for a load having a power factor of, say, 0.8 lagging, would lie between the curves for a purely resistive and purely inductive loads.

However, for a purely capacitive load, the *increasing* strength of the combined rotor/ armature fields *causes the terminal voltage to increase for increasing values of load current*.

Percentage voltage regulation

Alternators are designed to provide a **specified value of terminal voltage** when **supplying its rated current** at a **specified power factor** (typically either **unity** or at **0.8 lagging**).

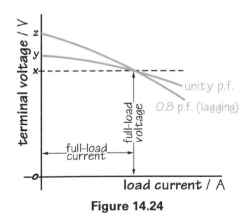

Figure 14.24

Voltage regulation is the *difference* between the alternator's **no-load terminal voltage** and its **rated full-load terminal voltage**.

Referring to Figure 14.24, distance *y–x* represents the voltage regulation for a *unity-power factor* load, while distance *z–x* represents the voltage regulation for a load of *power factor 0.8*.

So *there's a significant difference in the voltage regulation at different power factors* and, for this reason, voltage regulation is ***always*** quoted *at a specified power factor* – normally either **unity** or **0.8 lagging**.

It's more usual to describe voltage regulation in terms of **percentage voltage regulation**, which is defined as the difference between the alternators no-load voltage and its rated full-load voltage, *expressed as a percentage of its rated full-load voltage*, as follows:

$$\%\text{voltage regulation} = \frac{E_{no\,load} - E_{full\,load}}{E_{full\,load}} \times 100\%$$

The 'ideal' percentage voltage regulation would be zero percent.

Excitation systems

There are *three* things that determine the value of the e.m.f. induced into an alternator's armature. These are the:

▶ number of active conductors per phase.
▶ speed of the rotor.
▶ strength of the excitation field.

Any one of these could, in theory at least, be used to control the value of the e.m.f. induced into the armature.

However, in practice, we obviously cannot vary the *number of active conductors* as that number was fixed when the alternator was manufactured. And we cannot vary the *speed of the rotor* because this would then vary the frequency of the induced e.m.f., and this must not be allowed to vary beyond its statutory limits (50 Hz ±1% for the UK).

Generation companies regulate the daily average frequency so that electric clocks stay within several seconds of the correct time, as determined by International Atomic Time.

So this leaves us with *varying the strength of the excitation field* as the only way of controlling the value of the alternator's e.m.f.

In order to *create* the excitation flux within an alternator's magnetic circuit, **direct current** must flow through the rotor's field windings. This direct current is supplied by the machine's '**excitation system**'.

The **excitation system** must not only supply the direct current necessary to 'excite' the machine, but it must also incorporate some means of *varying the excitation current* in order to adjust the machine's output voltage – remember, we can't do this by varying the *speed* of the machine, because it must maintain a fixed output frequency that conforms to statutory requirements.

So let's examine the **block diagram** of a basic excitation system illustrated in Figure 14.25.

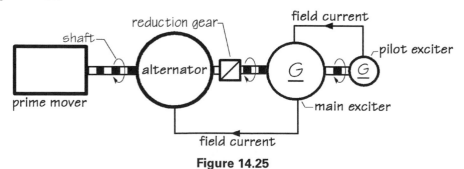

Figure 14.25

As well as driving the alternator's rotor, the machine's shaft also drives a small d.c. generator, called the '**main exciter**', together with another, even smaller, d.c. generator, called the '**pilot exciter**'.

For high-speed turbine/alternator sets, the main and pilot exciters are driven through a reduction gearbox, reducing their speed to 1000 or 750 rev/min, in order to avoid commutation problems.

The d.c. output from the **main exciter** is fed to the alternator's rotor via slip rings – thus allowing the large alternating voltage to be controlled by a relatively small direct voltage.

The **main exciter** is, itself, excited directly from the **pilot exciter** whose output is achieved through its rotor's residual magnetism – as described in the chapter on d.c. generators. Overall control of the alternator's excitation is achieved by adjusting the field current of the main exciter – as illustrated in Figure 14.26.

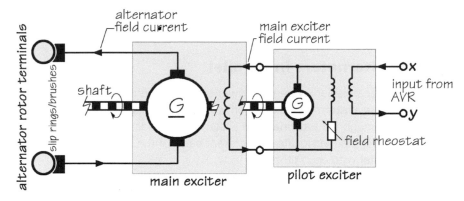

Figure 14.26

These days, to cut down on the amount of commutator/brush maintenance required, it is more common to use '**brushless' excitation**. With this method of excitation, the main exciter (d.c. generator) is replaced by an **a.c. generator** that provides a d.c. output via a built-in **rectifier**. This system enables the alternator's rotor to be connected *directly to the output from the rectifier* instead of via a commutator.

Automatic voltage regulator (AVR)

The device used to control an alternator's excitation system, in order to ensure that the terminal voltage remains within its statutory requirements, is termed an '**automatic voltage regulator**' ('**AVR**').

A detailed explanation of how an AVR operates is beyond the scope of this book. However, we *can* discuss the basic principle behind it.

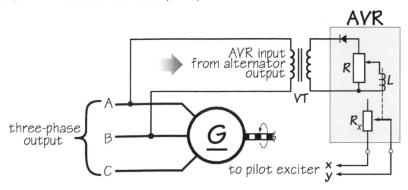

Figure 14.27

As we have already learnt, to control the output e.m.f. of an alternator, we need to be able to *adjust the field current that is supplied from its excitation system*. A basic circuit for doing this is shown in Figure 14.27.

To achieve this, the generator's phase voltage (in this case, phase A-B) is constantly monitored, and fed back to the input of the AVR, via a **voltage transformer** (**VT**). The output-voltage from the VT (which, of course, is proportional to the generator's output voltage) is rectified, and applied across a variable resistor, **R**. The resistance setting of this variable resistor establishes the **reference voltage** for the AVR. This reference voltage is then applied to the solonoid, **L**, which controls the variable resistor R_x via a 'floating' mechanical link. Variable resistor, R_x, in turn, varies the pilot exciter's field and, ultimately, the field of the alternator itself.

The magnetic strength of the solenoid determines the position of the floating link that controls the value of the variable resistor R_x. So, if the alternator's output e.m.f. should *increase* beyond the reference value, then the solenoid will act to *increase* the resistance of R_x and *reduce* the pilot exciter's output. On the other hand, if the alternator's output e.m.f. should *fall* below the reference value, then the solenoid will act to *decrease* the resistance of R_x and *increase* the pilot exciter's output.

Operating alternators in parallel

In an electricity transmission system, a power station's alternators each feed into a common **bus system** that is interconnected with other alternators at other power stations through the **national grid** system. Hundreds, or even thousands, of alternators in power stations located in different parts of the country are, therefore, all operating in **parallel** with each other.

> The term, '**bus**', is derived from the Latin word, 'omnibus', meaning 'to serve all', and describes a common point of interconnection for electrical equipment. In practical terms, this is achieved using sets of parallel copper bars (one per line, in the case of three-phase systems), called '**bus bars**', to which alternators, transformers, etc., are connected via circuit breakers. The term, 'bus' is frequently misspelled as 'buss'!

> The term, '**infinite bus**', is used to describe a power grid that is so large that the action of an individual alternator will have absolutely no effect on the overall operation of that grid in terms of its voltage and 'speed' (i.e. frequency). In other words, the 'infinite bus' system's voltage is constant and independent of any reactive power supplied or consumed, and its frequency is constant and independent of power flow.

Individual alternators have *no* influence on the overall behaviour of the bus system in terms of its voltage or 'speed' (frequency), and is termed an '**infinite bus**'.

Synchronising

The process of **paralleling** an individual alternator with another alternator, or with this infinite-bus system, is termed '**synchronising**'. Synchronising involves preparing the alternator so that the following conditions are met *before* it can be safely connected to the bus:

▶ corresponding line voltages are the same
▶ corresponding line voltages are in phase with each other
▶ corresponding speeds must be such that the frequencies are the same
▶ corresponding phase sequences are the same.

At this point, it might be a good idea to look at a simple **mechanical analogy** of this synchronising process.

So, imagine a huge mechanical system (equivalent to the 'infinite bus'), as shown in Figure 14.28, consisting of numerous mechanical drives all interconnected by a system of gear trains and running at a constant speed. Imagine, too, that access to this mechanical system is available via a constantly spinning toothed gear (labelled 'B' in the Figure 14.28).

Figure 14.28

Imagine, now, that we wish to connect an additional drive (equivalent to an individual alternator) to this mechanical system (equivalent to the infinite bus) by engaging, or 'synchronising', its gearwheel (**A**) with the system's gearwheel (**B**), so that it can contribute to driving that system. If the teeth of the two gears are properly synchronised when they are brought together, then those teeth will mesh and the additional drive will engage with the mechanical system smoothly.

However, several things *could* go wrong when we bring these gears together – as illustrated in Figures 14.29a–e:

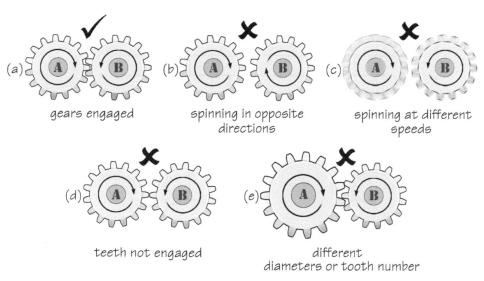

Figure 14.29

For example, if the two gears were spinning in *opposite directions* (Figure 14.29b) then catastrophic damage could result however gently we try to engage them.

If the gears were spinning in the *correct* directions, but at *different speeds* (Figure 14.29c), then damage could also result – particularly if the difference in their speeds were great.

If both gears were spinning at exactly the same speed, but their relative positions were such that their teeth are unable to mesh (Figure 14.29d), then they would be unable to engage.

Finally, if their diameters/tooth numbers didn't match then, again, they would simply be unable to engage.

Now let's see how this mechanical analogy relates to the requirements for synchronising an alternator.

Table 14.1

Mechanical analogy:	Electrical equivalent:
spinning in opposite directions:	the alternator's **phase sequence** is *opposite* to that of the bus system: i.e. 'A-C-B', rather than 'A-B-C'.
spinning at different speeds:	the alternator's output **frequency** is different to that of the bus system.
teeth not engaged:	the alternator's voltage is **out of phase** with that of the bus system.
different diameters/tooth count:	the alternator's **voltage** is *different* from that of the bus system.

In practice, the alternator's terminal line voltage should typically be within 5% of the busbar line voltage and any phase difference shouldn't exceed 5 electrical degrees. The alternator's frequency should, ideally, exceed that of the busbars by between 0.2 and 0.4 Hz when synchronisation takes place. It is unlikely that the phase sequence would ever change unless the alternator was to be driven *backwards* by its prime mover, which is highly improbable.

Figure 14.30 shows a set of busbars, with one alternator (G1) already connected and in sync with the bus system, and a second alternator (G2) being prepared to be paralleled with alternator G1 and the bus system. The requirement for the same voltage is satisfied when both voltmeters indicate the same value. The requirement for the alternator's line voltage to be in phase with that of the bus system is called '**synchronising**' and is aided by using a **synchroscope**.

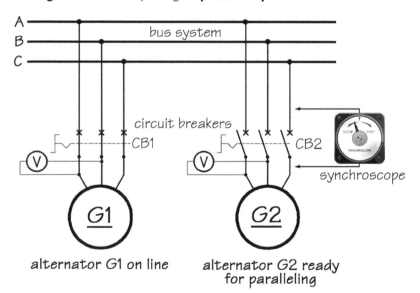

alternator G1 on line   alternator G2 ready for paralleling

Figure 14.30

The principle of operation of a synchroscope is beyond the scope of this book so, instead, let's examine an alternative, but far older, method of synchronising an alternator with the bus system: the '**cross-connected lamp method**'. In fact, this method is an excellent learning tool, as it explains exactly what is going on in relatively simple terms.

Figure 14.31 shows how these lamps are connected. In the circuit diagram, for the sake of simplicity, the lamps are shown connected directly to the line conductors but, in practice, for reasons of safety (we are dealing, after all, with high voltages!), the connections would be via **instrument transformers**.

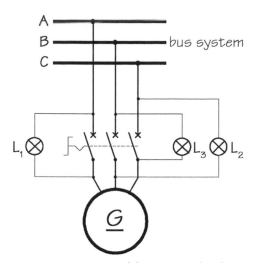

cross-connected lamp method

Figure 14.31

In this example, we can see that **lamp L₁** is connected between line A (bus) and line A (alternator), **lamp L₂** is connected between line C (bus) and line B (alternator), and **lamp L₃** is connected between line B (bus) and line C (alternator).

A potential difference appearing across any lamp will cause it to illuminate, and its brightness will be determined by the magnitude of that potential difference.

These lamps are installed on the control panel where they can be clearly visible from the circuit-breaker control position, and form the corners of an isosceles triangle, as illustrated in Figure 14.32. During the process of synchronising, these lamps illuminate one after the other and, thanks to our persistence of vision, give the impression of 'rotating' in either a clockwise or counterclockwise direction – depending on the speed of the alternator compared with that of the bus system.

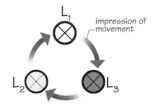

Figure 14.32

So let's examine the behaviour of these lamps whenever the oncoming alternator is running slightly 'faster' (i.e. generating a higher frequency) than the bus system and, therefore, with the alternator's phasor continuously *advancing* on the bus system's phasor, as illustrated in Figure 14.33 (where the alternator's phasors are shown in black and the bus's phasors are shown in blue):

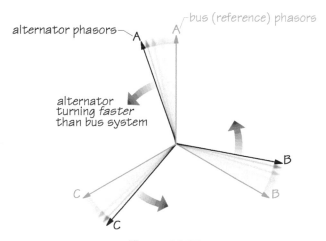

Figure 14.33

Now, let's see what effect this has on the lamps.

In Figure 14.34, we show two sets of phasors: one (coloured blue) representing the reference voltages (i.e. that of the bus system), and the other (coloured black) representing the alternator's voltages and three consecutive instants in time. Strictly speaking, of course, these phasors should be superimposed one exactly over the other (as in Figure 14.33) but, in order for us to be able to see what is going on, we've had to slightly displace them from each other.

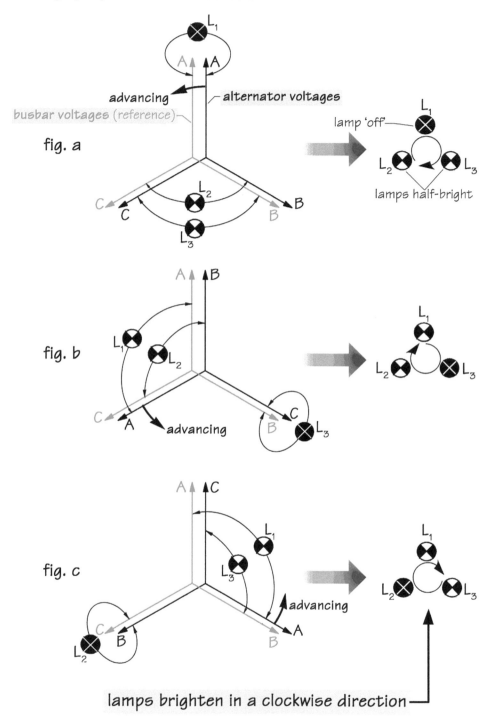

On-coming Alternator running 'faster' than Busbar

Figure 14.34

In Figure 14.34a, although the alternator's phasors are rotating somewhat faster than the bus's, they are shown at the instant when both sets of phasors happen to be exactly in phase. In Figure 14.34b, the alternator's phasors have moved on by

120 electrical degrees so, at that instant, phasor A of the alternator is now in phase with (but in the process of overtaking) phasor B of the bus system. Finally, in Figure 14.34c, the alternator's phasor A has moved on by a further 120 electrical degrees, and is shown at the instant when it happens to be in phase with (but in the process of overtaking) phasor C of the bus system.

By carefully examining each of these diagrams, we should be able to confirm which lamps are 'off' and which lamps are 'on', at each of these particular instants.

Now, if these lamps are arranged to form the corners of a triangle, the sequence in which their brightness changes will, thanks to our persistence of vision, give us the impression of a rotational movement in a *clockwise* direction.

The speed of this apparent rotational movement provided by the glowing lamps will, therefore, simulate the speed of the rotation of the alternator, with respect to that of the bus system. The slower the lamps' apparent rotation, the closer the speed of the alternator is to that of the bus system. And, if the lamps were to reverse sequence and appear to rotate in a *counterclockwise* direction, then they are indicating that the alternator is moving more *slowly* than the bus system.

When lamps no longer 'rotate', and lamp L_1 remains extinguished while lamps L_2 and L_3 remain equally bright, we know that the alternator and the bus system are (a) running at the same speed and (b) the voltages are in phase with each other. In other words, they are running in sync with each other.

The operator, therefore, can adjust the speed of the alternator's prime mover, while monitoring the lamps until they indicate that the alternator is running at exactly the *same* speed as the bus system and in phase. At that point, the alternator is synchronised with the bus system, and its circuit breaker can be closed.

As explained earlier, this 'cross-connected lamp' method of synchronising is very old and is no longer used these days, but it is a good learning tool for us to understand the theory behind synchronisation. This technique has long been superseded by an instrument called a '**synchroscope**', which may be analogue or digital – as illustrated in Figure 14.35. You could describe the digital synchroscope as being a modern-day equivalent of the cross-connected lamp system in the sense that its display is produced by the illusion of movement produced by numerous light-emitting diodes (LEDs).

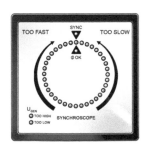 

digital synchroscope   analogue synchroscope

Figure 14.35

In either case, when the synchroscope's indicator rotates *clockwise*, it indicates that the alternator is turning *too fast*; when it turns counterclockwise, it indicates that the alternator is moving *too slowly*. Synchronisation is usually achieved as the indicator reaches the 12 o'clock position in the clockwise direction.

Procedure for synchronising

step 1: The alternator is set to hand control.
step 2: The alternator's excitation is adjusted until its voltmeter reading matches that of the bus system.

step 3: The speed of the alternator's prime mover is adjusted until the synchroscope's indicator rotates slowly in the clockwise direction (i.e. the alternator is running slightly 'faster' than the bus system).

step 4: When the synchroscope's indicator just passes through the 12 o'clock position, the main circuit breaker is closed, placing the alternator on line.

It should be pointed out that in most modern power stations, these days, synchronisation is usually achieved *automatically*, rather than manually, but this is well beyond the scope of this book.

When an alternator has been synchronised with an infinite bus system, its speed will normally remain 'locked' or 'tied' to that of the bus system until it is disconnected.

Behaviour of a synchronised alternator

Once an alternator has been synchronised to the infinite-bus system, its **speed** and its **terminal voltage** are *fixed*, so it's not immediately obvious *how* it is able to supply energy to the bus system when that system demands it.

To understand this, we must return to the earlier section on the effects of **armature reaction**. But before we do so, let's consider the behaviour of an alternator at a *very basic level*.

So, one way of looking at a **generator** is that it *converts* **torque** *into* **current** and, at some remote point down the transmission/distribution line, a **motor** then *converts that* **current** *back into* **torque**. So, in *extremely simple terms*, we *could* say that a generator provides a means of *transferring the torque* provided by its prime mover *to somewhere else where it can do useful work!*

> In simple terms, we could say that a generator provides a means of transferring the torque provided by its prime mover to somewhere else where it can do useful work!

This is an over-simplification, of course, but it's essentially true, and an easy concept to understand, and is useful analogy for what follows.

So, for the sake of simplicity, let's assume that the 'infinite bus' system behaves like a purely resistive load.

Let's remind ourselves of the combined, or resultant, flux for a purely resistive load, as illustrated in Figure 14.35.

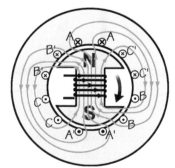

combined rotor and stator flux

Figure 14.35

Because the direction of the **armature flux** is such that it *opposes* the **excitation flux** emerging from the *leading* edge of the rotor, and *reinforces* the excitation flux emerging from the *trailing* edge of the rotor, the resultant flux is unsymmetrical – with the flux *weakened* at the leading edge and *strengthened* at the trailing edge.

The result of this is that the flux lines are 'stretched' at the trailing edge, as illustrated in Figure 14.35.

And, because flux lines behave somewhat like stretched 'rubber bands', the effect of this is to apply a *rearward-acting pull* or *torque* on the rotor.

This rearward pull *doesn't* slow down the rotor but, instead, causes the prime mover to compensate by exerting an *increased torque* compared with that which would be delivered when the generator is supplying a lower load.

Now, let's assume that the infinite bus 'demands' *more current* from the armature. *More* armature current means an *increase* in the armature flux, and an *increase* in the armature flux means an *increase* in the rearward pull on the rotor! Again, the rotor *doesn't* slow down but, to compensate, *the prime mover must further increase the amount of torque it applies to the rotor*.

So variations in load current result in the prime mover providing corresponding variations in torque. Going back to our very simple explanation, earlier, the synchronised alternator can be thought of as *converting torque into load current*.

If the infinite-bus system should feed current *back* into the alternator's armature windings, then the direction of the flux lines due to the armature windings will *reverse*, and *strengthen* the flux at the *leading* edge of the rotor and *weakening* the flux at the *trailing* edge of the rotor, resulting in a *forward-acting* torque being applied to the rotor. The machine will now be acting as a synchronous *motor*, and this condition is described as '**motoring**'.

Conclusion

Now that we've completed this chapter, we need to examine its **objectives** listed at its start. Placing a question mark at the end of each objective turns that objective into a **test item**. If we can answer those test items, then we've met the objectives of this chapter.

14

345

CHAPTER 15

Three-phase induction motors

Objectives

On completion of this chapter you should be able to:

1. list the three electromagnetic phenomena that contribute to the operation of all single- and three-phase a.c. motors.

2. describe the construction of a three-phase induction motor.

3. describe how a rotating magnetic field is set up within the stator of a three-phase induction motor.

4. describe how torque is developed by a three-phase induction motor's rotor.

5. explain what is meant by 'synchronous speed', when applied to a three-phase induction motor.

6. list the factors that determine the synchronous speed of a three-phase induction motor.

7. explain the term 'slip', when applied to a three-phase induction motor.

8. explain why an induction motor's rotor can never achieve synchronous speed.

9. explain the effect that rotor speed has upon the torque of an induction motor.

10. describe the effect of a varying rotor current and power factor on the torque of an induction motor.

11. interpret the torque/slip characteristics of different classes of induction motor.

12. explain the purpose of a wound-rotor induction motor and compare its behaviour with that of a squirrel-cage induction motor.

13. explain what is meant by a 'synchronous motor'.

14. describe the construction of a synchronous motor.

15. explain why synchronous motors cannot self-start, and describe methods of starting them.

16. identify applications for synchronous motors.

Introduction

While studying at his college in Carlstadt, Croatia, a then unknown Serbian electrical-engineering student, named **Nikola Tesla** (1856–1943), was perplexed by the severe arcing that he saw occurring at the commutator of a d.c. motor being demonstrated by his lecturer, a professor Poeschl. *'Surely,'* Tesla argued, *'it should be possible to construct a motor that would work without the need for a commutator?'* Poeschl, who considered Tesla to be *'. . . an ingenious man who often demonstrated (electrical) principles using apparatus of his own invention'*, took Tesla's comments very seriously and spent much of the following lecture explaining *why* alternating current simply couldn't be used to drive an electric motor. *'It is,'* he argued, *'an impossible idea!'*

Despite his professor's comments, Tesla remained convinced that it *was* possible, reasoning that *'if an alternating current is produced by circular motion then, surely, circular motion could be produced by an alternating current?'*

Electrical Science for Technicians. 978-1-138-84926-6 © Adrian Waygood.
Published by Taylor & Francis. All rights reserved.

And, of course, it turned out that he was correct! In an article published many years later, Tesla described how, in 1883, the idea for his a.c. motor came to him as he was taking a stroll through a park in Budapest: *'Like a flash of lightning, and in an instant, the truth was revealed. I drew, with a stick in the sand, the diagrams of my motor. A thousand secrets of nature, which I had stumbled upon accidentally!'*

What he had done was to use *two* separate alternating currents that were *out of phase with each other*, to create a *naturally rotating magnetic field* within the motor's stator, which then 'dragged' the rotor around with it. Tesla had invented what was to become known as an **induction motor** and, within five years of his inspiration, he was able to build a 400 W 'Tesla polyphase induction motor', selling the manufacturing rights to the American industrialist, George Westinghouse.

In 1885, working quite independently from Tesla, an Italian engineer, **Galileo Ferraris** (1847–1897), had also managed to design an induction motor but, unfortunately for Ferraris, it was Tesla who patented his design first and, therefore, was generally credited with the invention.

Galileo Ferraris
(1847-1897)

Nikola Tesla
(1856-1943)

Figure 15.1

Although the design of the induction motor is, these days, credited to both Tesla and Ferraris, the *concept* of a naturally rotating magnetic field was in fact described over half a century earlier by a French physicist, **François Arago** (1786–1853), which he was able to achieve by using coils controlled by manually operated switches. However, by using alternating current, Tesla and Ferraris were able to eliminate the need for using manually operated switches to create the rotating magnetic field necessary for their machines to work.

Around a third of the world's electrical energy is used to drive motors that are used to run compressors, pumps, fans, and a huge variety of other machinery. And, because alternating current is readily available, **a.c. induction motors** represent around 90% of the various types of motors in general use worldwide, and they vary in output power from a few tens of watts right up to over 30 MW.

The elegantly simple operation of alternating-current induction motors depends upon *three* electromagnetic principles:

▶ **Naturally rotating magnetic field**: By applying phase-displaced voltages to the stator's field windings, a naturally rotating magnetic field is created, radially, across the air gap between the stator and the rotor.
▶ **Generator action**: The rotating radial magnetic field then 'cuts' the rotor conductors, inducing voltages into them which, in turn, cause currents to flow through those conductors.
▶ **Motor action**: The currents flowing within the rotor conductors set up their own magnetic fields around those conductors, which then combine with the rotating magnetic field *to drag the rotor round in the same direction as the rotating field*.

General principle behind operation of induction motors

Before we examine the operation of a 'real' **induction motor**, let's first examine the *basic principle* behind its operation.

Imagine a horseshoe magnet, arranged horizontally, so that it can be rotated around its horizontal axis, using a hand-operated crank, as illustrated in Figure 15.2. Imagine, too, a rectangular closed-loop of wire, located between the poles of the magnet, sharing the same axis, but free to rotate quite independently from the magnet:

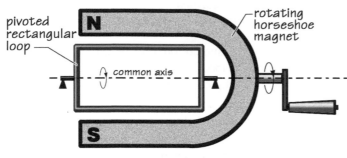

Figure 15.2

Now, let's change the direction in which we view this arrangement. The illustrations in Figure 15.3 represent the view when we look at it 'end on', from the left.

The basic 'dot/cross' convention that we learnt in the companion book, **An Introduction to Electrical Science**, apply here too. That is, current flowing *away* from us is represented by a *cross* (×), while current flowing *towards* us is represented by a *dot* (·). At the same time, the *direction* of a magnetic field is always taken as being *from north to south* (i.e. the direction a compass needle would point, if placed within that field).

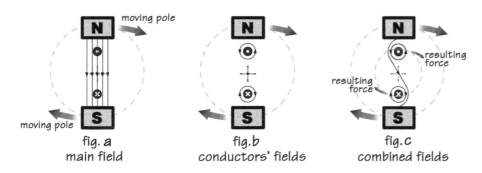

fig. a
main field

fig. b
conductors' fields

fig. c
combined fields

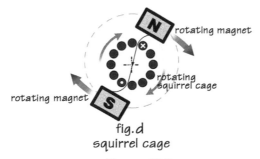

fig. d
squirrel cage

Figure 15.3

In **Figure 15.3a**, we've assumed that the horseshoe magnet is being hand-cranked in a *clockwise* direction, while the loop of wire is stationary. As the magnet rotates, the permanent magnetic field between its poles sweeps past, or 'cuts', the horizontal conductors of the stationary loop, inducing voltages into them.

At the top of the loop, the magnetic field cuts the conductor from left-to-right. If we apply Fleming's Right-Hand Rule* for generator action to this conductor, we see that the resulting induced voltage acts *towards* us – as illustrated by the 'dot'.

At the bottom of the loop, the same magnetic field cuts the conductor from right-to-left. If we apply Fleming's Right-Hand Rule* to this conductor, we see that the resulting induced voltage acts away from us – as illustrated by the 'cross'.

> *Remember, for **Fleming's Right-Hand Rule**, the thumb *always* indicates the direction of motion of *the **conductor** relative to the field* – so, if the main field moves clockwise, this is *equivalent* to the conductor moving counterclockwise.

The voltages induced into the top and bottom conductors of the loop results in a current that circulates around the loop of wire in the same direction as the induced voltages – i.e. *towards* us in the top conductor, and *away* from us in the bottom conductor.

In Figure 15.3b, we see the magnetic field around each conductor due to the currents passing through them. Applying the 'corkscrew rule' (or the 'right-hand grip rule') to each conductor will confirm that the resulting magnetic field set up around the top conductor acts in a *counterclockwise* direction, while the magnetic field around the bottom conductor acts in a *clockwise* direction. *For the purpose of clarity, we have temporarily not shown the main magnetic field due to the rotating magnet.*

Figure 15.3c shows the resulting shape of the *combined magnetic fields*. To keep the figure from getting too complicated, we have represented the main field with a just single line of magnetic flux but, of course, there are *many* lines of flux. This distortion to the main field clearly indicates that the top conductor will be subject to a force acting to the *right*, while the bottom conductor will be subject to a force acting to the *left* – in other words, the loop of wire is being subjected to a torque that is acting to rotate it *clockwise* or, if you like, the coil is being *dragged along in the same direction as the rotating magnet!*

Of course, the torque (turning force) acting on the loop will *only* occur just as the main magnetic field 'cuts' the coil's horizontal conductors (i.e. whenever the horseshoe magnet overtakes the loop), so the torque will *not* be constant; the coil will *only* move just as the horseshoe magnet passes it by.

So if, instead of a *single* loop of wire, we used a rotor comprising *numerous* loops of wire (**Figure 15.3d**), arranged to make it look like an exercise wheel for a pet gerbil or hamster (Figure 15.4) then there would be a more-or-less *continuous* torque acting to rotate the rotor, and it will continue run for as long as we continue to hand-crank the horseshoe magnet.

From the above explanation, it should be apparent that *the speed of the rotor can **never** quite match the speed at which the main field rotates*. If it were to do so, then the rotating main field would no longer 'cut' the rotor's conductors and, so, no voltages would be induced into them, and no currents would flow through them, and the torque on the rotor will completely disappear!

And this is a very important concept to understand about induction motors: *they can never quite run at the same speed as the rotating magnetic field*.

The speed of rotation of the main field is determined by the frequency of the three-phase supply and, so, is said to be rotating at '**synchronous speed**'. So the rotor can *never* run at synchronous speed.

For this reason, three-phase (and, indeed, single-phase) **induction motors** are classified as '**asynchronous machines**' (i.e. 'non-synchronous' machines), meaning that the rotor does not run at the same speed as the field.

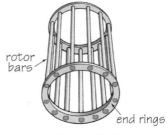

rotor bars

end rings

Figure 15.4

> The rotor cannot rotate at the same speed as the main field. If it were to do so, then the torque acting on the rotor would disappear!

The above explanation, then, describes the principle of operation of an '**induction motor**'. It's called an '*induction* motor' because, as we have seen, the current(s) circulating within the rotor is the result of voltages that are *induced* into its conductors by the rotating main field, and is *not* provided from some external source as it would be in the case of a d.c. motor. In this respect, *there are certain similarities between an induction motor and a transformer:* a theme we shall return to later!

Of course, the above explanation describes nothing more than a simple laboratory demonstration but, nevertheless, it is important because it demonstrates *the principle behind the operation of all induction motors.*

Surprisingly, perhaps, the operation of a three-phase induction motor is very much easier to explain than the operation of a single-phase induction motor, so we will start by examining the behaviour of a three-phase machine, and worry about single-phase motors in a later chapter.

The three-phase induction motor

Creation of a rotating magnetic field: the stator

As we learnt in the previous section, *the key requirement for the operation of an induction motor is the creation of a* **naturally rotating main field** and, in this section, we will learn just how this is achieved in a *real* induction motor, as opposed to the crank-operated laboratory demonstration we described above.

To understand this, we need to understand the construction of the machine's **stator**. However, *before* we describe the machine's **stator**, we need to clarify the terminology used to describe its **magnetic poles**.

A '**pole**' is a longitudinal section of the stator, surrounded by a distributed field winding inserted into longitudinal 'slots', machined into the inner surface of the stator.

The simplest three-phase induction motor is described as being a '**two-pole machine**', which means that it has *two poles* **per phase** or, if you prefer, one **pair** of poles per phase, giving it a total of *six* poles which are evenly distributed around the inner surface of the stator, with each pair located opposite each other (see Figure 15.5). More complicated machines have four, six, eight, ten, or twelve *poles per phase*.

> The **number of poles** specified for induction motor's stator is *alway* quoted on a *per phase* basis. So a 'two-pole' three-phase induction motor has a total of *six* poles.

The **stator** of any electrical machine is, as we know, its *stationary* part and, in the case of the a basic 'two-pole' machine, is manufactured from a stack of circular-shaped silicon-steel laminations, which are clamped together, and whose internal surface is machined to form three *pairs* of opposite-facing **poles**, with each pair displaced from each other by 120°. Three separate distributed **field windings** (one per phase) are then installed such that *half* of each phase's field winding is distributed around opposite-facing poles, and wound in such a direction that, when energised, those poles assume opposite magnetic polarity.

The field windings are then interconnected *either* in **delta** *or* in **star** ('**wye**'), and connected to an external three-phase supply – often, depending on its power output, via a **motor starter** circuit (described later).

> As the field windings represent a balanced load, there is no requirement for a neutral connection, in the case of star-connected field windings.

Figure 15.5 (grossly simplified, for the purpose of clarity) shows the basic construction of a **two-pole stator**, in this case having one *pair* of **opposite-facing poles** per phase (labelled **A–A'**, **B–B'**, and **C–C'**), with each pair being physically displaced from each other by 120°. For each phase, the field winding is equally distributed between its two poles, with each half-winding wound in such a direction that the exposed faces of opposite-facing poles always assume *opposite* magnetic polarity. The distance between adjacent poles of opposite polarity, is termed the '**pole pitch**' of the machine.

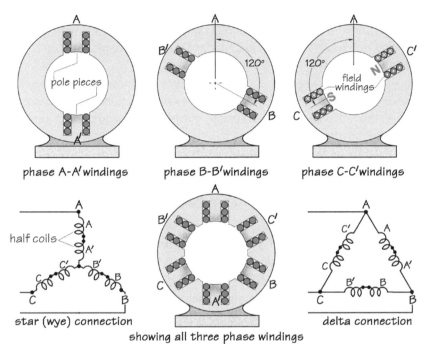

Figure 15.5

A three-phase motor's stator winding and rotor connections may be represented as shown in the simplified schematic diagram in Figure 15.6.

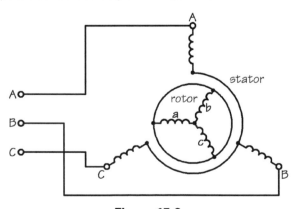

Figure 15.6

In this particular schematic diagram, the **stator windings**, whose line-terminals are marked, **A**, **B**, and **C**, are seen to be connected in *star (wye)*, but they could, instead, be connected in *delta*. However, the way in which they are connected makes absolutely no difference to the principle of operation of the motor.

The *simplified* cutaway diagram, shown in Figure 15.7, illustrates the main design features of a practical 'two-pole' stator. For the purpose of clarity, only *two* half-windings (corresponding to half-windings **A'** and **C** in Figure 15.6 above) are shown occupying the slots machined into the stator but, for this machine, there will, of course, be *six* half-poles, with those connected to the same phases facing each other (i.e. on opposite sides of the stator).

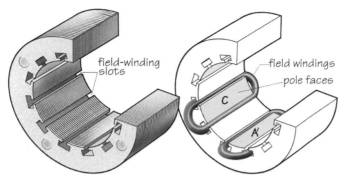

laminated silicon-steel stator

Figure 15.7

Actual stator winding arrangements are more complicated than this, often having a far more complex winding distribution arrangement, but this goes beyond the scope of this book.

When energised, the field windings will set up a *naturally rotating magnetic field*, as explained in Figure 15.8. For clarity, we have not shown the rotor and limited the number of lines of flux to just three.

Figure 15.8 shows the waveforms of the three-phase currents flowing in the machine's stator windings. Let's examine the directions of the resulting magnetic field at points 1–6 along this waveform.

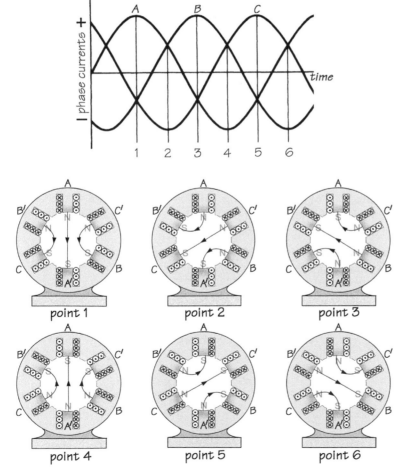

Figure 15.8

15

▶ At **point 1**, we'll assume that the maximum positive current through phase **A** makes the face of **pole A** magnetically *north*, so the face of the opposite **pole A'** must be *south*. At the same time, somewhat smaller currents are flowing in the *opposite* (negative) direction through the phases **B** and **C**. Because they are flowing in the opposite direction, the faces of **poles B** and **C** must both be *south*, so the faces of **poles B'** and **C'** must be *north*. The combined fluxes therefore, act vertically downwards (remember, the direction of flux is from *north to south*).

▶ **At point 2 on the graph**, the current in phase **A** has reduced in value, but is still flowing in the positive direction, while the current through phase **B** has reversed and is now flowing in the positive direction. At the same time, the current in phase **C** continues to flow in the *negative* direction, but has reached its maximum value. So, the faces of poles **A** and **B** will each be *north* together with the face of pole **C'**, while the faces of poles **A'**, **B'** and **C** must, therefore, all be *south*. So the combined fluxes will have now *rotated 60° clockwise*.

You can analyse the remaining positions for yourself. Just remember that whenever the phase current through a particular pole-winding is flowing in the *positive* direction, we are assuming that the face of that pole will be *north* and *vice versa*.

What should now be clear is that *the combined (or resultant) flux is continuously rotating, or whirling around within the stator in a clockwise direction*. In fact, if we were to hold a compass along the central axis of the motor, in place of its rotor, then we would observe its needle rapidly revolving as it tries to keep up with the rotating field.

What should *also* be clear is that, in this particular example, the *combined flux will make one complete revolution during one complete cycle of a.c. current*. So if, for example, this machine is connected to a 50 Hz mains supply, then the flux will rotate at 50 revolutions per second, or 3000 revolutions per minute.

This is called the machine's **synchronous speed** (symbol: N_s).

So, one factor that determines a three-phase machine's synchronous speed is the *frequency of the supply voltage* – so if the above machine was operated in, say, Canada, which has a 60 Hz mains supply, then it will have a synchronous speed of 60 revolutions per second, or 3600 rev/min.

The other factor that affects the machine's synchronous speed is the **number of poles** per phase. The machine illustrated above has just *two* poles (or one-*pair* of poles) per phase, *but most practical machines will have more than this*. So if, for example, a machine has *twice* as many pole-pairs per phase, then its synchronous speed would be *halved*.

This is because, for a two-pole machine, the field will rotate 360° mechanical degrees during one cycle but, for (say) a four-pole machine, the field will rotate only 180° mechanical degrees during one cycle.

The **synchronous speed** (symbol: N_s) of a three-phase motor is *the speed at which the field flux rotates*, and is given by the equation:

$$N_s = \frac{120f}{p}$$

where: N_s = synchronous speed (rev/min)

f = supply-frequency (hertz)

p = number of poles *per phase*

So, for a machine with *two* poles per phase, running from a 50 Hz supply, the synchronous speed will be:

$$N_s = \frac{120f}{p} = \frac{120 \times 50}{2} = 3000 \text{ rev/min}$$

So, for machines operating at frequencies of 50 Hz (e.g. Europe) and 60 Hz (e.g. North America), the following speeds result:

Table 15.1

Poles per phase	Sync speed at 50Hz/(rev/min)	Sync speed at 60 Hz/(rev/min)
2	3000	3600
4	1500	1800
6	1000	1200
8	750	900
10	600	720
12	500	600
16	375	450
20	300	360

If we want to achieve *different* synchronous speeds from those listed above, then we must be able to *vary the supply frequency*, which is only possible with **inverter-fed** motors (discussed briefly later).

The **standard direction** of an a.c. motor (three-phase or single-phase) is *counterclockwise*, when viewed from the *'front'* end – i.e. the end *opposite* the motor's driveshaft, as shown in Figure 15.9.

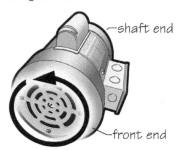

standard direction of rotation

Figure 15.9

The *direction* of rotation (see Figure 15.9) is determined by the **phase sequence** of the supply: **A–B–C**. By 'phase sequence', we mean the *order* in which the potential of each one reaches its peak value. For the sequence, **A–B–C**, the standard direction of rotation of the load, as viewed from the motor's front end, is *counterclockwise*. The direction of the rotating field and, therefore, the direction of rotation of the load, can easily be reversed by changing the phase sequence to **A–C–B** – which is achieved simply by *interchanging any pair of the three line conductors supplying the machine*.

Great care, therefore, needs to be taken whenever a three-phase induction motor is disconnected for maintenance, as it is *very* important that the supply leads are then reconnected in the correct phase sequence. Failure to do so will result in the load being driven in the wrong direction, and possibly causing damage to that load. Labelling the leads and motor terminals *prior* to disconnection is an obvious precaution, although the use of a **phase sequence indicator** (meter) might become necessary following any maintenance work downstream from the motor, for example at the service entry point to the building.

The rotor

Now that we've examined the stator and seen how a rotating magnetic field is created by the field windings, let's move on to find out how that rotating field causes the machine's **rotor** to turn.

First of all, in order to minimise the reluctance of the motor's magnetic circuit, it's necessary to minimise the axial length of the air gap (typically between 0.2 and

3.0 mm, depending on the power and physical size of the machine) by using a silicon-steel **rotor** ('silicon-steel' is a widely used misnomer for what is *actually* 'silicon-**iron**'). This is manufactured from a stack of stamped circular laminations, with a light layer of insulation between each lamination, in order to minimise eddy-current losses – as illustrated in Figure 15.10.

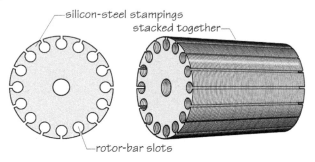

Figure 15.10

Circular slots (in this case, but other shapes are also used), stamped into the circumference of each lamination accommodate the rotor conductors while minimising the radial length of the air gap (i.e. the distance between the adjacent surfaces of the stator and rotor).

By 'burying' the rotor conductors inside the silicon-steel rotor, we are able to minimise the air gap, but we might be forgiven for assuming that the magnetic flux would then 'bypass' those conductors by taking the path of least magnetic reluctance around them. This is indeed the case but, for some reason that isn't apparent, this doesn't actually prevent voltages from being induced into the conductors. The actual reason, of course, is that the behaviour of lines of magnetic flux is simply a 'model' which, like most models, simply fails to explain all the phenomena associated with magnetic fields.

The rotor is manufactured in such a way that the number of longitudinal slots in the rotor is never quite the same as the number of slots in the stator to avoid points of zero torque, known as 'dead spots'. The same effect can be obtained by 'skewing' the rotor bars – as illustrated in Figure 15.13.

As can be seen in Figure 15.11, the design of the rotor conductors, or '**bars**', is known as a '**squirrel cage**'. This is a bit of a misnomer, as these types of rotor resemble 'exercise wheels', rather than 'cages', and the term dates back to when this type of motor was first developed, and *squirrels* were kept as family pets.

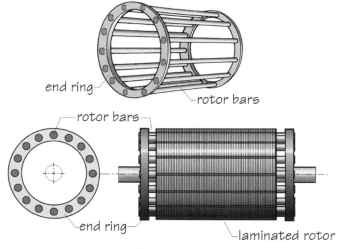

Figure 15.11

Rotor bars, roughly one pole pitch apart, form a 'turn', allowing the rotor currents to circulate along the bars, and around the end rings, as illustrated in Figure 15.12.

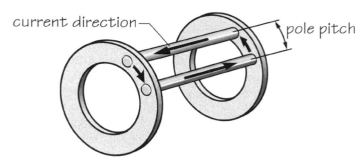

Figure 15.12

In many designs, the rotor bars are slightly 'skewed' (i.e. run at a slight angle along the length of the rotor), as illustrated in Figure 15.13, in order to smooth out any torque variations, to improve starting, and to reduce the 'growling' noise that sometimes occur with machines having horizontal bars.

Notice that *there are no external connections* to the squirrel cage because, as we shall see, the currents that circulate along the rotor bars (and around the end rings) are the result of voltages *induced* into them as they are 'cut' by the stator's rototating magnetic field (hence the name, 'induction' motor). This means that the motor is simple to manufacture and *practically maintenance-free* with, for example, no commutator brushes to wear out or replace – a *major* advantage compared with machines with commutators or slip rings.

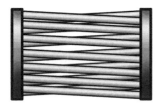

Figure 15.13

Although the rotor's copper (or aluminium) bars are uninsulated and laid (or, in some cases, poured as a *molten* metal) into the slots cut into the steel armature, practically *all* the rotor current flows through those bars because the silicon-steel rotor is a relatively poor conductor and, more importantly, is *laminated*, with each lamination lightly insulated from its neighbour, which prevents it from conducting current in the same direction as the bars. Effectively, therefore, *all* of the rotor current is confined within the rotor's bars and end rings.

The cross-section of the rotor bars may be circular, rectangular, or various other shapes, and may be either confined towards the surface or located deeper within the rotor because, as we shall learn, their design and location within the rotor affects the torque characteristics of the machine.

Another type of rotor is the '**wound rotor**' type, in which the rotor windings are connected, externally, via slip rings, to variable resistors, which enable the torque/speed characteristics to be adjusted. These days, this type of rotor is confined to very large induction motors, but we shall briefly discuss this type of machine later.

Now that we have learnt about the construction of the rotor, let's move on to see how the machine's **torque** is produced.

Creation of a torque on the rotor

In Figures 15.14 to 15.16, for the purpose of clarity, only *two* rotor bars (located on opposite sides of the rotor) are highlighted and, for the purpose of clarity, just three lines of magnetic flux are used to represent the rotating field set up by the stator's field windings.

When the field windings are first energised, the rotor is stationary, and its bars are 'cut' by the rotating magnetic field – as illustrated in Figure 15.14.

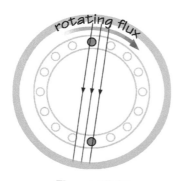

Figure 15.14

15

As the rotating magnetic field 'cuts' the rotor bars, it induces voltages into them which, in turn, cause circulating currents to flow along the bars and through the end-rings.

If we apply Fleming's Right-Hand Rule (for generator action) to each of the two highlighted rotor bars, we see that the induced voltages will cause (conventional) currents to flow along those bars in the directions shown in Figure 15.15, using the 'cross' and 'dot' notation – i.e. *towards* us in the top bar, and *away* from us in the bottom bar, circulating around the end-rings.

> **Important**! Remember, when applying Fleming's Right-Hand Rule, the thumb always indicates the <u>m</u>otion of the *conductor* relative to the field – i.e. as though the conductors were moving in the opposite direction to the field.

Figure 15.15

If we now temporarily ignore the rotating magnetic field, and concentrate, instead, on the current flowing along the rotor bars (Figure 15.15), we see that it will set up its own magnetic field around each of the bars. The directions of these fields can be determined by applying the 'corkscrew rule' (or the 'right-hand grip rule'): in this case, *counterclockwise* in the top bar and *clockwise* in the bottom bar.

The three magnetic fields – the *rotating field* and the *fields around the two individual rotor bars* – now *combine*, as shown in Figure 15.16, resulting in a force that acts on each of the rotor bars. These forces act *in the same direction as the rotating field* (in this case, clockwise) and create the torque that drives the rotor.

Figure 15.16

Another way of explaining this action is to say that the rotating flux *reinforces* the rotor flux at the *trailing* edge of each rotor bar, while *weakening* the rotor flux at the *leading* edge – resulting in a torque that pushes the bars in (this case) the clockwise direction.

Of course, the *actual* rotating field is distributed around the entire circumference of the air gap, and creates a torque on *all* of the rotor bars. But the resulting combined magnetic field far is too complicated to be represented in our diagram!

But what is, perhaps, not quite so obvious is that the **flux** of the rotating field is *not* distributed uniformly throughout the circumference of the air gap, causing its **flux density** to *vary sinusoidally* in magnitude and axial direction *as it travels, like a wave, around the air gap*.

This 'travelling flux wave' is difficult envisage or to illustrate for a circular air gap, but if we were to imagine that the air gap was *horizontal*, or flattened out, rather than circular, then the travelling flux wave could be represented as illustrated in Figure 15.17.

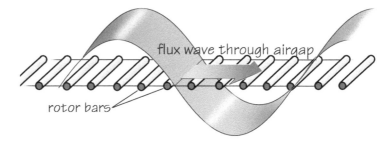

flux wave through airgap

rotor bars

Figure 15.17

This travelling 'flux wave' can also be represented as a sinusoidal variation in the rotating field's flux (and, therefore, its flux density), as shown in terms of the varying *closeness* of the lines of flux within the air gap, as shown in Figure 15.18.

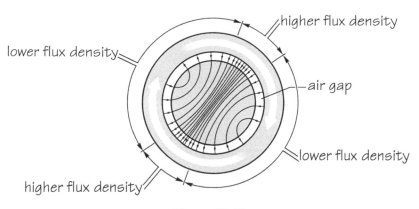

Figure 15.18

To summarise, the operation of the three-phase induction motor is brought about by:

▶ the stator's **rotating magnetic field** 'cutting' the rotor bars, which
▶ induce voltages into those bars, causing currents to flow through them (through '**generator action**'), which then
▶ set up their own magnetic fields that react (through '**motor action**') with the rotating field, dragging the rotor in the same direction as the rotating field.

One of the major advantages, then, of a three-phase induction motor, is that it is **self-starting**.

> **Important**! The above description explains the operation of a three-phase motor in terms of the behaviour and interaction of magnetic lines of force. But, once again, it is important to remember that the concept of '*magnetic lines of force*' is just a 'model' (a representation), proposed by Faraday, and intended to try and help us understand how the machine behaves. It should never be interpreted as what is *actually* happening inside the machine!

A more accurate (but, perhaps, less obvious) explanation would be that an induction motor behaves just like a three-phase **transformer**, with the *stator windings* being equivalent to the transformer's *primary windings*, and the *rotor bars* being equivalent to its *secondary windings*. The rotor currents are, therefore, the result of voltages induced into the rotor due to 'transformer action' (mutual induction). However, because the rotor is free to move, it rotates as the resulting secondary magnetic field tries to align itself with the rotating primary field.

Slip

It should be obvious, now, that *the rotor can never reach the same speed as the rotating field* (i.e. its '*synchronous speed*'). If it were to do so, then the rotating field would no longer 'cut' the rotor bars, no voltages would be induced into them, no rotor current would flow, and there would be no rotor field to react with the rotating field to produce the torque necessary to drive the rotor!

So an induction motor's rotor speed must *always be less than its synchronous speed*.

For this reason, induction motors are often termed '**asynchronous machines**', to distinguish them from '**synchronous machines**' – i.e. those types of a.c. motors whose rotors *do* rotate at synchronous speed.

The **difference** between a machine's synchronous speed and its rotor speed is usually expressed as a percentage of the synchronous speed, and is termed its **slip** (symbol: *s*). where:

15

$$s = \frac{(N_s - N)}{N_s}$$

where: s = slip
N_s = synchronous speed
N = rotor speed

Slip can be expressed either as a *per unit* value or as a *percentage* value.

So, when the motor is stationary ($N = 0$), the slip will be unity (1) or 100%. If it were possible for the rotor to turn at synchronous speed (i.e. the speed of the rotating field, $N = N_s$), then the slip would be 0 or 0%:

▶ **slip 0 (or 0%):** rotor at synchronous speed (impossible!)
▶ **slip 1 (or 100%):** stationary rotor

When the field windings are first energised, the rotor is stationary so its slip is *maximum* (unity or 100%) and, as the rotor's speed increases, its slip falls towards, *but never quite reaches*, zero.

Worked example 1

A four-pole three-phase induction motor is connected to a 50 Hz supply. Calculate (a) its synchronous speed, and (b) the speed of the rotor when the slip is 5%.

Solution

a. $N_s = \dfrac{120f}{p} = \dfrac{120 \times 50}{4} = 1500$ rev/min

b. $s = \dfrac{(N_s - N)}{N_s} \times 100\%$

$5 = \dfrac{(1500 - N)}{1500} \times 100$

$(1500 - N) = \dfrac{5 \times 1500}{100} = 75$

$-N = 75 - 1500 = -1425$

$N = 1424$ rev/min (**Answer b.**)

Performance of a three-phase induction motor

So, we can think of a three-phase induction motor as being *a unique type of* **transformer**, having a rotating, rather than fixed, secondary winding. We know that, for an ordinary transformer, the primary and secondary frequencies are *always* the same value but, for an induction motor, the 'secondary frequency' is *the frequency at which the rotating field cuts the rotor bars (the 'secondary'), which will depend upon the amount of slip*.

For example, at standstill (a slip of unity or 100%), the **rotor frequency** (i.e. the frequency of the voltages induced into the rotor bars) will be exactly the *same* as the supply frequency because it is behaving just like an ordinary transformer but, as the rotor speeds up and approaches synchronous speed (a slip approaching 0%), the frequency must be *reducing* because the rotating flux is cutting the rotor bars at a much lower rate. If it *were* possible for the rotor to run at synchronous speed, then the rotor frequency would, of course, be zero!

We can express this relationship between rotor frequency and slip as:

rotor frequency = slip × supply frequency

$$f_r = sf$$

Worked example 2

Calculate the frequency of the voltages induced into the rotor of a three-phase induction motor connected to (a) 50 Hz supply and (b) a 60 Hz supply, when the machine is operating with slips of 40% and 90%.

Solution

a. $f_r = sf = 0.4 \times 50 = 20$ Hz

$f_r = sf = 0.9 \times 50 = 45$ Hz (**Answer a.**)

b. $f_r = sf = 0.4 \times 60 = 24$ Hz

$f_r = sf = 0.9 \times 60 = 54$ Hz (**Answer b.**)

Torque development

We will recall, from ***An Introduction to Electrical Science***, that the **force** (F) acting on a current-carrying conductor located within a magnetic field is proportional to the product of the **flux density** (B) of the field, the magnitude of the **current** (I) flowing through that conductor, and the length of the conductor (l) within the field:

$$F = BIl$$

As the length of the rotor conductor and its radius of rotation are constants (set by the design of the motor), it should be clear that the torque acting on any rotor conductor must be proportional to the flux density of the field in which it is located as well as the current (I_r) flowing in that conductor:

$$T \propto BI_r \quad ---\text{(equation x)}$$

This expression is equally true whether the flux density and rotor current are *constant* or *sinusoidal* in nature.

As we have learnt, the **frequency** of the rotor current is maximum when the rotor is stationary (unity or 100% slip). But as the *rotor starts to turn* and the *slip starts to fall*, so too does the frequency of the voltages induced into the rotor bars.

While the **resistance** of each rotor conductor is, essentially, independent of frequency (if we ignore the 'skin effect') and, therefore, *remains more-or-less constant*, the conductor's **inductive reactance** (X_L) is proportional to the rotor frequency, as confirmed by the equation we know from our a.c. theory:

$$\text{i.e. } X_L = 2\pi f L$$

From which we can deduce that:

▶ at *lower* frequencies (higher rotor speeds), the main opposition to the rotor-current flow will be due to the *resistance* of the rotor bars but, as the frequency *increases*, so too does the **inductive reactance** of the bars.

▶ so, at *lower* frequencies the rotor currents are practically *in phase* with the induced voltages but, as the frequency *increases* (lower rotor speeds), the currents will increasingly *lag* behind those induced voltages.

These two situations are illustrated in Figure 15.19.

This situation might be clearer if, instead, we show the relationship between the maximum flux density and the maximum rotor current in the following way (Figure 15.20). We can compare this with the effect of **armature reaction**, which we described in the chapter on d.c. motors.

Note that the varying sizes of the 'cross' and 'dot' notation in Figure 15.20 represent the varying magnitudes of current flowing in the rotor conductors due to the sinusoidal nature of the voltages induced into those conductors.

15

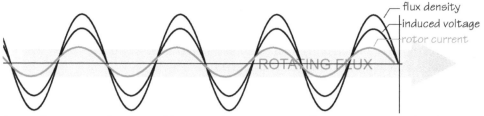

lower frequency (approaching synchronous speed) -current near in phase

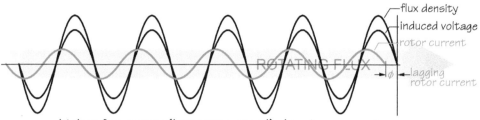

higher frequency (low rotor speed) -lagging current

Figure 15.19

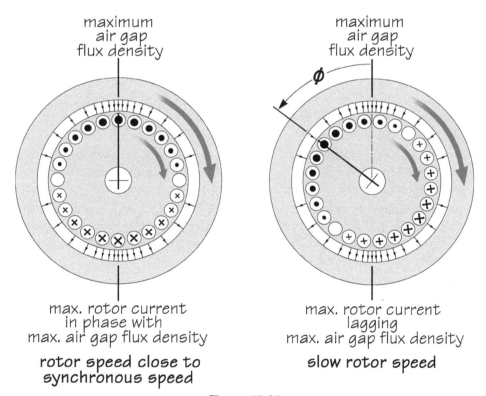

rotor speed close to
synchronous speed

slow rotor speed

Figure 15.20

From the equation ($F = BIl$), it should be obvious that the *maximum* force and, therefore, the **maximum torque** on the rotor, will occur *when the maximum flux density and the maximum rotor current coincide*. That is, *when they are both in phase with each other*.

But, at *slower* rotor speeds, the rotor currents *increasingly lag* the flux density, so not only do their maximum values *not* coincide with each other but, at some points around the circumference of the rotor, we even have a situation in which they even act in *opposite* directions during the waveform!

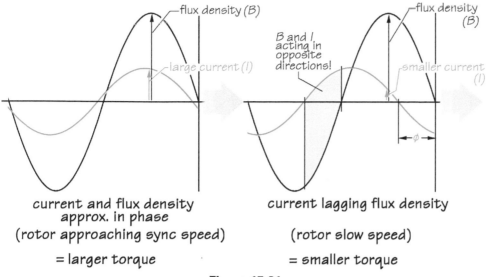

Figure 15.21

Despite this, however, the torque itself will continue to act in the *same direction*, but at a somewhat lower value than it would be if the rotor current and flux density were to be in phase with each other.

As we learnt when we were studying a.c. circuits, we express the phase relationship between sinusoidal quantities in terms of its **power factor**, which is numerically equal to the *cosine* of the phase angle (the angle by which the current, in this case, *lags* the induced voltages). When the current is in phase with the induced voltage, the phase angle is zero, so its power factor is 1, and if the phase angle lags by 90°, then the power factor is 0.

So, we can modify the basic expression for the force on a conductor, to reflect this, as follows:

$$F \propto BI_r \cos\phi \quad \text{therefore} \quad T \propto BI_r \cos\phi$$

Let's now see how the torque acting on the rotor is affected by variations in **rotor current** and **power factor**.

Effect of variations in rotor current on torque

The **current** in any particular rotor bar depends, of course, on the value of voltage induced into that bar which, in turn, depends upon the rate at which that bar is being cut by the rotating magnetic field. The maximum rate occurs when the rotor is stationary (100% slip), starts to fall off as the rotor picks up speed, and would reach zero if the rotor could achieve synchronous speed (0% slip).

So *maximum* rotor current occurs when the rotor is *stationary*. As the rotor's speed *increases*, the rotor current starts to fall. If we were to plot a graph of *current against speed*, then the resulting curve would look something like that shown in Figure 15.22.

As torque is proportional to current, we can conclude that, *if we temporarily ignore the effect of variations in power factor*, maximum torque occurs at start-up (unity or 100% slip), and falls towards zero as the rotor approaches synchronous speed (0% slip).

Effect of variations in power factor on torque

At standstill, with *maximum* phase displacement between the rotor current and the induced voltage, the rotor power factor is *minimum*. As the rotor starts to pick up speed, the rotor current starts to move into phase with the induced voltage, so the power factor starts to increase and will continue to do so as it approaches

15

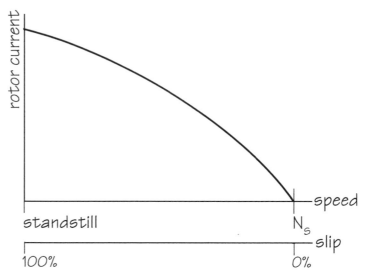

Figure 15.22

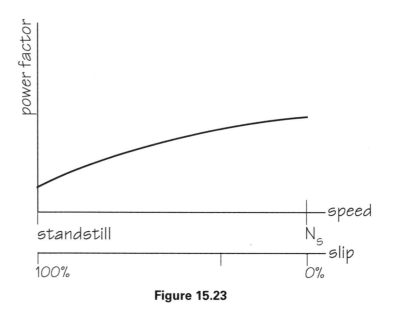

Figure 15.23

synchronous speed. So, if we were to plot a graph of *power factor against speed*, then the resulting curve would something like that shown in Figure 15.23.

As torque is proportional to power factor, we can conclude that, *if we were to temporarily ignore the effect of variations in rotor current*, then the minimum torque would occur at start-up (100% slip), and increases as the rotor approaches synchronous speed (0% slip).

However, since **torque** is proportional to *both* rotor current *and* power factor, if we were to combine the effects due to variations in both rotor current *and* power factor, against speed/slip, then the resulting curve of **torque** versus **speed** curve would look *something* like that shown by the broken line in Figure 15.24.

However, the *actual* shape of the **torque curve** depends on the design of the motor, with the value of slip at which maximum torque occurs being dependent upon the ratio of *rotor resistance* to *rotor reactance*.

The effect of different ratios of resistance to reactance is illustrated in Figure 15.25.

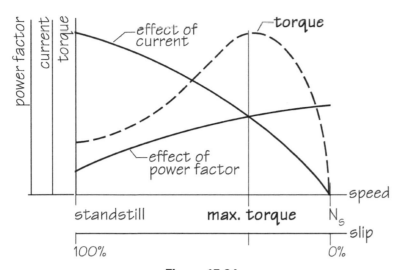

Figure 15.24

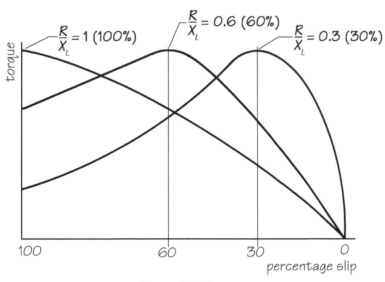

Figure 15.25

As we can see from Figure 15.25, the *maximum starting torque* occurs when the ratio of R to X_L is 1; in other words when the rotor's resistance equals its inductive reactance. This occurs when the maximum rotor current lags the maximum stator flux density by 45°.

These ratios are achieved in various ways, including through the use of 'double cage' rotors, in which there are *two layers* of rotor bars: a higher resistance outer layer, close to the surface of the rotor and a lower resistance inner layer, much deeper in the rotor. Alternative methods include shaping the cross-sections of the bars to exploit variations in their resistance at different frequencies, due to the resulting variations in the 'skin effect'. Typical examples of these various rotor-bar shapes and depths are illustrated in Figure 15.26, and relate to the torque/speed characteristics described later.

Figure 15.26

Further development of this particular topic is well beyond the scope of this chapter.

> The '**skin effect**' describes how alternating currents tend to travel towards the surface of a conductor, thus reducing the effective cross-sectional area, which raises the resistance of a conductor. This effect increases with frequency.

Regardless of the *shape* of the torque curve, the **operating region** of the torque curve (i.e. where the motor is driving its rated load), is always towards the *synchronous-speed end* of the speed/slip axis, with the machine's full-load speed occurring at around 5% slip or less, as shown by the illustrations in Figure 15.27.

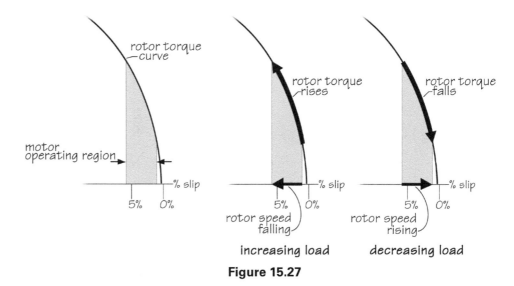

Figure 15.27

As illustrated in Figure 15.27, if the mechanical load should *increase*, then the rotor will *slow down*, causing its torque to *increase* to match the new load; if the mechanical load should *decrease*, then the rotor will *speed up*, causing its torque to *decrease* to match the new load. In this way, equilibrium is achieved with the rotor's torque constantly adjusting itself to match any changes in the load torque.

> In fact, at full load, there is very little variation in speed and, for this reason, induction motors are considered to be **constant-speed motors**. The term, 'constant speed' isn't intended to be taken literally; what it actually means is that any speed variation is *insignificant*.

For 50 Hz induction motors, the most common operating speeds are:

▶ 1425 rev/min (four-pole machine) – most common type
▶ 2850 rev/min (two-pole machine)
▶ 950 rev/min (for six-pole machine)
▶ 712 rev/min (for eight-pole machine) – for slow-speed applications

Induction motors are **classified** according to the *shape of their torque curves*. For example, the following illustration shows the shape of the torque curve for what is known as an **IEC Class N** motor (more on this, later), which represents the most commonly used design class of three-phase motor. Let's examine this curve from left to right, as illustrated in Figure 15.28.

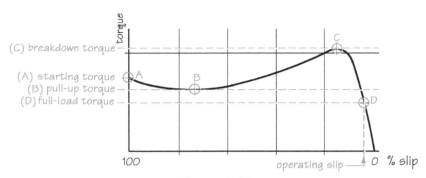

Figure 15.28

▶ **'Starting torque'**, also known as the **'locked-rotor torque'**, is the torque that the motor will develop as it *starts from rest* when its rated voltage and frequency are applied. *High* starting torques are important for *heavy* loads that are difficult to start, whereas lower starting torques are acceptable for light loads that are easy to start.

▶ **'Pull-up torque'**. With some designs, just as the motor starts to accelerate, its torque tends to fall somewhat before starting to increase again. The point of lowest torque in this region of the curve is termed the 'pull-up torque'.

▶ **'Breakdown torque'**. This is the highest torque achieved by the motor as it accelerates towards its operating speed. Continuous operation near breakdown torque is undesirable, as it will cause the machine to overheat. Furthermore, if an increasing load should cause the motor's speed to drop *below* this point (i.e. towards the left), then the rotor torque will rapidly reduce in value towards point **A**, which represents the **'starting torque'** (the minimum torque necessary to get the machine turning), causing it to stall.

▶ **'Full-load torque'**. This is the torque required for the machine to produce its rated power at full speed; this is also called the machine's **'rated torque'**. This is the torque that will match that of the load and, as already explained, occurs at, or below, 5% slip.

As already mentioned, the **shape** of the torque curve will actually vary according to the design of the induction motor and, in fact, induction motors are classified *according to the shapes of their torque curves*. This enables the end-user to be able to choose a motor that best matches the torque characteristics of the load it is expected to drive.

There are *two* main standards organisations that classify induction motors: **IEC** and **NEMA**. The **IEC (International Electrotechnical Commission)** is a European organisation that also sets standards internationally, while **NEMA (National Electrical Manfucaturers Association)** is a US organisation that primarily sets standards in North America. And, of course, there is a great deal of co-operation between the two organisations.

Figure 15.29 contrasts the torque curves for **NEMA Classes A–D**:

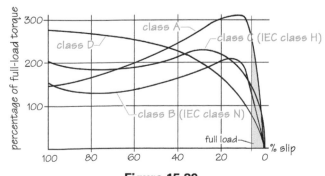

Figure 15.29

15

With the exception of **NEMA Class D** design, all of these motor types operate at around 5% slip or less (a Class D machine operates at a somewhat higher percentage slip).

Due to the steepness of those parts of the curves within the operating region, *all* three-phase induction motors are considered to be 'constant-speed' motors but, as the above curves also reveal, some speeds are 'more constant' than others! For example, the **NEMA Class C (IEC Class H)** motor has the steepest curve in the operating area, which means that it can react to changes in load with very small changes in speed, whereas a **NEMA Class D** motor has the shallowest curve in this region, meaning that it speed varies the most when it reacts to changes in load.

So it would make sense to choose a **NEMA Class C (IEC Class H)** motor if maintaining a constant speed for variations in load is important as it would be, for example, to drive a compressor.

At the opposite end of the torque curves, we see some significant variations in amount of starting torque provided by these motors, with the **NEMA Class D** providing the greatest, and making it the most suitable choice for driving high-inertia loads such as reciprocating pumps (i.e. a type of pump that has to raise liquids vertically by working *against* gravity when it lifts, yet which works *with* gravity when it falls).

The most widely used three-phase induction motor is the **IEC Class N (NEMA Class B)** design, although its maximum or breakdown torque is one of the lowest, which means that it cannot cope with heavy overloads. Its applications include driving fans, blowers, pumps, and machine tools.

Wound-rotor three-phase induction motor

Before the development of **variable frequency supplies**, speed control of three-phase induction motors was *only* achievable using relatively expensive '**wound rotor**' or '**slip-ring**' **induction motors**, in which the resistance of the rotor windings was controlled by external variable resistors, as shown in Figure 15.30.

These days, with the exception of some very large induction motors, wound-rotor motors are no longer being manufactured in any significant numbers but, of course, many such motors remain in service and, so, in this section, we will examine them only very briefly.

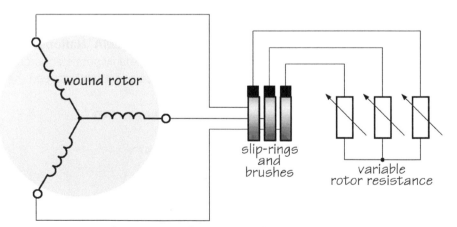

wound rotor with external variable resistors

Figure 15.30

In the previous section, we learnt that the *ratio of resistance to inductive reactance* of the squirrel-cage rotor bars affects the shape of the *torque/speed* characteristic curves. But, as the rotor itself is quite inaccessible, there is absolutely no way in which this ratio can be controlled, externally, in order to enable us to modify the behaviour of the motor.

This drawback can be overcome by using a '**wound-rotor**' or '**slip-ring**' **induction motor** of the type illustrated above.

With this type of machine, the rotor's slots accommodate a set of *three-phase distributed windings*, similar to the distributed stator windings. These rotor windings are connected in star (wye) with their three outer ends connected to a set of **slip rings** mounted on the shaft of the motor. Carbon brushes then connect the windings to a set of three external **variable resistors**.

These external variable resistors enable the effective resistance of the three rotor windings to be varied in order to:

▶ increase the starting torque, while reducing the starting current drawn from the supply, and to

▶ provide some degree of speed control.

If you refer back to Figure 15.29, you can see how the various **torque/speed characteristics** result in different slopes at the operating region of the motor which, in turn, results in variations in speed for a given torque.

Compared with squirrel-cage induction motors, wound-rotor induction motors:

▶ *provide excellent starting torque* for high-inertia loads,

▶ offer *variable speeds*, between 50–100% of full speed, and

▶ have *higher maintenance requirements*, due to brushes/slip ring wear/arcing.

Three-phase synchronous motors

We'll finish this chapter on three-phase motors by briefly examining the three-phase **synchronous motor**.

A synchronous motor's three-phase stator windings set up a magnetic field that naturally rotates at synchronous speed, just like a three-phase induction motor.

But, as the name suggests, this is a machine whose rotor, *unlike* that of an induction motor, runs at **synchronous speed** – i.e. at the same speed as the rotating field.

Synchronous motors vary in size from a just a few kilowatts up to several megawatts. These larger 'motors' are actually *alternators* run as motors.

'**Pumped-storage**' **hydroelectric power stations** (Figure 15.31) provide an example of where water-turbine driven alternators are run as motors.

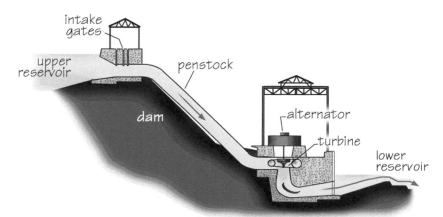

Pumped-Storage Hydroelectric Power Station

Figure 15.31

15

In a conventional hydroelectric power station, the water from a reservoir flows through the turbines, and is discharged downstream. A *pumped-storage system*, however, has *two* reservoirs:

▶ **Upper reservoir**. Like a conventional hydroelectric power station, a dam is used to create a reservoir. The water from this reservoir flows through water turbines and is discharged and stored in a lower reservoir.

▶ **Lower reservoir**. Using a reversible turbine, the power station *can pump water back to the upper reservoir* during off-peak periods – essentially, refilling the upper reservoir for re-use during periods of peak consumption.

The construction of a synchronous motor is essentially the same as that of a salient- or cylindrical-pole synchronous generator, which was described in the previous chapter. So a synchronous motor has a *wound rotor* fed, via slip rings, from a d.c. supply, and the magnetised rotor 'locks' onto the stator's rotating field and rotates at exactly the same speed.

However, the rotor *cannot lock onto the rotating field from standstill* as the torque developed simply isn't strong enough to overcome the **inertia** of the rotor. In this context, 'inertia' is defined as *the property of a body to persist in a state of rest*. We see the effect of inertia whenever we push-start a car whose battery has died: it's difficult to get the vehicle to move at first but, once it starts to move, it becomes easier to push.

A synchronous motor, therefore, must be provided with some means of driving its rotor up to speed, thus enabling it to lock onto the rotating field.

There are *two* ways of doing this.

The *first* way is to couple the rotor to an external induction motor, via its shaft. An induction motor used in this way is called a '**pony motor**'. Once the synchronous motor reaches synchronous speed, the pony motor is disengaged.

The *second* way is to incorporate a **squirrel-cage rotor** into the wound rotor itself. This enables the motor to behave like a three-phase induction motor at start-up but, as it approaches synchronous speed, the rotor's d.c.-fed windings take over and drive the rotor at synchronous speed. This technique is termed, '**asynchronous starting**'. At synchronous speed, of course, the squirrel-cage is running at *zero slip*, so an e.m.f. is no longer being induced into the bars and, no currents will circulate through them. So the torque provided by the squirrel-cage to start the machine disappears, and is replaced by that provided by the rotating field.

As already explained, in pumped-storage hydroelectric power stations, *alternators are used as motors* in order to pump water from a lower reservoir to an upper reservoir during off-peak periods. The major difficulty is *starting* the alternator so that it can be used as a motor, and this is achieved in one of *three* ways. These include using a 'pony motor' or by 'asynchronous starting', as already explained. The third method involves connecting two alternators back-to-back, applying excitation to both while at standstill, and then applying torque to one of them (the 'starting' alternator). The two machines then pull into synchronism at low speed and accelerate towards synchronous speed.

There is a limit to how much torque can be developed by a synchronous motor. Typically, if the torque is allowed to exceed around 150% of its continuous rated torque, then the rotor will be forced *out of synchronism*. We call this excessive torque the '**pull-out torque**' but, for all values of torque below this value, the rotor will run at synchronous speed.

Adjusting the rotor excitation when a synchronous motor is running under load will result in changes to the machine's operating power factor and, by using a *high* rotor excitation current, the motor can be made to operate with a *leading power factor*.

This is a particularly useful feature, which allows large synchronous motors to be used for **power-factor improvement**. Used for this purpose, these machines are known as '**synchronous capacitors**' (an older term was '**synchronous condenser**'). Synchronous motors used for power-factor improvement compensate for the lagging power factor resulting from induction motors operating at the same site thereby reducing the value of load current drawn from the supply system.

Let's finish by listing the main disadvantages of synchronous motors, which are:

▶ they are more expensive than induction motors (on a cost per kilowatt basis).
▶ a d.c. supply is required (usually provided by a small d.c. generator driven by the same shaft.
▶ they require some means of starting.

Conclusion

Now that we've completed this chapter, we need to examine its **objectives** listed at its start. Placing a question mark at the end of each objective turns that objective into a **test item**. If we can answer those test items, then we've met the objectives of this chapter.

15

CHAPTER 16

Single-phase motors

Objectives

On completion of this chapter you should be able to:

1. explain the difference between a 'pulsating' magnetic field, and a 'rotating' magnetic field.

2. in simple terms, describe how, once started, a single-phase motor will continue to operate within a pulsating magnetic field.

3. explain how a rotating magnetic field is generated by the stator windings of a 'capacitor-start', and 'split-phase', induction motors.

4. explain the principle of operation of a 'shaded pole' induction motor.

5. describe the operation of a 'universal motor'.

Introduction

As explained in the previous chapter, the principle of operation of a *three-phase* induction motor is somewhat easier to explain, and more intuitive to understand, than single-phase induction motors, but now that we have seen how a three-phase induction motor works, we can turn our attention to the operation of **single-phase induction motors**. In addition to this, we will also be briefly examining the **universal motor**.

Single-phase induction motors are less efficient than three-phase induction motors, and have been developed primarily for domestic and light-industrial applications where three-phase supplies are generally not available.

Although outputs of several kilowatts are possible, most single-phase induction motors are usually less than 1 kW. In North America such motors are widely known as '**fractional horsepower**' motors, because they are typically rated below one horsepower (where 1 horsepower = 0.746 kW).

Just as with a three-phase induction motor, a single-phase induction motor requires a **rotating magnetic field**, both to *start it* and to have it *run in the desired direction*.

Unfortunately, with a single-phase supply, it is *not* possible to obtain a *naturally* rotating magnetic field as easily as with a three-phase supply.

But, as we shall learn, once the rotor has *started* to run, a continuous torque *can* be achieved by using a *pulsating* (rather than rotating) magnetic field which *is* naturally generated by a single-phase a.c. supply. So, just as we did for three-phase induction motors, before examining 'real' single-phase induction motors, let's start by examining their basic *principle of operation*.

Electrical Science for Technicians. 978-1-138-84926-6 © Adrian Waygood.
Published by Taylor & Francis. All rights reserved.

Principle of operation of a single-phase induction motor

We'll do this by using a pivoted permanent magnet to demonstrate the behaviour of the magnetic field between opposite electromagnetic poles. This is *not* intended to show the operation of a *real* rotor, but merely *to demonstrate the behaviour of the magnetic field created by a single-phase induction motor's stator windings*.

So, in the following diagrams, a **permanent magnet** is pivoted at its centre of gravity, so that it is free to rotate between the faces of two **coils** that are connected to an alternating-current supply. The coils are wound in the same direction relative to each other, so that the poles facing each other are of *opposite* magnetic polarity.

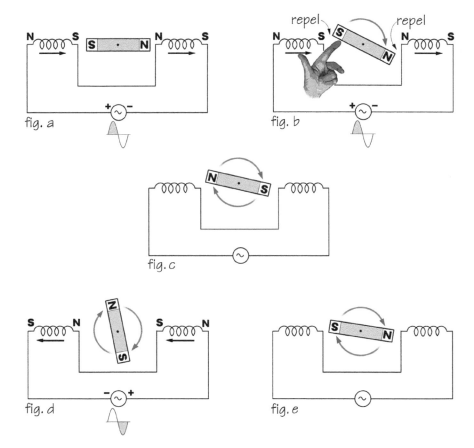

Figure 16.1

We'll assume that, during the first half-cycle of supply voltage, with the (conventional) current flowing clockwise around the circuit, the two coils assume the polarities shown in Figures 16.1a and b.

So, with the magnet initially lying along the horizontal axis relative to the two coils, as shown in Figure 16.1a, the repulsion between the inner faces of the two coils and the poles of the magnet acts *along the axial length* of the magnet, so there is no torque generated to cause the magnet to rotate.

However, if, as in Figure 16.1b, we give the magnet a slight nudge in, say, the clockwise direction, the magnet is no longer axial with the coils, and its poles will now be *repelled* by the coils' poles, causing the magnet to rotate clockwise. Had we, instead, given it a nudge in the *opposite* direction, then the magnet would have just as easily rotated counterclockwise.

In Figure 16.1c, the supply current passes through zero and the two coils demagnetise, but the momentum of the magnet causes it to continue to rotate *past* the horizontal position, to the position shown.

In Figure 16.1d, during the second half-cycle, the supply current now flows *counterclockwise* around the circuit, causing the magnetic polarities of the coils to reverse, making them once again *repel* the magnet's poles, so that the magnet continues to rotate clockwise, trying to align itself with the field with its north pole pointing towards the field's south pole.

In Figure 16.1e, the current again passes through zero and, once again, the momentum of the magnet allows it to rotate past the horizontal position. The magnet is now back to where it was in Figure 16.1b, and the whole process repeats itself.

So, once something causes the magnet to *start* rotating in the first place, it will then *continue* to rotate in that particular direction, due to a combination of the torque created by the interaction between the coils' and the magnet's magnetic fields, aided by the momentum of the magnet.

Unlike the magnetic field obtained from a three-phase supply, the stator field is *not* a rotating field but, rather, a *pulsating* field: in our example, it simply changes direction, horizontally, between the innermost faces of the two coils.

The magnet, therefore, is *not* subject to a continuous torque (as is a three-phase machine's rotor) but, rather, a *pulsating* torque. For most purposes, this is perfectly adequate for the types of load typically driven by single-phase machines . . . *once they get started!*

Of course, a *practical* single-phase induction motor doesn't use a *magnet* as its rotor; it uses a squirrel-cage rotor, just like that of a three-phase motor. So, you may wonder *how* it acquires the magnetic polarity it requires to react with the stator's pulsating magnetic field.

The answer is, just like the rotor of a three-phase induction motor discussed in the previous chapter, it is the equivalent of the *secondary winding of a transformer*. So, voltages are induced into it by the changing stator field, due to 'transformer action' and, in accordance with Lenz's Law, the direction of the resulting rotor currents must be such as to create a magnetic field that *opposes* the stator (primary) field, as shown in Figure 16.2:

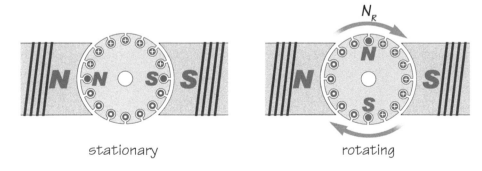

stationary   rotating

Figure 16.2

In Figure 16.2 (left), with the rotor stationary, the direction of the currents circulating along the rotor bars creates a field that must oppose the stator's field – we can confirm this by applying the 'right-hand rule' to the rotor currents. But as the resulting forces between the stator and rotor fields act straight through the centre of the rotor, no torque is developed, and the rotor cannot turn.

However, if something starts the rotor turning in a particular direction – as in Figure 16.2 (right) – it will *continue* to turn in that direction, just as the magnet did in our earlier example.

A more academic explanation of *why* the machine's rotor should continue to rotate with just a pulsating magnetic field is called the '**Double-Revolving Field Theory**'.

According to this theory, the horizontal pulsating field ($\Phi_{pulsating}$) that we have described can be *resolved* into two identical fields (Φ_1 and Φ_2) that rotate at a constant speed in *opposite* directions – as illustrated in Figure 16.3a–e.

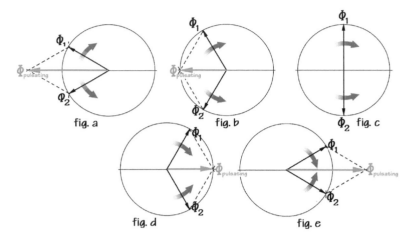

Figure 16.3

Each of these resolved fields apply its own torque to the machine's rotor in *exactly* the same way in which the rotating field in a three-phase motor applies torque to a rotor.

Assuming the rotor is started in a particular direction, the 'forward' torque will be greater than the 'backward' torque, and the resultant torque is considered to be 'positive' – so the motor will continue to accelerate in that forward direction. The final speed of the rotor, of course, cannot match the speed of rotation of the resultant resolved flux because, just like the rotor in a three-phase induction motor, **slip** is necessary for the rotor to continue rotating.

The problem with the motor we've described so far, of course, is that it requires some *external torque* applied to its rotor to get it turning in the first place, as well to get it turning in the desired direction. So just *how* do we get it to move in the first place, and in the desired direction?

Obviously, we can't just poke it with our finger, as we did with the magnet in our explanation of the basic principle of a single-phase machine.

The answer is through the use of a *second* pair of windings (2a and 2b), as shown in Figure 16.4.

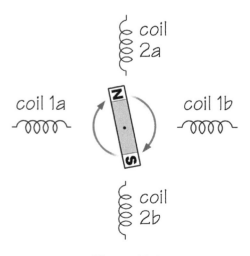

Figure 16.4

As well as their *physical locations*, the additional windings also differ from the original pair by drawing a current from the supply that is *out of phase* with the current supplying the original windings.

So, how do we obtain a current that is out of phase with the original windings' current?

Quite simply, by deliberately *making the reactance of each field winding different*, which results in two field currents that are out of phase with each other.

The most common method of doing this is through the use of a **capacitor**. As illustrated in Figure 16.5, field windings 1a and 1b are connected in series with each other *directly* across an a.c. supply, while field windings 2a and 2b are connected in series with each other *and with a capacitor*.

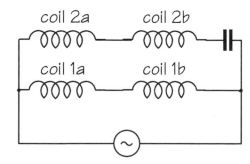

Figure 16.5

The value of capacitance of the capacitor inserted into the upper branch of our circuit must be such that the current flowing through that branch will *lead* the supply voltage by around 45°, while the natural inductance of the lower branch is such that it causes the current flowing in that branch to *lag* the supply voltage by around 45°.

So the two currents are displaced *from each other*, by around 90°.

The first 'polyphase' a.c. systems were, in fact, **two-phase systems**, generated by alternators that had their two armature coils physically displaced by 90°, resulting in two phase voltages that were displaced from each other by 90 electrical degrees, and the system was developed for driving early induction motors. So, the system we are describing, essentially, provides a way of simulating a two-phase a.c. system.

In this arrangement, the **R-L** stator winding is called the 'running winding', while the **R-C** stator winding is called the 'starting' (or, as we shall learn, the 'auxiliary winding').

How this pair of windings is able to establish a rotating magnetic field is illustrated in Figure 16.6. In this diagram, **R-R′** represents the **running** (resistive-inductive) **winding**, and **A-A′** represents the **starting** (resistive-capacitive) **winding**.

Reversing the direction of the rotating magnetic field and, therefore, reversing the direction of rotation of the rotor, is easily achieved by reversing the connections to the auxiliary winding.

Once the motor has started to move in the desired direction, the *starting winding (2a and 2b) can actually be disconnected*, and the machine will then continue to rotate as described earlier for the single pair of field windings – i.e. under the action of the pulsating field (as described in Figure 16.1). This disconnection is achieved by means of a centrifugal switch connected to the motor's shaft; when the rotor is up to speed, the centrifugal switch will open and disconnect the starting winding.

16

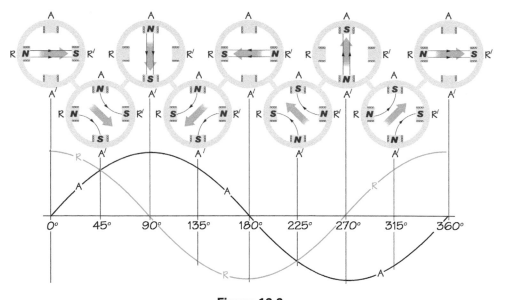

Figure 16.6

Although, in the past, it's been common practice to *disconnect* the starting winding once the machine is running, the modern tendency is *not* to do so but, instead, to leave it permanently connected. For this reason, these days, it's more appropriate to term this winding, the '**auxiliary winding**' rather than 'starting winding'.

This, then, in simple terms is the *principle of operation of a single-phase a.c. motor*. Now let's move on to look at specific *types* of single-phase motor.

Capacitor-start induction motors

As explained above, once a machine's rotor has been started in the desired direction, the starting (auxiliary) winding can be disconnected. This is often what happens with a **capacitor-start** induction motor.

With a capacitor-start motor, a second set of field windings, in series with a capacitor, is used to provide sufficient torque to start the motor in the desired direction. Once the motor reaches about 80% of its full speed, a centrifugal switch attached to the machine's shaft (or, in some cases, a timer switch) opens and disconnects the starting winding. In the simplified diagram, below, each coil represents a *pair* of field windings.

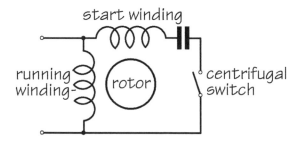

Figure 16.7

As already explained, the modern tendency is *not* to disconnect the starting winding. So why was it usual to *disconnect* the capacitor circuit in the first place? Why not simply leave it connected – allowing it to continue to provide, when connected, a *rotating*, rather than *pulsating*, field? The answer is that the capacitive circuit contributes to energy losses in the motor (bearing in mind that single-phase motors are already between 50% and 25% as efficient as a three-phase motor).

Capacitor-start motors are easily identified because of the large capacitor usually strapped to the outside of the machine, or by the external metal housing (Figure 16.8) in which the capacitor is secured onto the frame. These types of induction motor are commonly used in household applications, such as refrigerators.

capacitor housing

Single-Phase, Capacitor-Start, Induction Motor

Figure 16.8

Capacitor start/capacitor-run induction motors

A variation on the single-phase 'capacitor-start' induction motor, is the '**capacitor-start/capacitor run' induction motor** in which the necessary starting torque is provided by means of a *large* capacitance, but a *smaller* capacitance is used to improve its running torque performance when running, by maintaining the rotating field – while minimising the losses caused by the capacitors.

This is achieved by placing *two* capacitors in parallel (initially providing a higher capacitance: connecting two capacitors in parallel *increases* the circuit's capacitance) and, then, removing one of them once the machine is running. Again, this is achieved using a centrifugal switch, as shown in the circuit in Figure 16.9.

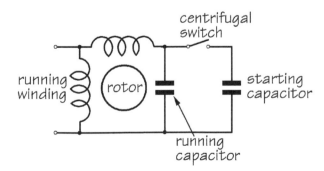

centrifugal
switch

running
winding

rotor

starting
capacitor

running
capacitor

Figure 16.9

Split-phase induction motors

A variation of the capacitor-start induction motor is the '**split-phase' induction motor**.

Like the capacitor-start motor, two separate field windings, connected in parallel with each other, are used. However, the necessary phase difference is achieved by manufacturing the main (or 'running') winding with a low resistance/high inductance, and the auxiliary winding with a high resistance/low inductance.

16

When connected to the supply, the current through the main winding will be more in phase with the supply voltage than the current through the auxiliary winding. In other words, the current through the main winding will lag the current through the auxiliary winding, resulting in a rotating magnetic field.

Like the capacitor-start motor, a centrifugal switch will disconnect the auxiliary winding once the machine has reached its operating speed.

Shaded-pole induction motor

A 'shaded-pole' induction motor is a self-starting squirrel-cage motor that uses a somewhat inefficient, yet rather ingenious, method of creating a rotating magnetic field, through the use of what is termed a 'shaded pole'. This term refers to a short-circuited turn of thick copper wire, placed around a small auxiliary pole created by machining a slot into part of each of the machine's main pole pieces – as shown in the *four-pole* machine in Figure 16.10.

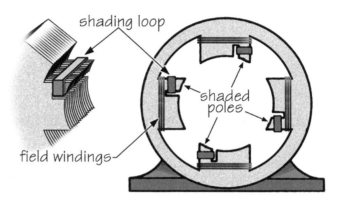

Figure 16.10

Let's examine how this works, by examining the behaviour of an individual pole (Figure 16.11):

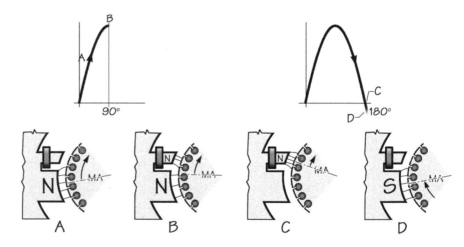

Figure 16.11

Point A: As the field current *increases*, a voltage is induced (by transformer action) into the shading ring in such a direction that the resulting current circulating in the ring (in accordance with Lenz's Law) *opposes* the main field, weakening the flux in the shaded pole, causing the flux to crowd, or shift, towards the unshaded part of the pole. The magnetic axis (labelled '**MA**') of the resulting field, therefore, lies roughly along the centre of the unshaded pole.

Point B: As the field current approaches its peak value, its rate of change falls towards zero, and no longer induces a voltage into the shading ring so its opposition to the main field disappears. As a result the flux now redistributes itself uniformly across *both* poles, with the magnetic axis moving clockwise to roughly centre itself between them.

Point C: After passing through its peak value, the field current now decreases rapidly towards zero, causing the main flux to *decrease* towards zero. At the same time, the voltage induced into the shading ring reverses direction, and now acts to *maintain* the flux in the shaded pole. As the field current reaches **point C** (zero), the flux in the main pole disappears, while the flux in the shaded pole reaches its maximum. So the magnetic axis now moves, clockwise, to the centre of the shaded pole.

Point D: With the direction of the field current now reversed, the action repeats itself, as described above, but with the polarities of the poles reversed.

With this action happening at *each* of the machine's poles, the result is a rotating magnetic field within the air gap between the stator and rotor. The direction of rotation is always *towards the shaded pole*, which should become clear by studying the sequence illustrated in Figure 16.12.

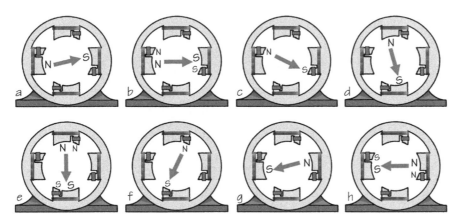

Figure 16.12

Shaded-pole motors are inexpensive and reliable, easy to construct, and rugged, but they suffer from a low starting torque, and are relatively inefficient. Figure 16.13 shows a typical two-pole shaded-pole motor.

So shaded-pole induction motors can be considered to be an inexpensive solution to the problem of starting a single-phase motor. However, because they have a relatively low starting torque, their use is restricted to light load applications such as toys, small fans, ventilators, etc.

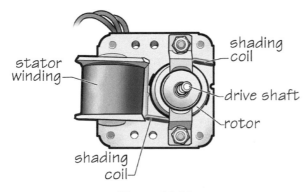

Figure 16.13

381

Universal motor

The single-phase induction motors described above are *fixed-speed* induction motors – i.e. once constructed, their speeds *cannot* be controlled (except with high-frequency inverters, which are generally not economically viable for fractional-kilowatt induction motors). Furthermore, due to **slip**, they are limited to speeds that are *below* synchronous speed (which, as we have learnt, for a given number of poles, is determined by the frequency of the supply).

As the name suggests, **universal motors** are motors that can work on either a.c. or d.c. These are machines that use *separately excited rotor windings* which, of course, require commutators and brushes to supply current to their rotors. In other words, they are *not* induction motors.

The universal motor is a slightly modified version of a d.c. **series motor** – i.e. a d.c. motor whose field winding is in series with the supply.

The modification involves reducing the number of turns in the field winding, and increasing the number of turns in the armature winding, to ensure that there is no phase-displacement between their magnetic fields – thus ensuring the directions of those fields reverse at the same time. Of course, when the current through the field-winding *and* the current through the armature winding *are both reversed at the same time*, the direction of the resulting torque remains unchanged.

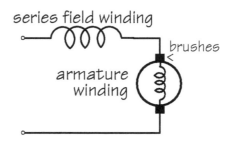

Figure 16.14

Unlike induction motors, the speeds of these 'universal' motors are not limited by the mains' frequency (synchronous speed), and they can run as fast as 30 000 rev/min but, more typically, between 8000 and 12 000 rev/min. However, because the sparking at the commutators is rather worse than for d.c., such machines are limited to low-power applications – typically below 1 kW. Such high-speed/low-power applications include vacuum cleaners, food blenders/liquidisers, and power tools such as drills, routers, etc.

Conclusion

Now that we've completed this chapter, we need to examine its **objectives** listed at its start. Placing a question mark at the end of each objective turns that objective into a **test item**. If we can answer those test items, then we've met the objectives of this chapter.

CHAPTER 17

A.C. motor starters

Objectives

On completion of this chapter you should be able to:

1. explain the purpose of using an a.c. motor starter.

2. describe the principle of operation of a direct-on-line starter.

3. given a schematic diagram of a direct-on-line starter and its control circuit, identify and state the purpose of each of its components.

4. describe what is necessary to reverse the direction of rotation of a three-phase induction motor.

5. explain the basic principle of a star-delta motor starter.

6. given a schematic diagram of a star-delta motor starter, identify and state the purpose of each of its components.

7. explain the basic principle of operation of an autotransformer motor starter.

8. given a schematic diagram of an autotransformer motor starter, identify and state the purpose of each of its components.

Introduction

When we apply the rated supply voltage to a stationary induction motor, the machine will initially develop around *twice* its full-load torque, and draw a starting current of up to *ten times* its rated load current. This initial '**starting**', or '**inrush**', current quickly falls as the machine accelerates towards its near synchronous speed.

With larger motors, this high starting current can have a detrimental effect not only on the machine itself, but also upon any other equipment or consumers connected to the supply system. Accordingly, any *direct* connection to the supply is normally restricted to machines of around 4 kW or less. Machines *larger* than this usually need to be started using some method that will **restrict the initial inrush current**.

Once running, an *overloaded* motor will draw a continuous excessively large supply current causing it to overheat, which will possibly cause permanent damage to its insulation. Such currents are called, 'overload currents', and motors must also be provided with some method of **overload current protection**.

It's also important that whenever a motor runs down following, say, a power interruption, it should *not* then be allowed to restart by itself when the power is eventually restored, so safeguards to prevent this from happening (called '**no-volt' protection**) must also be incorporated into the circuits that control the motor.

Electrical Science for Technicians. 978-1-138-84926-6 © Adrian Waygood.
Published by Taylor & Francis. All rights reserved.

So, the *three* requirements for starting and running an induction motor are:

(a) where necessary, restricting its starting current,

(b) providing overload current protection, and

(c) providing no-volt (self-starting) protection.

Devices for doing this are generically termed '**motor starters**', and are the subject of this chapter.

There are *two* ways of **starting** an induction motor. These are:

▶ full-voltage starting ('direct-on-line' starting)
▶ reduced-voltage starting, including:
  ▶ star-delta starting
  ▶ autotransformer starting

Although beyond the scope of this chapter, it's worth noting that solid-state starter/controllers are now widely used and provide both reduced-voltage starting, together with stepless speed control.

Full-voltage (direct-on-line) starter

'**Full-voltage**', or '**direct-on-line**' **(DoL)** starting can normally *only* be applied to motors rated up to about 4 kW because, beyond this rating, the resulting high inrush current may overheat the machine's field windings. In addition to this, a high starting current may have a detrimental effect on the behaviour of the supply system. This is because a high starting current causes a voltage drop along the supply cable, and a reduction in the voltage appearing across the motor's terminals. This causes the machine to compensate by drawing even more current which, in turn, causes an even greater voltage drop! The resulting, cumulative, reduction in the supply voltage can adversely affect other consumers or loads connected to the same supply system.

The most obvious adverse effect caused by motors starting up is termed '**voltage flicker**', and describes how other consumers supplied from the same electrical system are affected by the sudden and temporary drop in supply voltage. For example, an incandescent lamp's power output is determined by its voltage, and a small reduction in that voltage will cause a proportionally much larger drop in the lamp's power output. If the voltage drop is large enough, it might also cause some types of equipment (e.g. computer systems) to turn off and reset.

The main component of a **direct-on-line starter** system is a heavy-duty electromagnetically operated switch, called a '**contactor**'. A contactor is, essentially, a heavy-duty relay, whose contacts are capable of making and breaking a heavy current without deteriorating due to the resulting arcing.

A DoL starter also incorporates some means of overload current protection. A typical example of a three-phase motor starter is illustrated in Figure 17.1.

Figure 17.2 shows a schematic diagram of the internal connections of a typical **three-phase direct-on-line starter** of the type illustrated in Figure 17.1. The heavy lines represent the three-phase lines and main contacts; the thinner lines show the internal operating coil's control circuit and auxiliary contacts.

The starter has three heavy-duty **main contacts**, one per line, together with a light-current **auxiliary contact**. Each of these contacts (main *and* auxiliary) is mechanically interlocked, so that they *all* operate together when electromagnetically activated by the starter's **operating coil**, a strong electromagnet that slams the contacts open or closed. As we shall learn, the function of the auxiliary contact is to provide a '**latching**' or '**hold-in**' feature for the external control circuit – this ensures that the main contacts remain closed once the 'start button' is released by the operator.

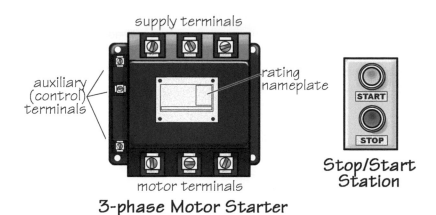

3-phase Motor Starter

Stop/Start Station

Figure 17.1

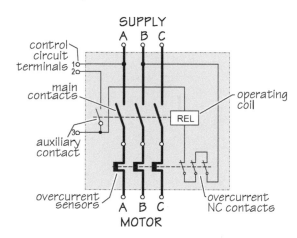

Figure 17.2

In *series* with the main contacts are **overload current protection sensors**. These are temperature-sensitive (bi-metallic or 'solder pot' operated) devices – in effect, 'heaters' – each of which, when subjected to a prolonged overload current, will cause one or other of the three associated normally closed **overcurrent contacts**, connected in series with the operating coil, to open and de-energise the operating coil, causing the main contacts to trip open.

It's important to understand that the overload current *sensors* themselves don't 'open'; they are simply **heaters** which, when a certain temperature is reached, cause one or other of the normally closed overcurrent contacts to open, which then de-energises the operating coil.

As well as providing heavy-duty terminals (**A-B-C**) for connecting the supply to the motor, the motor starter also provides three smaller terminals (marked: **1**, **2**, and **3**), in Figure 17.2, for connecting to the external (low current) **control circuit**.

In the schematic diagrams that follow, you should be aware of the following conventions which, in fact, apply to *all* schematic diagrams:

▶ *all* switches and contacts are always shown in their *de-energised* state: i.e. the positions they assume when the control circuit is de-energised
▶ the terms, '**normally open**' (**NO**) and '**normally closed**' (**NC**), describe the position of switches or contacts when the control circuit is de-energised
▶ all contacts are identified by letters or numerals that identify which relay coil causes them to operate (e.g. contact 'A' is operated by coil 'A', etc.), and several contacts may be operated by the *same* coil

17

385

▶ in schematic diagrams, switches and contacts are not necessarily drawn anywhere near the coil that controls them, whereas they are *actually* adjacent to, and may be physically connected by, a mechanical linkage.

You should also be aware of the ISO **circuit symbols** shown in Table 17.1, which are used in the schematic diagrams that follow.

Table 17.1

Device	Normally open	Normally closed
pushbutton switches:		
operating coil:		
operating coil and its contacts:		
overcurrent protective sensor:		

(**Note.** North American schematics use a circle to represent the operating coil, and a pair of parallel lines – similar to ISO symbols for capacitors – to represent contacts.)

Direct-on-line start/stop control circuit

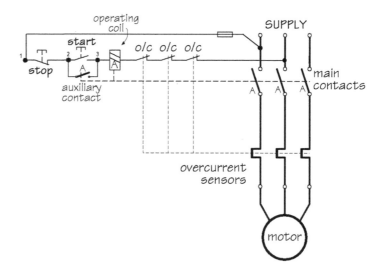

Figure 17.3

(**Note:** In this schematic, the operating coil is labelled 'A'; when energised, this coil will operate *all* contacts also labelled 'A', as it is physically connected to each by means of a mechanical linkage from its magnetic armature: N/O contacts will close and N/C contacts will open.)

The **direct on-line** three-phase starter, illustrated in Figure 17.3, consists of three **main contacts**, one per line, and one **auxiliary contact**, controlled by an **operating coil**. In series with each of the main contacts is a thermally operated **overload**

current sensor; in the event of a prolonged overload current, e.g. due to a stalled motor, any one of the three overcurrent sensors will cause its corresponding normally closed contact in the control circuit to *open*, de-energise the control coil, trip the main contacts, and disconnect the motor from the supply. In the event of an overload current sensor operating to disconnect the machine that sensor must be manually reset in order to reclose their associated NC contacts.

The operating coil is supplied between any pair of line conductors (in Figure 17.2, between lines A and B), but in some starters, a **transformer** is used to reduce the line voltage applied to the coil and its control circuit.

In *schematic* diagrams, the various components are not necessarily shown in their *actual* locations – e.g. in Figure 17.3 the operating coil, auxiliary contacts, and overload contacts *appear* to be located well away from the starter itself, whereas they are *actually* located within the starter's enclosure and are only accessible via the terminals marked 1, 2, and 3.

When the NO '**start**' pushbutton switch (located between terminals 2 and 3) is pushed closed, the control circuit is energised, and current flows through the starter's control coil (A), which closes the three NO **main contacts** (energising the motor) together with the NO **auxiliary contact**.

When the NO **auxiliary contact** closes, it short-circuits the NO **start** pushbutton switch and keeps the operating coil energised after that pushbutton has been released by the operator. This arrangement is termed a '**latching**' or '**hold-in**' circuit. This arrangement also provides 'no-volts' protection – preventing the motor from self-starting, following a loss and subsequent restoration of the mains-supply voltage, which would be *very* dangerous.

The NC **stop** pushbutton switch is in series with the control circuit. When this switch is pushed, it opens, and the control coil is de-energised, causing the three NO **main contacts** to open, de-energising the motor. At the same time, the NO **auxiliary contact** also opens so, when the NC '**stop**' pushbutton switch is released by the operator, the control circuit remains de-energised.

The three NC **overcurrent (O/C) contacts**, connected in series with the control coil, provide protection against any overcurrent. If the machine draws a prolonged overcurrent (for example, if the motor should stall under load), then any or all of the overload sensors in the main line will act to trip open their corresponding NC **overcurrent contacts**. These are connected in *series* with the operating coil and, so, will de-energise the control circuit, opening the NO **main contacts** and disconnecting the motor.

Any number of additional stop/start stations can be added to the control circuit, to provide remote stop/start locations, or to provide additional emergency stop pushbuttons. Such additional NC **stop** pushbutton switches must be connected in *series* with the existing stop pushbutton, and any additional NO **start** pushbutton switches must be connected in *parallel* with the existing start pushbutton.

Direct-on-line forward/reverse control circuit

The basic direct-on-line starter, described above, can be replaced with a direct-on-line **forward/reverse motor controller**, which will enable a three-phase motor to be started and stopped *in either direction*.

You will recall that, in order to reverse a three-phase induction motor, all that is necessary is to interchange any two line conductors – thus supplying the machine with a *negative phase sequence* (A-C-B-A) of voltages.

17

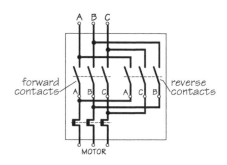

Figure 17.4

The diagram in Figure 17.4 shows the main contact circuits of a forward/reverse motor controller. As we can see, there are *two* independent sets of main contacts. The left-hand set provides the motor with a positive phase sequence, A-B-C-A, supply; the right-hand set has lines B and C interchanged, providing the motor with a negative phase sequence, A-C-B-A. Only one set of overload current sensors is needed. The starter also incorporates operating coils and auxiliary contacts but, for the sake of clarity, these are not shown in the illustration.

The control circuitry for this forward/reverse starter is beyond the scope of this chapter, but would include both electrical *and* mechanical **interlock** arrangements to prevent both sets of main contacts from closing at the same time – if this were allowed to happen, then a dead-short would be applied to the supply!

Pressing a NO '**forward**' pushbutton switch will close the main 'forward' contacts. Pressing a NO '**reverse**' pushbutton switch will trip the main forward contacts *before* closing the main 'reverse' contacts. These two sets of main contacts are *mechanically* interlocked, to prevent them from closing at the same time. The 'forward' and 'reverse' pushbuttons are *electrically* interlocked to prevent them *both* from operating if pressed at the same time.

Control of this starter can be made more and more sophisticated, by adding more and more features to the control circuit. Such features include enabling the machine to be **incrementally rotated** (termed '**jogging**') in order to accurately position a load, or brought to an **immediate stop** (termed '**plugging**').

Reduced-voltage ('star-delta') starter

For motors rated at around 4 kW and above, some method must be employed to *limit* the starting current in order to avoid overheating the stator windings and having a detrimental effect on other consumers supplied from the same system. The '**star-delta**' **starter**, illustrated in Figure 17.5, is one of several methods of achieving this.

A three-phase motor's field windings are connected either in **star** ('wye') or in **delta**. These configurations may be achieved internally or externally. Provided these configurations are accessible *externally* (for example, inside a terminal box attached to the motor), then the motor can be easily changed from one configuration to the other.

If we refer to the chapter dealing with three-phase systems, we will recall that, for a star-connected system, the phase voltage is the line voltage, divided by the square-root of three, or **58%** times the line voltage.

A **star-delta starter** normally operates with the motor running with its field windings connected in delta. On starting, however, they are initially connected in **star**, thus subjecting them to just 58% of their rated voltage, thus limiting the inrush line current to *one-third* of its rated current, i.e.:

$$\text{starting current} = \left(\frac{1}{\sqrt{3}} \times \frac{1}{\sqrt{3}}\right) = \frac{1}{3} \times \text{rated current.}$$

This is because, for a star-connected motor, the line current is the reciprocal of root-3 times the rated (delta-connected) winding, and the applied voltage has been reduced by the same amount.

When the machine is running, the starter reconnects the motor's field winding in delta.

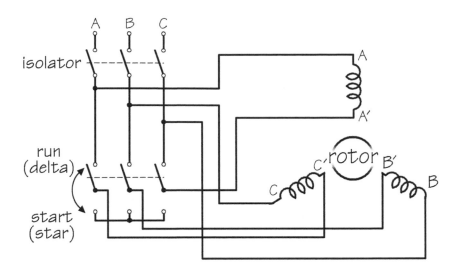

Figure 17.5

In Figure 17.5, the outermost end of each field winding (A, B, and C) is permanently connected to the incoming lines. The innermost end of each field (A', B', and C') winding is connected to the common terminals of the two-way switches labelled '**start**'/'**run**'.

In the '**start**' position, the inner ends of each field winding (A', B', and C') are connected together – so the field windings are connected in **star**.

In the '**run**' position, the inner end of each field winding is connected to the outer end of their adjacent field windings (A' to B; B' to C; and C' to A) – so the field windings are connected in **delta**.

For simplicity, *overcurrent protection devices are not included in the schematic.*

Reduced-voltage (autotransformer) starter

Another method of restricting a motor's starting current is by using an **autotransformer starter**.

An autotransformer starter uses *three* **variable output autotransformers** (one per line) to reduce the initial voltage applied to, and therefore current drawn by, the motor. Unlike the 'star-delta' starter, where the starting-current is fixed at *one-third* of the rated current, the autotransformer started can be adjusted to *any* desired value.

In Figure 17.6, with each rotary switch turned to its upright position, the output from each autotransformer is applied to the field windings.

The output from each autotransformer depends upon the position of its slider. As they are moved (together) upwards, the output-voltage increases.

When the rotating switches are turned to their downward positions, the full supply-voltage is applied to the field windings.

For simplicity, *overcurrent protection devices are not included in the schematic.*

17

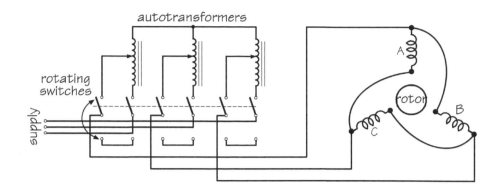

autotransformers

rotating switches

supply

A

rotor

B

C

Figure 17.6

How a 'latching' (or 'hold-in') circuit works

A very important feature in all motor-control circuits is the '**latching**' or '**hold-in**' feature, which keeps the contactor's operating coil energised *after* the '**start**' pushbutton switch has been released.

In fact this basic feature is incorporated into a great many other control circuits, so an understanding of how it works is a fundamental requirement for understanding the operation of any control circuit.

In Figure 17.7, a load (for simplicity, represented by a lamp but, in the case of a motor starter, this would be a motor) is supplied through a NO (normally open) **start** pushbutton switch. When the **start** switch is closed, current will flow through the load; when the switch is released, the current will stop flowing.

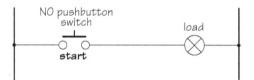

NO pushbutton switch

load

start

Figure 17.7

Now let's modify the circuit, to incorporate the 'latching' or 'hold-in' feature. In the modified circuit, shown in Figure 17.8, a relay's operating coil is connected in *parallel* with the load, and its **NO** (normally open) **contact** is connected in parallel with the pushbutton switch. Remember that, in a schematic diagram, it's *not* necessary to show the relay's contacts physically next to its operating coil (where they are *actually* located). In the following diagram, the relay's contacts are indicated by the broken line connecting them to their operating coil.

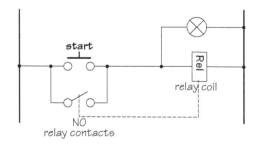

start

Rel

relay coil

NO
relay contacts

Figure 17.8

Now, let's see what happens when we close the 'start' pushbutton switch.

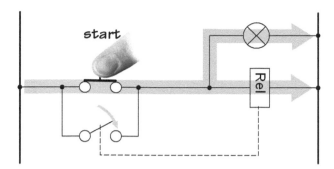

Figure 17.9

When the pushbutton switch is closed (Figure 17.9), current flows through both the load *and* through the relay's operating coil (because they are in parallel). As soon as the relay coil is energised, it closes its NO contact – enabling current can flow through both the pushbutton switch *and* the closed NO contact (Figure 17.10).

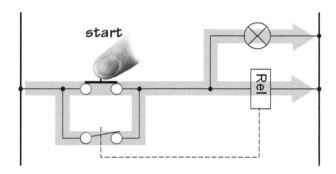

Figure 17.10

So, what happens when the pushbutton switch is released?

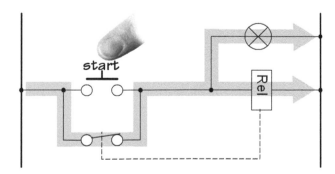

Figure 17.11

As we can see in Figure 17.11, although the pushbutton switch re-opens, current is *still* able to flow through the closed relay contacts. This current not only continues to supply the load, but it *also* keeps the operating coil energised which *holds its contact closed* – maintaining the supply.

This, then, is the principle behind the '**latching**' or '**hold-in**' circuit, which is widely used in a great many control circuits.

Now that the load is continuously energised, *how do we switch it off?*

17

The answer is shown in Figure 17.12: by incorporating a NC (normally closed) **stop** pushbutton switch *in series* with the NO **start** pushbutton switch.

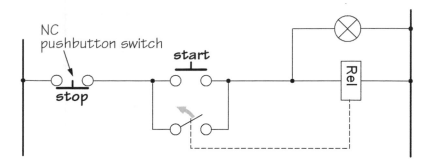

Figure 17.12

Pushing the NC **stop** pushbutton switch will *break* the circuit. As well as disconnecting the load, this *also de-energises the relay's operating coil*, allowing its contact to reopen. When the NC **stop** pushbutton switch is released, the circuit will *not* re-energise because both the NO **start** pushbutton *and* the relay's NO contacts are open – so there is *always* break in the circuit.

Conclusion

Now that we've completed this chapter, we need to examine its **objectives** listed at its start. Placing a question mark at the end of each objective turns that objective into a **test item**. If we can answer those test items, then we've met the objectives of this chapter.

CHAPTER 18

Electric lighting

Objectives

On completion of this chapter you should be able to:

1. explain what is meant by the term, 'visible light'.

2. explain the relationship between 'white' light and the 'colour spectrum'.

3. explain the relationship between the colour of visible light and the temperature of its source.

4. compare the explanations of visible light in terms of (a) electromagnetic energy, and (b) a stream of photons.

5. explain the difference between measuring visible light using (a) radiometry, and (b) photometry.

6. explain the difference between (a) a continuous colour spectrum, and (b) bright-line emission spectrum, and how they relate to different types of lamp.

7. explain the propagation of visible light in terms of its *source*, *flow*, *arrival*, and *reflection*, identifying the quantities used to measure each, the relationship between each, and the relevant SI units of measurement.

8. explain the 'inverse-square rule' and 'cosine rule' as they relate to the propagation of light.

9. solve simple lighting design problems, using the 'lumen formula' method.

10. explain the term, '*luminous efficacy*', and how it relates to different lamp types.

11. outline the difference between *incandescent*, *gas-discharge*, *arc*, and *solid-state* categories of lighting, giving at least one example of each.

12. describe the construction and principle of operation of a tungsten incandescent lamp.

13. describe the construction and principle of operation of a tungsten-halogen incandescent lamp – in particular, describe the halogen-regenerative cycle.

14. describe the general principle of operation of all gas-discharge lamps.

15. sketch the basic circuit diagram of a standard fluorescent lamp, and identify and describe the function of each of its major components.

16. describe the starting cycle for a standard fluorescent lamp.

17. describe methods of overcoming the potential hazards of a fluorescent lamp's 'stroboscopic effect'.

18. compare the advantages/disadvantages of the compact fluorescent lamp (CFL) and the tungsten (GLS) lamp it is intended to replace.

19. describe the construction and principle of operation of the

    (a) high-pressure mercury-vapour lamp

    (b) low-pressure sodium-discharge (LPS), or SOX, lamp

    (c) high-pressure sodium-discharge (HPS or SON) lamp

    (d) high-voltage luminous-discharge ('neon' display) lamp.

Electrical Science for Technicians. 978-1-138-84926-6 © Adrian Waygood. Published by Taylor & Francis. All rights reserved.

Introduction

Electric lighting was the very first benefit that most people experienced following the introduction of electricity distribution systems: first as *street lighting*, then as *lighting in theatres, hotels* (Figure 18.1) *and shops* and, eventually, *in the home.*

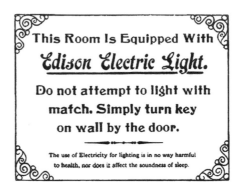

Figure 18.1

Although the incandescent lamp was widely available by the 1900s, only a very small percentage of British residents could afford to have their homes wired for electricity, so the majority of homes continued to be lit by gas, paraffin or candles well into the 1930s. In fact, it was still possible to come across some British homes without electricity well into the 1940s and, in some rare cases, even into the 1950s and early 1960s.

Today, electric lighting is such a normal part of our everyday lives that we hardly give it any thought; in fact, it's been said that most of us are only aware of electric lighting when it stops working!

We use artificial lighting to:

- provide a safe environment and help us find our way around
- extend the length of the working day
- improve our working conditions and our productivity
- reduce crime
- display objects
- attract attention.

What is light?

So what exactly *is* light?

> The **Illuminating Engineering Society of North America** defines **light** as *'radiant energy that is capable of exciting the retina and producing a visual sensation'.*

In other words, **light** is defined in terms of **human physiology**: that is, the ability of our eyes to convert the radiant energy received by the retina into electrical signals that our brain then interprets as images.

If we were to search for other definitions, we would find that they practically all explain light in terms of *its effect on the human eye*, which is the reason, of course, why it's generally referred to as '**visible light**' in order to distinguish it from other forms of radiant energy, such as **infrared** and **ultraviolet** light, which cannot be seen by humans despite being visible to some reptiles and insects.

But what do we mean by '**radiant energy**'? In simple terms, radiant energy is energy in transit from one place to another by means of **'electromagnetic waves'** without the need for any medium to support it, and is a combination of alternating magnetic and electric fields. For example, energy from our sun reaches the earth, through the vast expanse of space, as radiant energy.

Radiant energy is not confined to just 'light'. The *entire* spectrum of radiant energy is illustrated in Figure 18.2, and is classified according to either its **wavelength** (expressed in **metres**) or its **frequency** (expressed in **hertz**), from 'long radio waves' at one extreme to 'gamma rays' at the other. Some, like X-rays and gamma rays, are extremely harmful to the human body.

As we can also see from Figure 18.2, 'visible light' represents just a tiny range of the wavelengths or frequencies within the overall radiant energy 'spectrum'.

In the science we are traditionally taught at school (called 'classical science'), while *some* of the ways in which visible light behaves can be adequately explained if it is considered to be an **electromagnetic wave**, there are *other* types of behaviour that can only be explained if light is considered, instead, to be a **stream of individual energy 'packets'** we call '**photons**'!

So *which* of these explanations is correct?

The answer is that *both* these descriptions are simply 'analogies' or 'models' that scientists use to try to explain the behaviour of visible light in understandable terms – with one analogy best describing *some* types of behaviour, while the other analogy best describes *other* types of behaviour.

For example, when we discuss visible light in terms of its **colour**, we usually need to explain it in terms of its wavelength or frequency – so we are treating it as a form of 'electromagnetic energy'. On the other hand, when we describe, say, the **reflection** or **refraction** of light, it is more understandable if we treat it as a 'beam of photons'.

It seems, to the layman at least, that *neither* model can adequately describe *all* the ways in which visible light behaves! We simply have to accept that there is no single understandable explanation for *all* the ways in which light behaves. To paraphrase the American physicist, Richard Feynman (1918–1988):

> *If we choose to examine a phenomenon which is impossible to explain . . . (then) it contains only mystery, and we cannot necessarily make that mystery go away with a simple explanation of how it works.*

So, scientists will have to continue to use whichever analogy best describes a particular behaviour in order to describe visible light to the layman. And this is what we will be doing in this chapter.

Visible light

Whenever we talk about **'visible' light**, we are describing that tiny part of the complete **electromagnetic spectrum** that can be perceived by the human eye – i.e. the range of electromagnetic wavelengths or frequencies sandwiched between those of infrared ('*below* red') and ultraviolet ('*above* violet'). This is illustrated in Figure 18.2.

In air, the **frequency** range of visible light is approximately **430–790 THz** (terahertz). Expressed in terms of their wavelengths, this corresponds from about 700 nm down to about 390 nm (nanometres).

The great physicist, Sir Isaac Newton (1642–1727), is credited with discovering that what we consider to be simply 'white light' is actually not white at all, but *a continuum of colours* that we call the '**colour spectrum**'. He was able to demonstrate this by means of a simple experiment in which he first passed a beam of white light through a prism, which *separated* the colours present in that white light (as illustrated in his own hand, from his scientific papers, in Figure 18.3) and then, using a second prism, was able to *recombine* those colours to form white light once again – thus proving that the prism itself wasn't simply 'colouring' the beam.

18

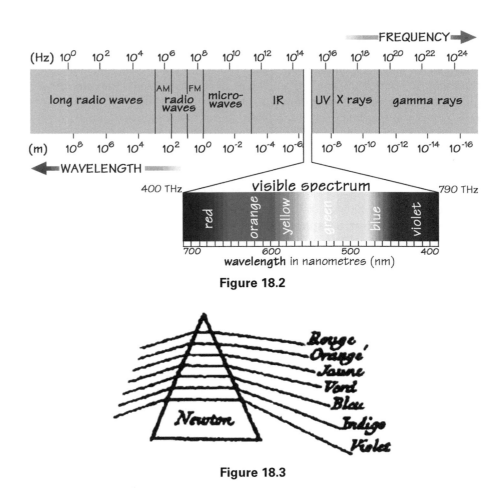

Figure 18.2

Figure 18.3

We often see this effect in nature, after a rain shower for example, when moisture suspended in the atmosphere separates 'white' sunlight to form a **rainbow**.

The reason why 'white' light separates in this way is because rays of visible light of different wavelengths are 'refracted' (deflected) through different angles as they pass through a glass prism and, so, present themselves as this spread or 'spectrum' of different colours.

Our main source of natural light is, of course, our sun whose surface temperature is about 6000 K (approximately 5726°C). And it is *temperature* that determines the levels of radiant energy at the various frequencies throughout the entire electromagnetic spectrum and, in particular, the various colours we perceive.

We have all, at some time or another, watched iron being worked in a foundry. As its temperature gradually increases, the first thing we notice is that the iron will start to glow a dull red. As its temperature continues to increase, this colour changes from red to white and, if the temperature is raised high enough, eventually to blue.

Colour, therefore, is defined in terms of the *temperature* of a source of visible light, expressed on the absolute or **kelvin scale**. This '**colour temperature**' scale ranges from around 1000 K (a candle flame) to around 10 000 K (deep-blue northern sky). Understanding colour temperature is *very* important in a number of different fields ranging from photography and publishing to astrophysics.

The colour spectrum comprises, in order of decreasing wavelength (or increasing frequency): red, orange, yellow, green, blue, indigo, and violet. Of course, the various shades of these individual colours merge one into the other, sometimes making it rather difficult for us to tell the individual colours apart – in particular, most of us seem unable to distinguish indigo from blue. The approximate wavelengths of these colours are:

Table 18.1

Colour	Wavelength / nm
Red	620 – 750
Orange	590 – 620
Yellow	570 – 590
Green	495 – 570
Blue/indigo	450 – 495
Violet	380 – 450

The back of the human eye, called the 'retina', has *two* types of light-sensitive nerve endings called 'cones' and 'rods', which are connected to the brain via the optic nerve.

The retina's **cones** respond to visible light between around 400 nm and 700 nm, but are not equally sensitive to *all* the colours within this range: being *least* sensitive to the colours at opposite ends of the spectrum and *most* sensitive to light with a wavelength of around 555 nm, which corresponds to the **green** part of the spectrum – as illustrated in Figure 18.4.

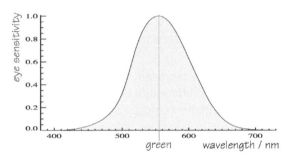

Figure 18.4

The retina's **rods**, on the other hand, are highly sensitive *but are only effective at low lighting levels* and *provide only monochromatic vision*. So, as the level of lighting falls, the eye's maximum sensitivity falls from around 555 nm to around 540 nm, causing the curve shown in Figure 18.4 to shift slightly to its left. However, as the rods' sensitivity takes over from cones well below the values of lighting discussed in this chapter, we can safely ignore this shift.

The human eye is a complex mechanism which, unlike artificial sensors, constantly adjusts itself to compensate for changes in the colour of light sources.

Those of us who use digital cameras, for example, are aware that we may have to change its 'white balance' setting whenever we wish to photograph objects under artificial lighting (particularly under gas discharge lighting, such as fluorescent lamps), otherwise the resulting reflected colours captured by the camera will be false (e.g. too 'green' under fluorescents or too 'orange' under incandescents).

Before the introduction of digital cameras, uncompensated ('daylight') colour transparency film would record subjects lit by tungsten lamps with an orange 'colour shift', and subjects lit by fluorescent lamps with a green 'colour shift'. To compensate, it was necessary either to purchase film specifically designed for use under artificial light, or to use appropriate colour-correction filters over the camera's lens. Digital cameras allow us to do this electronically instead.

To a large extent, the human eye can automatically compensate for this for itself; in most cases enabling us to see the 'correct' reflected colours of objects, regardless of the colour emitted by the lighting source.

18

All of this leads us rather neatly on to the subject of how we *measure* electromagnetic energy and, in particular, visible light. There are, in fact, *two* systems of measurement used in this field, called '**radiometry**' and '**photometry**'.

▶ **Radiometry** describes the system of measurement used for *all* radiant energy, both 'visible' and 'invisible'. In other words, it treats the *entire* electromagnetic spectrum equally, regardless of wavelength. In practice, however, it is usually restricted to measurements of those wavelengths that make up the *infrared*, *visible*, and *ultraviolet* spectrum. The SI units of measurement used in radiometry, such as the watt, are already familiar to us and relate to other SI units that are common to all other branches of science and engineering.

▶ **Photometry**, on the other hand, describes a system of measurement used for only that radiant energy *within the visible spectrum*, and in such a way that it takes into account human physiology, i.e. *adjusted to match the sensitivity curve of the human eye*. The SI units of measurement used to measure visible light, such as the **lumen** (**lm**), therefore, are unique to photometry and don't relate directly to other SI units.

For example, in radiometry, the *rate* at which an incandescent lamp emits its radiant energy is measured in **watts**, a unit with which we are already familiar and which is used for measuring the rate of transfer of *all* forms of energy, whereas, in photometry, the equivalent *rate* at which the same lamp emits only visible light is measured in **lumens**, which takes into account the sensitivity of the human eye, making that particular unit unique to photometry.

Because radiometry and photometry represent two completely different systems of measurement, using completely different SI units, there is *no* simple 'conversion factor' between radiometric and photometric units, such as the watt and the lumen.

For example, while one watt (radiometric) of 'green' light at 555 nm (to which the eye is most sensitive) is equivalent to about 683 lumens (photometric), this conversion factor changes significantly for other wavelengths of visible light: for example, one watt (radiometric) of 'red' light at 700 nm is equivalent to just 2.7 lumens (photometric)!

And, for *infrared* or *ultraviolet* 'light', one watt of radiant energy, of course, equates to *zero* lumens of visible light, because infrared and ultraviolet energy are not visible to the human eye!

Because photometric units vary according to the sensitivity of the human eye to light of different frequencies, SI photometric units simply do not relate directly to other SI units of measurement – so we *cannot* directly convert one to the other.

About 80% of the electrical energy supplied to a typical incandescent GLS lamp is emitted as radiant energy, with the remainder lost as heat through conduction to its light fitting or luminaire. Of the total radiant energy emitted, visible light accounts for just around 5%.

So, the *output power* of a 100 W incandescent GLS lamp, for example, will be about 80 W (expressed in radiometric units). This is made up of mainly *infrared radiation* from the lamp's filament and the hot gas surrounding it, with the *visible light* accounting for just around 5 W (expressed in radiometric units), which corresponds to around 960 lm (expressed in photometric units) for the wavelength at which the eye is most sensitive (green light).

Photometry, therefore, is important for measuring light *as it is perceived by humans*, i.e. 'visible' light in the 360–830 nm region, and adjusted to match the sensitivity curve shown in Figure 18.4, and is the method preferred by **lighting designers**: those who design outdoor and indoor lighting systems for the purpose of providing all forms of lighting, including residential, commercial, industrial, and theatrical lighting.

Light as a stream of photons

In the companion book, **An Introduction to Electrical Science**, we learnt that electrons move around the nucleus of an atom in separate and distinct elliptical orbits, we call 'shells'. The energy level of individual electrons determines which shell they occupy, with those having the greatest energy levels occupying those shells furthest from the nucleus.

We also learnt that electrons cannot occupy the space *between* shells because their energy levels *cannot* change gradually, but *only* in discrete amounts, or 'packets' of energy, that we call '**quanta**' (plural for 'quantum'). For an electron to move from an inner shell to an outer shell, therefore, it must *gain* a discrete **quantum** of energy; for an electron to move from an outer shell to an inner shell, then it must give up or *lose* a quantum of energy.

In the book based on his BBC-television series, **The Ascent of Man**, physicist Jacob Bronowski, explains that, while the inside of an atom is invisible to us, *'there is a window in it, a "stained-glass window": the spectrum of the atom'*. What he is describing is the fact that *every* element has its own, unique, colour spectrum; but one that is *not* continuous like the spectrum that Newton was able to extract from white light (see Figure 18.3) but, instead, one made up of *a number of individual bright coloured lines* (termed a '**bright-line emission spectrum**') that characterise each element, and which accounts for the unique range of colours emitted by different types of gas-discharge lamp.

Whenever an electron changes its energy level and moves between shells, it either *emits* or *absorbs* **photons**, whose energy level therefore must be *exactly* equal to the difference in energy between those shells.

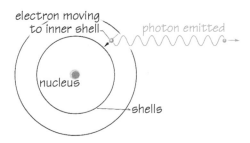

Figure 18.5

So when, for example, the single electron in a hydrogen atom jumps from an outer shell to an inner shell, it releases a photon (Figure 18.5). The combined photon emissions from billions-upon-billions of hydrogen atoms can be viewed, using a spectroscope, as four coloured 'spectrum lines': a red line, a **cyan** (blue-green) line, a **blue** line and a **violet** line (the '*stained-glass window*', described by Dr Bronowski).

The red line results from those hydrogen atoms whose electron has jumped from the third shell to the second; the cyan line from those whose electron has jumped from the fourth shell to the second, the blue line from those whose electron has jumped from the fifth shell to the second and, finally, the violet line from those whose electron has jumped from the sixth shell to the second.

This bright-line emission, then, represents the unique 'spectral fingerprint' for hydrogen gas, and is illustrated in Figure 18.6. Other gases each have their own, unique, 'spectral fingerprint' so, by examining its bright-line emission spectrum, *any* gas can easily be identified.

This explains why different types of **gas-discharge lamp** (which we will be examining, shortly) emit light of specific colour combinations unique to the gas used in that lamp. For example, a sodium lamp (the yellow lamp, widely used to

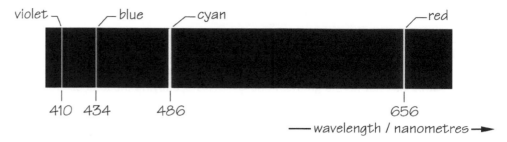

violet blue cyan red

410 434 486 656

— wavelength / nanometres →

Figure 18.6

illuminate motorways) has a bright-line spectrum consisting mainly of two yellow lines that are very close together (at 589.0 nm and 589.6 nm, respectively). These two lines are so close together, that we consider the light from a sodium lamp to be 'monochromatic' – i.e. having one wavelength.

Examining a bright line emission spectrum, such as that of hydrogen illustrated above, using a high-definition spectroscope will reveal that, in some cases, these coloured lines are not single lines at all but, in fact, are made up of several lines, very close together, that merge into each other. Furthermore, by using appropriate sensors, emission lines in the non-visible (e.g. infrared, ultraviolet, etc.) spectrum can also be detected.

So what exactly is a photon? Well, a **photon** is considered to be an *elementary particle* (just like an electron, a proton, or neutron), but one that:

▶ has no mass,
▶ is electrically neutral,
▶ represents exactly one quantum of energy,
▶ represents all forms of electromagnetic radiation, and
▶ travels in straight lines at a speed of approximately 300-million metres per second in a vacuum.

For our purposes, therefore, a '**beam**' of light can be considered to be a 'stream of photons'.

In everyday life, we probably use the terms 'ray' and 'beam' interchangeably. Scientifically, however, they are *different*.

▶ A '**ray**' is the *direction* of the path taken by visible light, and is represented in a diagram as a straight line with an arrow superimposed over it.
▶ A '**beam**' is made up of numerous rays that may be diverging, converging, or parallel with each other.

Rather in the same way as we represent electric fields using imaginary lines of force, called 'electric flux', we can think of the rays of light that form a beam as being '**luminous flux**', expressed in **lumens (lm)**.

Lighting and, in particular, **photometry**, is a special area of study within the electrotechnology industry and, so, the purpose of this section is to provide us with an overview of the subject; *not* to turn us into experts in artificial-lighting design, but *to help us converse intelligently with lighting designers*.

Propagation of light

The reason we see an object or a surface is because visible light is *reflected* from it towards our eyes.

It's convenient to think of the **propagation of light** (i.e. the way in which light travels) in terms of *four* stages (Figures 18.7 and 18.8), that is the:

1. **source of light**, in terms of its *luminous intensity*.
2. **flow of light**, in terms of its *luminous flux*.
3. **arrival of light**, in terms of its *illuminance*.
4. **reflection of light**, in terms of its *luminance*.

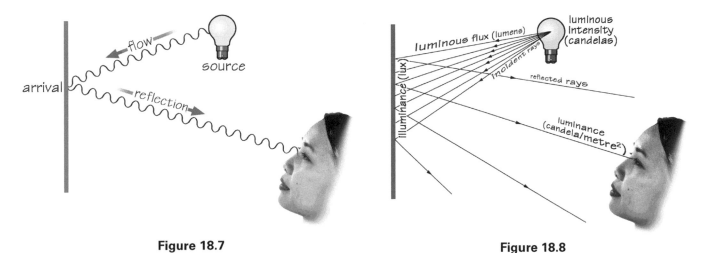

Figure 18.7   **Figure 18.8**

Despite the logical sequence described above, it is actually very much easier for us to understand this particular topic if we temporarily deal with it *out of sequence* and start, instead, by learning what we mean by the '**flow of light**' (**luminous flux**) rather than starting with **luminous intensity** which, otherwise, would be a relatively difficult concept to grasp.

Flow of light (*luminous flux*)

So, let's start out of sequence by examining the 'flow of light'. Visible light radiant energy 'flows' outward in *all* directions from its source. We can represent this as a 'beam' made up of straight 'rays' of light energy (photons) that radiate away ('diverge') from a light source in rather the same way as we can represent an electric field as lines of electric flux that radiate away from a positive electric charge.

The term, '**luminous flux**' (symbol: Φ), is used to describe the *rate* at which this light energy (as photons) leaves its source (e.g. a lamp). The SI unit of measurement for luminous flux, in the photometric system of measurement, is the **lumen** (symbol: **lm**). Luminous flux, therefore, is the photometric equivalent of 'power', and the lumen is the photometric equivalent of the watt.

The **lumen** is an SI derived unit, defined as '*the flux emitted in a solid angle of one steradian by a point source having a uniform luminous intensity of one candela*'.

To understand this definition, we must now move onwards or, rather, *backwards* (!), in the sequence of light propagation, and examine the 'source of light' in terms of its **luminous intensity**.

Source of light (*luminous intensity*)

Now that we understand what we mean by luminous flux and the lumen, it becomes very much easier to understand what we mean by the **luminous intensity** (symbol: *I*) of a 'source of light'.

In non-scientific terms, we could simply describe **luminous intensity** as being the '**brightness**' of any object which emits luminous flux, such as the sun or a lamp.

18

401

The problem with this description, unfortunately, is that 'brightness' is *subjective*. For example, to the human eye, the same object may appear 'bright' when judged against a *dark* background, but not particularly bright, or even invisible, when judged against a *light* background. This, for example, is why we can easily see the stars at night while they are not visible in daylight! So what we need is a more *objective* method of judging 'brightness', and this is what we mean when we describe 'luminous intensity'.

A 'source' of light, such as an incandescent lamp, then, is described in terms of its **luminous intensity**, which is a measure of *its ability to emit light in a particular direction*. In the photometric system of measurement, the SI unit of measurement of luminous intensity, is the **candela** (symbol: **cd**) which is 'equivalent' to (but *not* 'defined' as) 'luminous flux per steradian' (equivalent to a 'lumen per steradian').

In order to understand this explanation, we need to know what a '**steradian**' is and, in order to understand *that*, we need to remind ourselves what a '**radian**' is!

A '**radian**' (symbol: **rad**) is the SI unit of angular measure, defined as *'the angle subtended at the centre of a circle, by an arc whose length is equal to the radius of that circle'*, as shown in Figure 18.9.

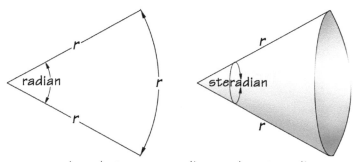

comparison between a radian and a steradian

Figure 18.9

There are **2 π** radians in 360°, so one radian is equivalent to very nearly 57.3°.

As shown in the right-hand illustration, a '**steradian**' (symbol: **sr**) can be thought of as a 'conical' radian, and is defined as *'the solid angle (or 'cone') subtended at the centre of a unit sphere, by a unit area on its surface'*.

So, objects having a *low* luminous intensity will have a *low* luminous flux per steradian, compared to objects having a higher luminous intensity, as should be made clear in Figure 18.10.

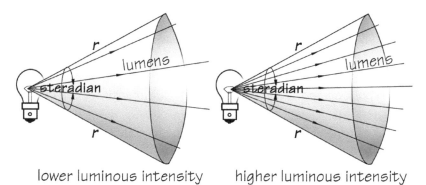

lower luminous intensity   higher luminous intensity

Figure 18.10

Although we have described luminous intensity as being 'equivalent' to 'luminous flux per steradian', it would be *wrong* to think that we can 'define' the **candela** as a 'lumen per steradian'. This is because the candela is an SI *base* unit, whereas the lumen is a *derived* unit, and we *cannot* define a base unit in terms of derived units!

This is rather like the situation with the **ampere**, the base unit for electric current, which is *defined* in terms of the force between a pair of current-carrying conductors, but which is *equivalent* to a coulomb per second.

The **luminous intensity** of *any* light source must be measured against a 'standard' source of artificial light. And, for a light source to be a 'standard', scientists must be able to reproduce it accurately whenever and wherever it is needed for the purpose of calibrating measuring instruments, etc.

Originally, this standard was quite literally based on the luminous intensity of a **candle**, and its unit of measurement was called a 'candlepower'. Of course, the 'candle' used at that time was not just *any* old candle that we might find stuffed in the back of a drawer at home, but one defined as a 'standard candle', which was carefully manufactured from the wax found in the head cavities of the sperm whale (called *'spermaceti'* wax), having a mass of exactly one-sixth of a pound (about 76 g), and one which would burn at a steady rate of 120 grains (a little under 65 mg) per hour!

These days, of course, it is possible to select a far more consistent and precise source to replace the 'standard candle'. This is based on the light emitted by a luminous radiator whose temperature is the same as that of platinum as it solidifies from its liquid state (1773°C). The details of this are well beyond the scope of this book, but can be replicated in a standards laboratory and this allows us to define the candela as follows:

> The **candela** (**cd**) is defined as *'the luminous intensity, in a given direction, of a source that emits monochromatic radiation of frequency 540 ×10^{12} hertz, and which has a radiant intensity in that direction of 1/683 watt per steradian'.*

The candela, of course, is a *photometric* unit of measurement, which is why its definition specifies the frequency of radiation as being 540 THz: that being the frequency of the visible-light spectrum to which the human eye is most sensitive under low-light conditions.

> The figure of 1/683 used in this definition was introduced so that the current definition of the candela would numerically match that of its previous definition.

This definition, then, enables scientists from all over the world to reproduce a standard light source having a luminous intensity of *exactly* one candela, thus allowing it to be used to calibrate instruments designed to measure luminous intensity.

At the beginning of this section, we described luminous intensity of a light source as being a measure of its ability *to emit light in a particular direction*. This is because 'real' lamps *don't* necessarily emit light equally in *all* directions, either because of the design and shape of the lamp itself, because of its 'luminaire' (light fitting), or because of a combination of the two.

> **'Luminaire'** is the term used by lighting engineers to describe a lamp fitting that distributes, filters, or transforms the light emitted by a lamp or lamps, and that includes all the items necessary for fixing and protecting those lamps, and for connecting them to the supply.

Because lamp/luminaire combinations don't necessarily distribute light equally in all directions, the luminous intensity of a specific type of lamp/luminaire, when viewed from various directions, is often represented using a '**polar diagram**'.

In polar diagrams, such as the one illustrated Figure 18.11, the concentric circles represent different values of luminous intensity, expressed in candelas, while the 'spokes' represent various angles measured relative to the perpendicular. The

18

grey area represents the luminous intensity for a particular type of lamp/luminaire combination, measured at various angles.

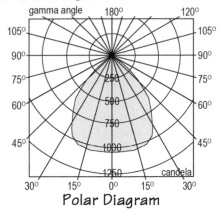

Polar Diagram

Figure 18.11

In the above example, the pattern of the polar diagram (which we have highlighted in grey) tells us that most of the luminous flux from the lamp/luminaire it represents is directed, symmetrically downwards, and within 45° either side of the perpendicular. The distances from the centre of the diagram, along the 'spokes', to the edges of the grey area represent the luminous intensity, in candelas, viewed from various angles. So, for example, at zero degrees (i.e. immediately below the lamp), the **luminous intensity** would be seen as being a little over 1000 cd, whereas at, say, 30° from the perpendicular, in both directions, the luminous intensity would be seen as being about 875 cd while, at 45°, it will be seen to have fallen off to less than 400 cd.

For an ordinary household incandescent lamp suspended from the ceiling, a single polar diagram is likely to apply to *any* vertical plane around the vertical axis of that lamp as the luminous intensity distribution is likely to be *symmetrical*.

But for other lamp/luminaire combinations, such as the fluorescent luminaire illustrated in Figure 18.12, which is long and thin, a polar diagram representing the vertical plane *along* the axis of the luminaire will be quite different from the polar diagram representing the vertical plane along its *transverse* axis. This type of distribution we call *asymmetrical*.

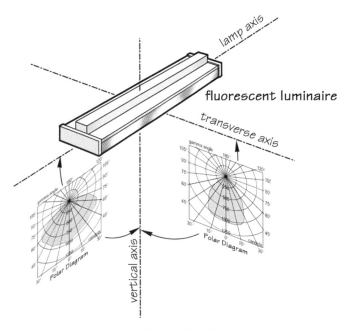

Figure 18.12

The luminous intensity distribution for some types of luminaire can be so complex that a large number of polar diagrams would be required to represent the various vertical planes, so other forms of representation are used instead. This, however, is a topic outside the scope of this book.

Arrival of light (*illuminance*)

Now let's turn our attention to the next step in the sequence of the propagation of light: the '**arrival of light**', or '**illuminance**'.

Light *arriving* at a surface is called '**incident light**', and we measure it in terms of its **illuminance** (symbol: *E*), which is defined as *'the luminous flux per unit area'* (or, in simpler terms, its 'luminous flux density') *perpendicular* to the surface. So we measure illuminance in lumens per square metre which, in SI, using the photometric system of measurement, is given the special name: **lux** (symbol: **lx**).

Illuminance is subject to what is termed the '**inverse-square law**', which means, for example, that if we were to *double* the distance between a surface and a point-source of light, then the level of illuminance will *fall to one-quarter*. We will deal with this topic in more detail later in this chapter. When incident light is *not* perpendicular to a surface that it strikes, then illuminance is also subject to the '**cosine law**', which we shall also deal with later.

We measure illuminance using an instrument called an **illuminance meter** (those of us with an interest in photography would recognise this as being equivalent to an 'incident light meter', which is held immediately in front of the subject to measure the light falling *on* that subject). A multi-range digital version of an illuminance meter is shown in Figure 18.13. When measuring general levels of illuminance in a room, the translucent-dome sensor is held horizontally about 850 mm above the ground, while taking care not to cast any shadow over that sensor.

Figure 18.13

The human eye is able to adapt to an enormous range of illuminance levels. For example, those of us with normal eyesight are quite able to read documents from around 15 lx all the way up to around 150 000 lx. Despite this, artificial lighting should always be designed to enable *the greatest task efficiency at the lowest level of illumination*.

Research into determining appropriate levels of illuminance for various types of task started in the 1930s once it had been realised that there was a relationship between those levels and the ability of people to perform visual tasks efficiently – thus establishing a link between illuminance and productivity at work!

By the mid-1970s, for example, a level of 500 lx had been established for general office lighting in the UK. Interestingly, the corresponding standard in the United States at that time was very much higher at 2000 lx due, in no small part, to the significant differences that existed at that time between our countries' attitudes to energy consumption and wastage. Since then, thanks to a general recognition of the need to save energy, US levels have fallen whereas the UK levels have remained about the same.

In the United Kingdom, the **Chartered Institute of Building Services Engineers (CIBSE)** produces a *Code for Interior Lighting*, which specifies minimum illuminance levels for residences, offices, shopping malls, schools, hospitals, theatres, etc., based on the requirements of British Standard **BS EN 12464-1 (1211)**. There are also various other standards for outdoor lighting, such as street lighting, outdoor-arena lighting, signage, etc.

The following table gives a glimpse of the numerous and highly detailed tables supplied in the Code, described above.

18

Table 18.2

Task	Example:	Illuminance / lux:
Casual vision	Storage spaces, etc.	100
Reading and writing	Offices, classrooms, etc.	500
Fine assembly	Component assembly	1000
Minute assembly	Watchmaking	3000

Incidentally, these recommended levels of illuminance must be maintained *throughout the entire life of the lighting installation* and, so, must allow for any eventual reduction of light reaching the working plane due to the ageing of the lamps or of any reflecting decor (e.g. walls), as well as to the accumulation of dust and grime on luminaires and on any other reflecting surfaces.

Reflection (*luminance*)

Let's now move on to examine the *final* stage in the propagation of light: the **reflection of light**.

Regardless of the level of illuminance of the incident light arriving on a surface, what the eye *actually* sees is the light *reflected* from that surface. This reflected light can be considered to be a *secondary light source*. Unlike luminous intensity, though, this secondary light source is *not* from a 'point source' but, rather, is reflected from a reflective *surface* and, therefore, dependent on the *area* of that surface. We call this secondary light source, '**luminance**' (symbol: **L**), and it is defined as *the luminous intensity per unit area*.

From this definition, it follows that **luminance** is measured in **candelas per square metre** (symbol: **cd/m²**).

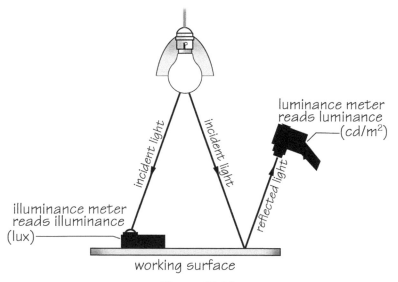

luminance meter reads luminance (cd/m²)

incident light

incident light

reflected light

illuminance meter reads illuminance (lux)

working surface

Figure 18.14

In simple terms, what we are describing is how 'bright' that surface appears to the human eye but, once again, we must remind ourselves that 'brightness' is very subjective, whereas '**luminance**' is absolute and can be measured accurately using an instrument called a **luminance meter** an example of which is shown in Figure 18.15, (those of us with an interest in photography would recognise this as being equivalent to a 'reflective' light meter, which is pointed at the subject to measure the light reflected *from* that subject, as in Figure 18.14).

Figure 18.15

Illuminance and luminance meters have photo resistive sensors that are actually sensitive to a wider range of visible light than is the human eye. These sensors are, therefore, fitted with an optical filter that simulates the characteristics of the human eye by restricting the measured light towards the centre (green) of the visible-light spectrum.

Luminance depends upon the **reflection factor** of an opaque surface. This is because some of the incident luminous flux will be *absorbed* by that surface, while the rest is *reflected*. For example, imagine a situation in which exactly the same amount of light falls on both a dark surface and bright surface. The *illuminance* will be exactly the same in each case, but the *luminance* of the brighter surface will be far higher than that of the darker surface.

If *all* the incident light is reflected, then the reflection factor will be unity or 100%, whereas if all the incident light is absorbed, then the reflection factor will be zero. This is summarised in the following equation:

$$\text{reflected luminance} = \text{incident luminance} \times \text{reflection factor}$$

Summary of light-propagation terms and units

Table 18.3

A **source** of light . . .	is described in terms of its **luminous intensity** (*I*), and is . . .	measured in **candelas (cd)**.
The **flow** of light . . .	is described in terms of the rate at which it leaves its source, is called **luminous flux** (symbol: Φ), and is . . .	measured in **lumens (lm)**.
Light arriving on a surface . . .	is called **incident light**, is described in terms of its **illuminance** (*E*), and is . . .	measured in lumens per square metre, or **lux (lx)**.
Light **reflected** from a surface and received by the eye . . .	is described in terms of the **intensity of reflected light**, is described in terms of its **luminance** (*L*), and is . . .	measured in **candelas per square metre (cd/m²)**.

Not surprisingly, people often confuse the terms 'illuminance' and 'luminance' so, to help us remember which is which, we can use the following mnemonic:

▶ **IL**uminance = **I**ncoming **L**ight (light arriving at a surface)

▶ **L**uminance = light **L**eaving a surface

18

The inverse-square rule and cosine rule

Earlier, we said that **illuminance** (luminous flux density) is subject to the 'inverse-square law', which was established by the Swiss physicist, Johann-Heinrich Lambert (1728–1777), in the mid-eighteenth century.

To understand this, let's consider the illuminance at a surface of a one-metre square area at a distance of one metre from a **point source** of one candela is one lux (a lumen per square metre). If that distance is *doubled*, to 2 m, then the same amount of luminous flux will cover an area that is *four* times as great, i.e. 4 m² – so the illuminance will have fallen to *one quarter*.

So, *the illuminance at a perpendicular surface is inversely proportional to the **square** of its distance from the source of luminous intensity.*

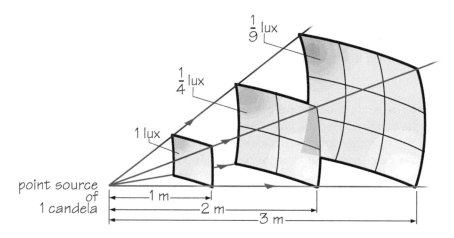

Figure 18.16

This is called the '**inverse-square rule**', where:

$$E = \frac{I}{d^2}$$

where: | E = illuminance (in lux)
I = luminous intensity (in candelas)
d = distance (in metres)

So, for Figure 18.16, applying the inverse-square rule for a luminous intensity of, say, 1 cd at 1 m, 2 m, and 3 m distances, we have:

$$E = \frac{I}{d^2} = \frac{1}{1^2} = 1\,\text{lx} \quad \Big| \quad E = \frac{I}{d^2} = \frac{1}{2^2} = \frac{1}{4}\,\text{lx} \quad \Big| \quad E = \frac{I}{d^2} = \frac{1}{3^2} = \frac{1}{9}\,\text{lx}$$

Of course, we would normally express these answers as decimals, *not* as fractions.

The **inverse-square law**, as expressed above, applies when the surface being illuminated is *perpendicular* (at right angles) to the incident light that strikes it, which is the reason why the surfaces are shown as being curved rather than being flat. If a flat surface is inclined at some angle, θ, (Figure 18.17) to the incident light that strikes it, then the surface will offer an even greater surface area for the same amount of flux, *so its illuminance (luminous flux density) will be reduced* – as explained below.

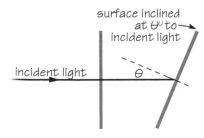

Figure 18.17

The inverse-square rule, therefore, can be modified to:

$$E = \frac{I}{d^2}\cos\theta$$

where:

E = illuminance (in lux)

I = luminous intensity (in candelas)

d = distance (in metres)

θ = angle between surface and incident light

This is generally known as the '**Cosine Rule**' for illumination.

Worked example 1

A street lamp, having a point source luminous intensity of 2000 cd, is suspended 5 m above a road – as shown in Figure 18.18. Calculate the illuminance (a) directly below the lamp, and (b) at a distance of 2 m to one side, and (c) 4 m to one side.

Solution

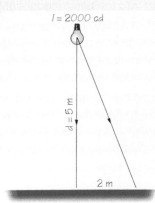

Figure 18.18

(a) The illuminance directly below the lamp:

$$E = \frac{I}{d^2}\cos\theta = \frac{2000}{5^2}\cos 90° = \frac{2000}{25}\times 1 = 800 \text{ lx } \textbf{(Answer a.)}$$

(b) To calculate the illuminance 2 m to one side, we need to know the *distance* from the lamp to that point (i.e. the hypotenuse of the triangle illustrated above), and the *cosine of angle* at which the incident light arrives at that point. To calculate the distance, we can apply Pythagoras' Theorem:

$$x = \sqrt{5^2 + 2^2} = \sqrt{25+4} = \sqrt{29} = 5.4 \text{ m}$$

18

$$\cos\theta = \frac{adjacent}{hypotenuse} = \frac{5}{5.4} = 0.93$$

Inserting these into the inverse-square rule equation:

$$E = \frac{I}{d^2}\cos\theta = \frac{2000}{5.4^2} \times 0.93 = 63.8 \text{ lx } \textbf{(Answer b.)}$$

(c) To calculate the illuminance 4 m to one side, again we need to know the *distance* from the lamp to that point, and the *cosine of angle* at which the incident light arrives at that point. To calculate the distance, we can apply Pythagoras' Theorem:

$$x = \sqrt{5^2 + 4^2} = \sqrt{25 + 16} = \sqrt{41} = 6.4 \text{ m}$$

$$\cos\theta = \frac{adjacent}{hypotenuse} = \frac{5}{6.4} = 0.78$$

Inserting these into the inverse-square rule equation:

$$E = \frac{I}{d^2}\cos\theta = \frac{2000}{6.4^2} \times 0.78 = 38.1 \text{ lx } \textbf{(Answer c.)}$$

Lighting design

Lighting design is a specialist subject within the electrotechnology industry, requiring extensive formal training. It also represents a wide range of sub-specialties, with engineers and technicians specialising in interior design, theatrical lighting, outdoor-area illumination, highway illumination, illuminated signage, etc.

So it is *not* the intention of this chapter to teach us how to become lighting designers but, rather, to give us an insight into the basic indoor-lighting design process.

In the previous worked example, we saw how the value of illuminance falls off very rapidly in any direction away from a point immediately beneath that luminaire. However, by using a number of suitably spaced luminaires, it is possible to achieve any desired average level of illuminance across any surface area.

Using what is sometimes termed the '**lumen method**' or '**lumen formula**' (see equation 1, below), it is possible to determine how many lamps will be necessary to achieve a particular average illuminance across a working surface and, from this, determine how those lamps should be spaced to achieve this.

Lumen method

We'll start by assuming we want to illuminate a working surface having an area of **A** square metres, in order to give us an average illuminance of **E** lux.

From this information, we can determine the 'design' (i.e. the 'required') amount of luminous flux that must fall on the working surface, expressed in lumens. Remember, 'lux' is simply another name for a 'lumen per square metre', so if we multiply the *area* to be illuminated *(A)* by the desired *illuminance (E)*, then we can determine the amount of *luminous flux* necessary to achieve this:

$$\text{design luminous flux} = AE \quad ---\text{(equation 1)}$$

If we apply simple dimensional analysis to this equation, we will see *why* this equation results in a luminous-flux output expressed in lumens:

$$\text{design luminous flux} = AE\left[(\text{metres}^2 \times \text{lux}) = \left(\text{metres}^2 \times \frac{\text{lumens}}{\text{metres}^2}\right) = \text{lumens}\right]$$

The actual amount of luminous flux emitted by the lamps themselves, however, must be somewhat *greater* than the amount actually needed to illuminate the working surface, because some of the lamps' luminous flux will be absorbed by their luminaires as well as by the reflective surfaces, such as walls, curtains, etc., that surround the working surface.

The ratio between the *design* luminous flux that must fall *on* the working surface, and the *actual* luminous flux that must be emitted *by* the lamps, is called the '**utilisation factor**', so we can modify equation 1 as follows:

$$\text{lamps' total luminous flux} = \frac{\text{design luminous flux}}{\text{(utilisation factor)}} = \frac{AE}{(UF)}$$

Utilisation factor figures are available from data tables and graphs provided by lighting-manufacturers' publications, and can vary from around 0.03 to around 0.75, depending on the type of luminaire employed, the nature of the reflective surroundings, etc. An utilisation factor of, say, 0.4, means, for example, that the luminous flux produced by the lamp must *exceed* the required design luminous flux by the reciprocal of 0.4 – i.e. by a factor of 250%.

It's also necessary to allow for any reduction in luminous flux emitted by the lamp due to (1) any fall in its output over the lifespan of those lamps (a fluorescent lamp, for example, can lose up to half its output as it approaches the end of its lifespan), (2) the accumulation of dust and dirt, etc., not just on the luminaires that enclose the lamps, but on any reflective surfaces, such as surrounding walls, etc. We do this by applying what is called a '**light-loss factor**'.

Light-loss factor figures are also available from data tables and graphs provided by lighting-manufacturers' data publications, and typically range from around 0.6 to around 0.85, depending on the type of lamp the cleanliness of environment involved, and by the cleaning regime employed in the building. So, for a maintenance factor of, say, 0.65, it is necessary to increase the luminous flux falling on the working area by the reciprocal of 0.65 – i.e. (1/0.65 = 1.54) 154%.

By applying *both* the utilisation factor (UF) and light-loss factor (LLF) to equation 1, we have:

$$\text{lamps' total luminous flux} = \frac{AE}{(UF) \times (LLF)}$$

Finally, in order to determine the number of luminaires required to provide the design luminous flux at the work surface, we need to divide the above equation by rated luminous flux *per lamp* (Φ_{lamp}) which, of course, depends on the type of lamp selected:

$$n = \frac{\text{lamps' total luminous flux}}{\text{luminous flux per lamp}} = \left(\frac{AE}{(UF) \times (LLF)} \right) \div \Phi_{lamp}$$

$$n = \frac{AE}{(\Phi_{lamp}) \times (UF) \times (LLF)} \quad ---\text{(equation 2)}$$

Worked example 2

An office measures 10 × 5 m with a ceiling height of 4 m, and requires an average design illuminance of 500 lx. Assuming an utilisation factor of 0.8 and a light-loss factor of 0.6, how many lamps are necessary to achieve the design illuminance, if each lamp provides 6700 lm when new?

18

Solution

$$n = \frac{AE}{\left(\Phi_{\text{lamp}}\right) \times (UF) \times (LLF)} \quad ---\text{(equation 2)}$$

$$= \frac{(10 \times 5) \times 500}{6700 \times 0.8 \times 0.6}$$

$$\approx 7.77 \text{ lamps}$$

We must, of course, *round up* this answer to **8 lamps (Answer)**.

Planning the layout

Obviously, once we've determined how many lamps are required to produce the required illuminance at the working surface, it's then necessary to determine how to *distribute* these lamps in such a way that they provide an approximately uniform illuminance across the entire working surface.

If the lamps were suspended too far apart, for example, then the resulting variation in illuminance might be unacceptable. The rule of thumb is that the minimum value of illuminance falling on the working area should not fall below 80% of the average design level.

To achieve this, the layout of lamps should conform to the manufacturers' published maximum **spacing-to-height** ratio (S/h_m), which is typically between **1** and **1.5**.

Worked example 3

Assuming the height of the working surface in the previous worked example is 0.85 m above ground level, sketch a plan view of the optimum placement of the eight lamps if they are to be suspended by 0.5 m. Assume a spacing-to-height ratio of 1.

Solution

Step 1: Determine the height of the lamps above the working surface. This will be the height of the room, *less* the height by which the lamps are suspended from the ceiling and *less* the height of the working surface. That is:

$$= (\text{height of room}) - (\text{drop of lamps}) - (\text{height of work surface})$$

$$= 4 - 0.5 - 0.85$$

$$= 2.65 \text{ m}$$

Step 2: With a spacing-to-height ratio of just 1, the lamp spacing should be *no greater than* $1 \times 2.65 = 2.65 \text{ m}$.

Step 3: Sketch a plan (Figure 18.19) of the lamp layout that satisfies the above figures:

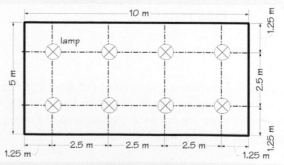

Figure 18.19

Luminous efficacy

As we've learnt, comparatively little of the electrical energy supplied *to* a lamp is actually converted into visible light emitted *from* that lamp. In the case of an ordinary household incandescent (GLS) lamp, for example, it could be argued that visible light is simply the desired by-product of the *heating effect* of an electric current.

However, the heat produced by lamps is not necessarily 'wasted' because, in cold weather, it can contribute to the general heating of a building. In fact, some commercial lighting fixtures (luminaires) actually incorporate 'heat recovery' systems so that this heat can better contribute to the general heating of a building.

> An argument frequently used by those who have campaigned to ban incandescent lamps is that the heat they produce is 'wasted' whereas, in fact, the heat is *not* wasted at all – particularly in cold weather, when it becomes a useful source of additional heating for a room.

The total input power to a lamp is, of course, measured in **watts**, whereas the output power of the visible light (i.e. its 'luminous flux') is measured in **lumens**.

So how can we realistically compare the *input power* of a lamp with the *photometric power of the luminous flux it emits?* The answer is to use what is known as '**luminous efficacy of a source**'.

> '**Efficacy**' is *not* the same thing as '**efficiency**'. We could say that 'efficacy' is a measure of *whether or not we have **achieved** a desired result*, whereas 'efficiency' is a measure of the **waste** *that results when achieving that result*.
>
> 'Efficacy' has units of measurements because we're comparing lumens with watts (hence: 'lumens per watt'), whereas 'efficiency' is dimensionless because we're comparing watts with watts!

The **luminous efficacy of a source** is '*the ratio of the luminous flux (measured in lumens) emitted by a luminous source, to its input power (measured in watts)*', and is measured in **lumens per watt** (symbol: **lm/W**).

$$\text{luminous efficacy} = \frac{\text{luminous flux output}}{\text{power input}}$$

As we learnt earlier, the eye is most sensitive to light in the 'green' part of the spectrum, so luminous efficacy is normally measured at this wavelength (around 555 nm). At this wavelength, the (theoretical) maximum luminous efficacy is 683 lm/W. However, in practice, luminous efficacy is nearly always *low* compared to this theoretical figure – except for those lamps (e.g. low-pressure sodium) that actually produce light near the green part of the colour spectrum.

For example, if the total luminous flux for a 230 V, 60 W, incandescent GLS lamp is 710 lm, then its luminous efficacy will be:

$$\text{luminous efficacy} = \frac{\text{luminous flux}}{\text{power input}} = \frac{710}{60} \simeq 12 \, \text{lm/W}$$

Generally, incandescent lamps typically have luminous efficacies between just 5 and 15 lm/W, depending on their type and operating voltage (the lower the voltage, the dimmer and more 'orange' the light will become). This figure improves for fluorescent lamps (between 50 and 100 lm/W), and improves considerably for other types of discharge lamp (e.g. low-pressure sodium: 183–200 lm/W) whose output wavelength peak around the green part of the colour spectrum.

18

Lumens are (or will be!) the 'new watts'!

Until recently, when the main source of residential lighting was the incandescent GLS lamp, it made sense for us to compare the light they produced (i.e. their luminous flux) in terms of their rated power or 'wattage'. It was reasonable to suppose, for example, that the output from a 200 W lamp would be roughly twice that from a 100 W lamp, and so on.

But, because different types of lamp have different luminous efficacies, it has become impractical to compare the luminous flux produced by one type of lamp with another type simply in terms of their rated 'wattage'. For example, if we want to replace a 100 W incandescent GLS lamp with a CFL (compact fluorescent lamp), what 'wattage' CFL should we select to ensure that it achieves the same amount of luminous flux?

To overcome this problem, ever since their introduction, CFL packaging has been labelled along the following lines: '**21 W – equivalent to 100 W (incandescent lamp)**', etc.

But with the emergence of completely new types of energy-efficient lighting technology, a more common-sense approach would be to ignore a lamp's rated wattage altogether and, instead, to rate *all* lamps according to the number of *lumens* they emit. This means that we can then easily compare the output of *any* type of lamps, regardless of the technology involved, thus enabling us to compare like with like.

(Of course, we will *still* need to know the rated power of a lamp in order to be able to calculate its electrical loading, but this value will no longer have any bearing on its luminous flux output.)

And this is the direction in which legislation within the EU, North America, and elsewhere, is heading with regard to the labelling of packaging for lamps.

As part of the European Union, the UK is obliged to implement measures contained in the **1998 EU Directive23**, which makes energy labelling of household lamps mandatory throughout the European Union. It requires *all* lamps sold in the domestic lighting market to carry an **energy label** that is similar to the 'A to G' energy labels used for other household appliances such as refrigerators and washing machines. And this scale must apply across *all* types of household lamps, regardless of their type (in some other countries, at present, this labelling is only done *within product classes*, so CFLs may have one efficiency rating system and incandescent lamps quite another!).

Under this EU labelling system, CFLs are typically rated as Class **A** or **B**, halogen lamps as Class **C** or **D**, and incandescent lamps as **E**, **F**, or **G**.

And, finally, *it is a requirement that all labels specify the number of lumens produced by the lamp*.

For example, the EU label, illustrated in Figure 18.20, has been taken from the packaging for a **5 W** (input) **CFL** providing a luminous flux of **200 lm** with an energy rating of '**A**':

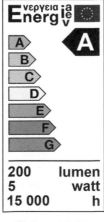

Figure 18.20

(The '**15 000 h**' value illustrated in Figure 18.20 energy label expresses the number of hours the lamp is estimated to operate before failing. Typically, this figure is between 1000 h for incandescent lamps and up to 15 000 h for the best-quality CFLs. This figure is based on the lamp operating for 3 h per day but, for CFLs in particular, it also depends on the number of times it is switched on and off. However, a report, published by *Which* magazine, in 2014, suggests that these figures are misleadingly optimistic!)

Typical values of efficacy for different types of lamp are shown below.

Table 18.4

Type of lamp:	Efficacy / (lm/W):
GLS incandescent	5–15
Tungsten halogen	12–35
Mercury vapour	40–60
Fluorescent	50–100
Compact fluorescent lamp	40–65
Metal halide	50–100
High-pressure sodium	40–60
LED	20–120

Worked example 4

If the efficacy of a particular brand of incandescent GLS lamp is 12 lm/W, what is the value of luminous flux produced by the lamp if its input power is 100 W?

Solution

$$\text{luminous flux} = \text{power input} \times \text{efficacy}$$
$$= 100 \times 12$$
$$= 1200 \text{ lm (Answer)}$$

Worked example 5

If the efficacy of a particular brand of CFL lamp is 60 lm/W, what must be the minimum input power of this lamp in order to provide the same luminous flux as the incandescent GLS lamp in the previous question?

Solution

$$\text{power input} = \frac{\text{luminous flux}}{\text{efficacy}}$$
$$= \frac{1200}{60}$$
$$= 20 \text{ W (Answer)}$$

Types of lamp

In the electrotechnology industry, we do *not* use the word, 'bulb' (we *plant* 'bulbs' in the ground whenever we want to grow daffodils!); the correct word is '**lamp**'. And 'electric lights' are more properly termed '**electric lamps**'. Similarly, 'light fittings' are termed '**luminaires**'.

Electric lighting fall into *four* general categories. These are:

▶ **incandescent** – e.g. tungsten (GLS) and tungsten-halogen
▶ **gas-discharge** – e.g. fluorescent, sodium, etc.
▶ **carbon arc** – now obsolescent
▶ **solid-state** – e.g. light-emitting diodes

18

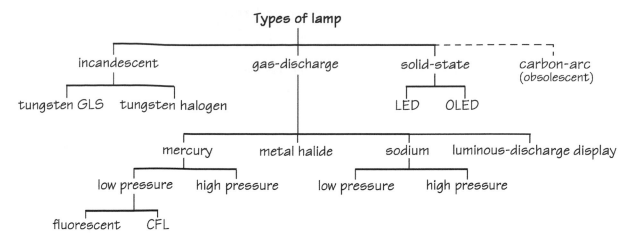

Figure 18.21

Incandescent, gas-discharge, and solid-state lighting are dealt with in great detail on this book's associated website; please go to: www.routledge.com/9781138849266.

Conclusion

Now that we've completed this chapter, we need to examine its **objectives** listed at its start. Placing a question mark at the end of each objective turns that objective into a **test item**. If we can answer those test items, then we've met the objectives of this chapter.

Index

Index